Siberian Industrial Days International Forum 2020

IOP Conference Series: Materials Science and Engineering
Volume 944

Tomsk, Russia
15 - 16 May 2020

ISBN: 978-1-7138-3196-9
ISSN: 1757-8981

Printed from e-media with permission by:

Curran Associates, Inc.
57 Morehouse Lane
Red Hook, NY 12571

Some format issues inherent in the e-media version may also appear in this print version.

Institute of Physics
Dirac House, Temple Back
Bristol BS1 6BE UK

Phone: 44 1 17 929 7481
Fax: 44 1 17 920 0979

techtracking@iop.org

Additional copies of this publication are available from:

Curran Associates, Inc.
57 Morehouse Lane
Red Hook, NY 12571 USA
Phone: 845-758-0400
Fax: 845-758-2633
Email: curran@proceedings.com
Web: www.proceedings.com

TABLE OF CONTENTS

IOP Conf. Series: Materials Science and Engineering **944** (2020) 011001 doi:10.1088/1757-899X/944/1/011001

PREFACE

We are proud to introduce the proceedings of the second Siberian Industrial Days international forum. The event was organized by the Siberian Design Center of Tomsk State University and held semi virtually on May 15-16. The coorganizers of the event are Co-organizers are Belgorod State Technical University n.a. Shukhov (Belgorod) and Saint Petersburg Art and Industry Academy n.a. Stieglitz (St Petersburg). The event is dedicated mostly to such topics as design, aesthetics, and arts. However, among the course of the event several dozen presentations had a scientific value so we made a decision to publish this volume.

The volume is comprised of 38 selected academic papers discussing various topics of industrial design. Most of the papers devoted to architectural design, computer graphics and its applications as well as the design of consumer goods. All materials were reviewed by the editorial team and discussed within the virtual meetings. Authors had their time to take into account not only remarks of referees but also feedback from the colleagues and attendees. The revised and edited papers were proofread by linguists of Tomsk State University to ensure the material's clarity.

Despite the fact that this year event took a part in the middle of the CoVid-19 pandemic the attendance was relatively high thanks to virtual participation. Several design projects and a dozen of researches were successfully presented virtually via BigBlueButton service. There were approximately 140 active participants in all meetings, including those dedicated to arts and business. Academic meetings in the course of the forum had almost 30 unique attendees excluding committee members.

We would like to express our highest appreciation to the organizers and patrons of the forum, multidisciplinary editorial team, and IOP Publishing for the opportunity to reach the global design community with these proceedings.

With all the best,

On behalf of the organizers

Victor A. Klimenko, *Director of Siberian Design Center, National Research Tomsk State University*

Natalia I. Kozhukhova, *associate professor, Belgorod State Technical University*

Peer review declaration

All papers published in this volume of *IOP Conference Series: Materials Science and Engineering* have been peer reviewed through processes administered by the Editors. Reviews were conducted by expert referees to the professional and scientific standards expected of a proceedings journal published by IOP Publishing.

- **Type of peer review:** Open peer review by the multidisciplinary editorial team.
- **Conference submission management system:** submissions were received and handled via e-mail and social media (vkontakte.ru)
- **Number of submissions received:** 113 (including non-academic manuscripts and art projects)
- **Number of submissions sent for review:** 45
- **Number of submissions accepted:** 38
- **Acceptance Rate (Number of Submissions Accepted / Number of Submissions Received X 100):** 40%
- **Average number of reviews per paper:** 2
- **Total number of reviewers involved:** 7
- **Any additional info on review process:** editorial team was comprised of the program committee and representatives of the co-organizers
- **Contact person for queries:** Natalia I. Kozhukhova, secretary of editorial board, associate professor, Department of Materials Science of Belgorod State Technical University n.a. Shukhov, kozhuhovanata@yandex.ru

IOP Conf. Series: Materials Science and Engineering **944** (2020) 012001 doi:10.1088/1757-899X/944/1/012001

City-planning transformation of the functional and planning structure of districts near a railway station in the largest megalopolises (international experience)

E A Akhmedova[1], **I V Kuznetsov**[1]

[1] Samara State Technical University, Molodogvardeyskaya Street, 194, 443001, Samara, Russia

E-mail: dir_inst_arch@bk.ru

Abstract. The article deals with the peculiarities of functioning, renovation and redevelopment of city districts proximately adjacent to the territories of railway stations and roads. Both decaying, degenerating districts and successfully developing ones are of interest. The functional, compositional and planning principles of these districts' development with their consecutive inclusion into the full-fledged city life are revealed by analysing foreign and Russian experience. The issue of city districts near a railway station's transformation becomes actual not only in Russia, but in many cities around the world as well. It is conditional upon the fact that to a greater extent railway infrastructure objects formerly situated outside a city begin to contact closely with the urban area in the process of urban sprawl. It is impossible not to consider their influence on planning, functional as well as architectural and stylistic solution of the infrastructure. The **aim** of the research is to detect the main compositional, planning and functional peculiarities and principles of the site development situated in city districts near a railway station. The following **tasks** were to be solved to achieve the abovementioned aim: to distinguish the examples of various countries' cities, the structure of which includes railway infrastructure objects actively; to study briefly the district history that can influence the further urban development of the site; to analyse the chosen cities' city-planning situations; to find out the main functions of buildings and housing as well as the urban realm elements present in districts near a railway station; to study the functionality of buildings, constructions and urban spaces that appeared in urban realm in different historic periods; to define planning and functional interconnections between the different elements of city tissue in districts near a railway station.

1. Introduction

On the rebound of industrial advance in many countries of the world such as Great Britain, Germany, Switzerland, Austria and others a new transport means, i.e. the railway one, appeared in the XVIIIth century. The first railway road appeared in Russia in the 1830s. It joined Saint Petersburg, Tsarskoe Selo (currently the town of Pushkin) and Pavlovsk. The active formation of the Russian Empire railway network began in the not-too-distant future, in the second half of the XIXth century. In such a way not only the new typology of public buildings, i.e. the railway station, but new city districts with their functional and planning specifity appeared. The districts that are proximately adjacent to the railway stations and road are meant. The railway road was the transportation means of large volumes of cargo and passengers' movement; and it is characteristic for many modern cities formed in the period of the railway transport development that the main railway stations and rail stops are situated in the city's central districts while they could be built on the city outskirts originally. The urban area development did not stop in the course of time, and the objects of the railway infrastructure turned out to be included in the city tissue. Railway service buildings and constructions, the maintenance of equipment, communal and warehouse as well as general-use objects and worker's neighborhood for the railway workers were placed along the lines [1, 2]. Very often the absence of strict planning structure, randomness, squatter development and other peculiarities characterizing these territories as undeveloped, uncomfortable for living [3] and mainly as territories most commonly cut from the main city tissue either physically or socially (or in both ways) were characteristic for the development of similar sites. The situation with the implementation of railway road objects in the city structure is characteristic for many European, American as well as Russian cities. However, the question of their literate inclusion in the city tissue is hardly solved everywhere; this fact explains the actuality of this question and its importance for many cities and their dwellers.

The main planning constraints related to the railway road and influencing the character of the adjacent site development are the easement area [4] and sanitary protection zone [5]. If the easement area includes only the objects of the railway road transport and engineering infrastructure, then sanitary protection zone allows making buildings and housing not belonging to the railway road on the territories occupied by it. However, the list of such ones is restricted by the governing documents. In particular, it is not allowed to place apartment blocks, landscape and recreation zones in the sanitary protection zone. Strict hygiene and other requirements limit the function package that the site development adjacent to the railway road can have. However, it should be noted that the regulatory environment of various states is different, that is why it is acceptable to place housing construction closely to the railway road in some of them.

The problem of the research is the civil site development stagnation of many city districts near a railway station and the lack of strict functional programme to develop them.

The **aim** of the present research is to reveal the main compositional, planning and functional peculiarities and principles of city districts near a railway station's effective development.

The following tasks should be solved to reach this set aim:

- to study the examples of cities where the railway road and its infrastructure objects are actively included in the city tissue of historic downtowns;
- to study briefly the history of a district near a railway station, the facts preceding the district's genesis, formation and development;
- to analyze the city-planning situations of the chosen cities;
- to look at the functionality of buildings and housing as well as the elements of urban realm present in districts near a railway station;
- to distinguish compositional, planning and functional interconnections of city tissue different elements in districts near a railway station.

The largest cities' districts near the railway stations are the **research objects**.

The research subject is the world experience of districts near a railway station's renovation as exemplified by large European cities.

In the course of the research the following hypothesis was developed, however, it does not pretend to be the conclusion, but it is directed at the more attentive treatment of the existing problem and the development of the logical and justified work system. The **hypothesis** lies in the fact that city districts near a railway station have their characteristic peculiarities of functional, compositional and planning structure formation due to various hygiene limitations that is reflected in the picture of the existing and perspective site development, that is why the core of the site development functional programme of such districts must consist of the unique public objects capable to attract investors as the capital influx and citizens as common users of the territory. In particular, public facilities, to be more precise, the research, development, production and managerial and business ones such as university campuses, technoparks, head-quarters of large industrial and entrepreneurial companies, waste recycling plants

IOP Conf. Series: Materials Science and Engineering **944** (2020) 012001 doi:10.1088/1757-899X/944/1/012001

and other objects studied in detail in the present article, are a spectacular example of the site development of the territories adjacent to the railway lines.

2. Materials and methods

In the course of the research such **methods** as comparative analysis, unit factor graphoanalytical method and architectural and spatial modeling were used, due to them the set tasks were solved and the aim of the research was reached.

For the more objective view of the site development problem and city districts near a railway station's development the world experience in solving this problem was considered.

Basel, Switzerland. The main part of the railway road in Basel is in the structure of restrained urban conditions. There are two main railway stations in the city. One of them, Basel Badischer Bahnhof, is situated on the right bank of the Rhine in the so called Small Basel. The other one, Bahnhof Basel SBB, is situated on the opposite bank of the river in Big Basel.

The buildings of chemical plant "Syngenta International AG", the complex of trade and exhibition centres consisting of the Basler Weinmesse, Hall 1, Messe Basel and Hall 3, hotel Hyperion Hotel Basel, Musical Theatre Basel, multi-storeyed garage Parkhaus Badischer Bahnhof AG of 300 metres long are in the structure of the districts near railway station Basel Badischer Bahnhof.

By contrast with the previous one the district near railway station Bahnhof Basel SBB is interesting by objects situated not simply in the pedestrian proximity from the railway road, but at the distance of not more than 100 metres and sometimes proximately adjacent to the easement area of the railway road. The city zoo is to the west of the railway station. The covered market Markthalle Basel built in 1929-1930 and reconstructed in 2006-2012 when the fourteen-storeyed tower was added to it among the other objects is just at a short distance from the railway station. The conference centre Viaduktstrasse built in 1885–1994 according to the project of bureau Diener & Diener Architekten is situated on the territory between the zoo and market. The twenty-four-storeyed tower of Bank for International Settlements (BIS) built in the 1970s is in the accessibility for pedestrians from the railway station. It is planned to finish the building of the nineteen-storeyed hotel Mövenpick Hotel Basel in front of the tower. Eastward, the architectural objects, Peter Merian Haus and Jacob Burckhardt Haus, built upon the project by Swiss architect Hans Zwimpfer (bureau ffbk Architekten) are on the elongated plot of land along the railway road in the place of the former locomotive shed and other buildings belonging to the railway road. Architectural-planning and voluminous-spatial solution of these buildings are similar. They are stretched along the railway road and face it with their perpendicular blocks and its opposite side, i.e. Nauenstraße St. The buildings occupy the offices of different companies and firms, commercial facilities as well as Hochschule für Wirtschaft FHNW in Western Switzerland and the university library of business and economics. The high-rise administrative office building Grosspeter Tower made in 2013–2018 is to the east of these constructions. Apart from the offices a hotel is there.

All the above mentioned objects are situated on the one side of the railway road. The other side contains buildings that are of interest as well. They are the following ones: office and commercial complex with nursing home for the elderly Südpark Basel, twenty-four-storeyed office and residential tower Meret Oppenheim Tower, Central Signal Box, museum Schaulager. Jacques Herzog and Pierre de Meuron who were awarded the Pritzker Architecture Prize are the authors of all the four described objects.

Zürich, Switzerland. The main "railway band" of Zürich is on the left bank of the river Limmat rooting down to the Lake of Zürich. It splits the city from the north-west to the south-east where on the bank of the river it ends with the city's main railway station. The Swiss National Museum with the new flanking building erected in 2012–2015 on the project design of architect bureau Christ & Gantenbein is near the key railway hub of whole Switzerland. From the southern side the railway road limits Zürich industrial quarter confined by the river from the north. The part of this district is known as "fashionable quarter" Escher Wyss. Swiss industrial machinery manufacturing plant from which the quarter got its name was situated on this territory from the beginning of the XIXth century and up to the middle of the XXth one. From the second half of the XXth century, after the closing of the plant, this district has been actively overbuilt by civic buildings. The theatre Schauspielhaus is in the former shipbuilding factory workshop from 2000. It neighbours the multiservice unit civic centre "Plus 5" during the building of which in 2001–2004 some elements of a foundry shop constructed in 1898 were preserved. The complex includes the showroom, business-centre with study rooms, offices of law firms as well as shops and catering enterprises. One of the faces of «Plus 5» looks at Turbinenplatz, the same square is faced with the building of Zürich technopark being one of the largest ones in Europe. It was built in 1989–1993 and became one of the first large community buildings erected in line with the redevelopment of this former industrial area. The premises of technopark have innovative structure, they are universal and suitable for the work of specialists in different areas of science, production operations and art, moreover, universities and research and development centres can lead their educational activities here. Several large hotels are in the accessibility for pedestrians. The railway road in Zürich does not have an exclusion zone as such. Large groceries, the fringe theatre "CLOSE-Theatre", the offices of architectural and design bureaus and other companies as well as housing accommodations and recreation sites such as, for example, Pfingstweiden Park and Maaghof-Park are almost close to it. The bridges transverse to the railway road with multilevel interchanges connect the districts of the city dissected by the road and the river.

Bern, Switzerland. The railway road in Bern goes through its very centre. The railway station of the city being the second in size in Switzerland after Zürich is situated in the historic downtown. The University of Bern, the Church of the Holy Spirit and a bit further from it Federal Palace on Bundesplatz are in the accessibility for pedestrians from the railway station. Here the Federal Council and the Federal Assembly of Switzerland, the main governing authorities of the state, have their sittings. Further to the east Bern Minster is situated. The railway lines in the district of the central station get under the ground because of the peculiarities of local topography and town surveying. It allowed utilizing the space that could be occupied by the easement area of the railway road. Open parking areas and a bus terminal are here. The urban area closely approaches the territory occupied by the railway road as in Zürich and Basel. Along with this, the typology of site development is various: both community and residence buildings are present. In some places the buildings are separated from the railway road only by the street way and green planting zone. In some cases the buildings face the railway road directly. Most commonly, these are community buildings. The repair depot is in one and a half kilometres to the west from the central station, and one more large railway station is slightly further. The Bremgarten cemetery currently representing parkland in the densely populated area of the city is near it. The cemetery borders the territory of the railway road. The group of university hospital buildings is nearby. The medical museum of Bern faces the railway road. The Bremgarten cemetery borders Bern's urban area. Parkland Glasbrunnen is northwards. Manufacturing and scientific companies such as recycling centre and the complex of university buildings are between the parkland and the railway road.

Eindhoven, the Netherlands. The factory of electric light bulbs known currently as "Philips" was opened in the Dutch city of Eindhoven in the end of the XIXth century. The company moved its headquarters to Amsterdam and its main production units to Austro-Asiatic countries in the late XXth century. After that the large industrial zones of the city present in its very centre against the background of the loss of local economic mainstay fell to desolation. Several quarters of district Strijp (Strijp-R (part of quarter Zwaanstraat, also known as Strijp-III), Strijp-S, Strijp-T (also known as Strijp-II and De Beuk) and Witte Dame are meant. All these former industrial zones are proximately adjacent to the railway road; the main railway station of the city is in quarter Stationsplein neighbouring the abovementioned quarters from the east.

District Witte Dame is the closest one to the city centre; it occupies the territory of approximately 15 hectares. The former factory building that after the reconstruction of 1995–1998 unites the design academy, public library and other various non-governmental organizations (bank departments, shops, public catering establishments, etc.) is on the territory of this district. The building of "Lichttoren" (the Tower of Light) is here as well in the heptagonal tower on the roof of which the testing of lamps produced by “Philips” factories was performed. After the reconstruction the building was geared to housing accommodations and public duties. Mall “La Piazza” built in 1999-2004 on the project design by Massimiliano Fuksas as the broadening of the existing building of department store “Bijenkorf Building” made in 1966-1969 on the project design by Italian architect Gio Ponti and Dutch architect Theo Boosten is near Witte Dame.

Quarter Strijp-S is on the fringe of Eindhoven centre and occupies 27 hectares. In 2001 the specialists of company West 8 devised the master plan of this territory’s development that meant the combination of new site development with the existing industrial buildings and housing. The central element of the planning concept became the landscaped boulevard Torenallee 60 metres wide going through the quarter in diagonal. A spacious paved square where different mass events are held and a big landscaped field also used for various public events are opposite the buildings placed along Torenallee. More than 4000 residential units were built in the course of almost 20 years of project installation in the quarter; and industrial buildings were modified to perform public duties: colleges, company offices, shopping centres, public catering establishments, museums of contemporary art. One of the buildings that catches the eye most of all is the cultural centre “Klokgebouw” (House with Watches) placed along the railway lines; different creative firms take offices on lease here.

Quarter Strijp-R is situated slightly further from the centre than all the above described quarters. The working-out of master plan on the transformation of the former industrial quarter into the residential quarters was done by architect bureau Diederen Dirrix and Buro Lubbers. In the new project the desire to preserve the identity of the place was taken into consideration. The old foot bridge, elevated platform, pump work for pipe bridge and security booth were preserved. The complex of the former workshop producing radio and television details is occupied by the company of the Dutch designer Piet Hein Eek. Workshops, a showroom, a restaurant and many other things are present here. This building was the first one to be geared to modern usage on this territory that played its positive role in the popularization of the quarter.

Vienna, Austria. The railway road goes along the historic districts in Vienna as in many European cities. The district of Althangrund and the ones adjacent to it give outstanding interest. Wien Spittelau Bahnhof that is a large multilevel commuting hub is situated here. Combustion plant Spittelau built in 1988-1997 on the project design by architect and artist Friedensreich Hundertwasser is near it. The federal bureau of criminal police is near. Further south the large university complex of the University of Vienna and the Vienna University of Technology is, Kurt Gödel Research Centre for Mathematical Logic (KGRC), pharmaceutical centre, geocentre, institute of regional planning and other structural scientific-research and training units of universities are here. The edifice by Dame Zaha Mohammad Hadid – residential compound Spittelau Viaducts the work over which was run from 1994 to 2006 is on the narrow plot of lands along the Donaukanal (Danube Canal). The object is situated on the currently non-functioning viaduct which was used for the overhead line of the city subway. The plot is lodged between the working overhead line of the city subway and canal front. The Jewish Cemetery in Währing (Jüdischer Friedhof Währing) neighbouring Währing park and the territory of the train shed are in 200 metres westwards the described district. The station of the above-ground subway is further south.

Oslo, Norway. The central railway station of Oslo is placed in the geometric downtown of the city, almost on the shore of the Oslofjord inlet that makes almost the whole historic, social and business centre of the city the district near the railway station. From the coastline and the station it is separated by the narrow strip of social and business site development and sea port. The underground station, tram stops are nearby. The southern part of district Gamle-Oslo including the waterside and the district of Sentrum adjacent to it from the north-west are of interest. The railway road is crossed by the highway and foot bridges, the most interesting of them is the bicycle and foot bridge “Acrobat” (in Norwegian “Akrobaten”) built on project design L2 Arkitekter and opened in 2011 and automobile and foot bridge Nordenga opened the same year.

Two most high-rise buildings of the city stand from the northern part of the railway road: thirty-seven-storeyed building of the hotel “Radisson Blu Plaza Hotel”, opposite which is multi-purpose indoor arena "Oslo Spektrum" and the twenty-six-storeyed office building "Biskop Gunnerus Gate 14". These buildings were made in the second half of the XIXth century.

The former sea port urban areas began to be settled comparatively recently from the southern side of the railway road in the district of Bjorvika ("Bjørvika" in Norman). In this way, for example, the cultural and business centre "Barcode Project" began to be formed in the 2000s along the railway road separated from it only by automobile and foot street. Architect bureaus A-Lab, MVRDV, DARK Architects, Snøhetta and Solheim + Jacobsen Arkitekter were engaged in the work on designing the quarters. All the buildings of the complex are the office and business ones. The first floors face the street with trading space and one of the buildings has housing accommodations. This is the KLP Building built upon the project Solheim + Jacobsen Arkitekter (currently AART Architects), where the last 10 storeys are occupied by 53 flats with balconies and access to the outdoor patio. All the buildings in the complex are at ground level and united between themselves by public space.

Apart from the quarter "Barcode Project" the waterside zone separated by the railway road is being developed with other modern objects. The opera theatre erected in 2003-2007 upon the project of bureau Snøhetta is distinguished particularly.

To the south-east from the described district in the immediate proximity from the railway road historic monuments in the ruined condition are: St. Mary's Church, the Church of Saint Clement and St. Hallvard's Cathedral. The old city cemetery is to the south.

3. Results

As can be seen from the active processes of engineering and construction in city districts near a railway station, the theme of land use planning and these territories' management is interesting to different participants of the construction procedure: citizens, builders, architects. This shows the topicality of the current theme, but at the same time it sets many tasks before the specialists to be solved, so that the districts under consideration (i.e. city districts near a railway station) functioned harmonically in the city system. The value of these territories is understood to a great extent; however, renovation brings its results most commonly only in big and the biggest cities, moreover, mainly in the European, American and Asian ones.

The Russian experience of districts near a railway station's renovation as well as former industrial, communal and warehouse and general-use territories is less successful than the foreign one. Community objects and urban open spaces having potential to attract citizens and developers there and revive these territories appear on such plots as occasional spots not very frequently.

On the basis of the analysed material it is clear that most commonly districts near a railway station are built with unique objects capable to attract citizens there, to make them socially important and interesting both for the rest of citizens and investments. These can be large research and development complexes and technoparks, university academic campuses and libraries, trade and exhibition centres and theatres as well as large administrative and office and business objects such as, for example, federal ministries, courts and company headquarters. All the above mentioned functions can be included in the abstract model of city districts near a railway station's development. Moreover, while designing on the territories adjacent to the railway lines it is necessary to adhere to the principles of transport security, environmental protection and landscape-aesthetic harmonization [6].

4. Discussion

The theme of city districts near a railway station's development is interesting to many architects. This is proved both by theoretical manuscripts and applied work, i.e. numerous designing and construction.

In that way Le Corbusier says in his article "The New Spirit in Architecture" written in 1924: "… a human being reveals themselves through the order; what unveils in front of your eyes when you leave Paris by railway road if not the largest reduction of things to order and not the fight with nature? Fight to submit it to themselves, to classify, to bring comfort to themselves, in brief to establish themselves in the human world different from the hostile natural environment, in our world where geometry reigns? The human being works way through only by geometry. The rails are absolute parallels, the railroad embankment is the realisation of geometrical drawing; bridges, viaducts, sluices, channels, all this city and near-city creation developing in a separate countryside shows that when a person acts when he/she wants to show his/her will, he/she unavoidably becomes a geometrician and creates with geometry" [7, p. 28]. The genius of modernism not only allows to understand the city-planning value of territories near a railway station, but shows the aesthetic importance of the railway road which in this or that form can be transmitted into architecture (Table 1).

Table 1 (beginning). The analysis of functional and planning structure of city districts near a railway station (as exemplified by foreign experience)

CITY, COUNTRY — **BASEL, SWITSERLAND**

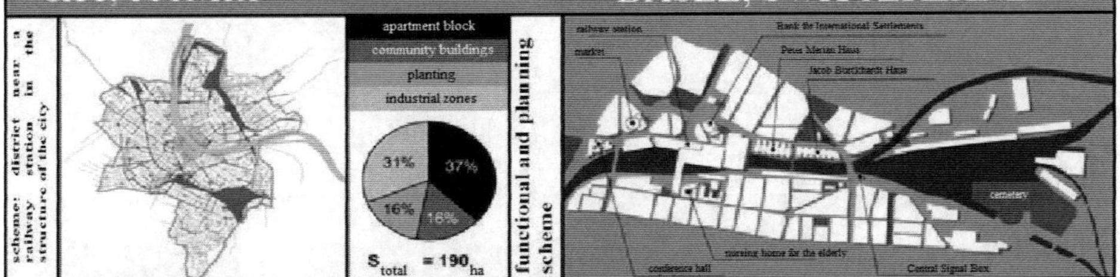

FUNCTIONS ESTABLISHED IN A DISTRICT NEAR A RAILWAY STATION historically (before XX century)	FUNCTIONS IMPLEMENTING IN THE DISTRICTS NEAR A RAILWAY STATION in contemporary times (XX-XXI centuries)	EXAMPLES OF CIVIL SITE DEVELOPMENT in contemporary times	COMPOSITIONAL AND PLANNING STRUCTURE OF A DISTRICT NEAR A RAILWAY STATION
historic apartment blocks	domestic buildings and hotels	Meret Oppenheim Tower	linear and linear-junctional
	objects of human services	Südpark Basel	
zoological garden	science education objects	Peter Merian Haus	
	art and culture objects	museum Schaulager	
cemetery	trade enterprises	covered market Markthalle	
	managerial and business objects	Bank for International Settlements (BIS)	

CITY, COUNTRY — **Zürich, SWITSERLAND**

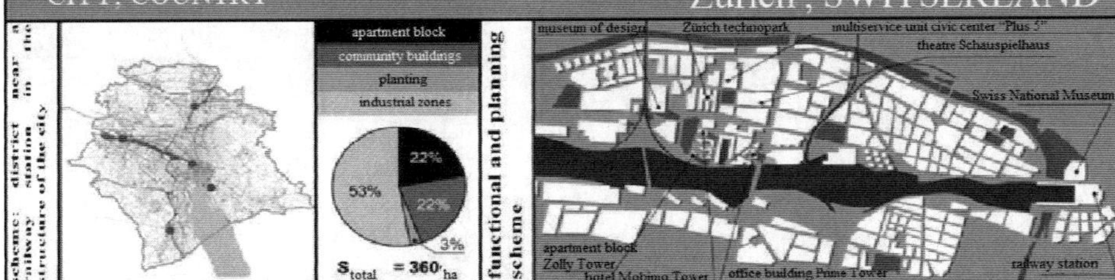

FUNCTIONS ESTABLISHED IN A DISTRICT NEAR A RAILWAY STATION historically (before XX century)	FUNCTIONS IMPLEMENTING IN THE DISTRICTS NEAR A RAILWAY STATION in contemporary times (XX-XXI centuries)	EXAMPLES OF CIVIL SITE DEVELOPMENT in contemporary times	COMPOSITIONAL AND PLANNING STRUCTURE OF A DISTRICT NEAR A RAILWAY STATION
	domestic buildings and hotels	Zolly Tower, Mobimo Tower	linear-dispersed
	science education objects	Zürich technopark	
industrial zone, machinery manufacturing plant producing turbines "Escher Wyss"	art and culture objects	centre "Plus 5", theatre Schauspielhaus	
	trade enterprises	mall "Migros"	
	managerial and business objects	Prime Tower	

CITY, COUNTRY — **BERN, SWITSERLAND**

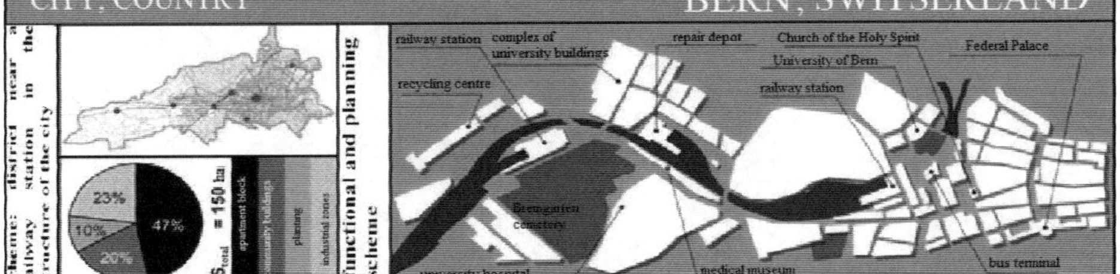

FUNCTIONS ESTABLISHED IN A DISTRICT NEAR A RAILWAY STATION historically (before XX century)	FUNCTIONS IMPLEMENTING IN THE DISTRICTS NEAR A RAILWAY STATION in contemporary times (XX-XXI centuries)	EXAMPLES OF CIVIL SITE DEVELOPMENT in contemporary times	COMPOSITIONAL AND PLANNING STRUCTURE OF A DISTRICT NEAR A RAILWAY STATION
historic apartment blocks	domestic buildings and hotels	residence buildings of middle height	linear-dispersed
	science education objects	the University of Bern	
	art and culture objects	medical museum	
Bremgarten cemetery	trade enterprises	shops	
	managerial and business objects	Federal Palace	

Table 1 (end). The analysis of functional and planning structure of city districts near a railway station (as exemplified by foreign experience)

CITY, COUNTRY — **Eindhoven, the Netherlands**

scheme: district near a railway station in the structure of the city

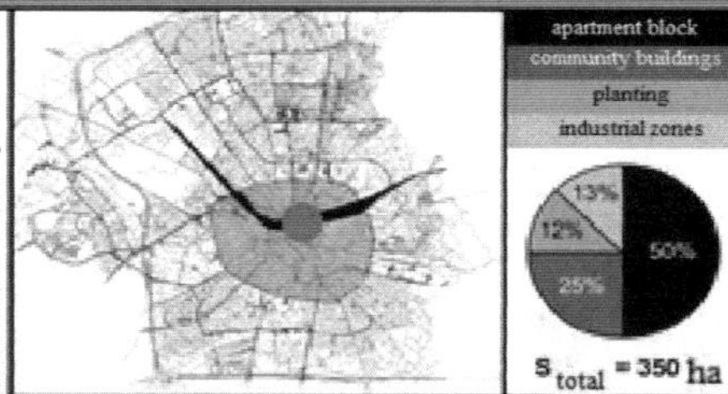

functional and planning scheme

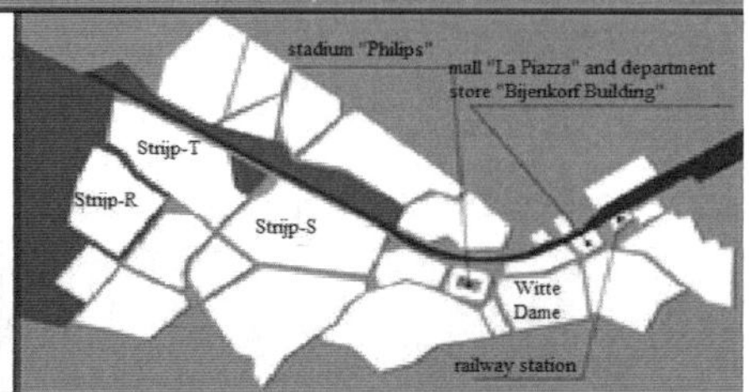

FUNCTIONS ESTABLISHED IN A DISTRICT NEAR A RAILWAY STATION historically (before XX century)	FUNCTIONS IMPLEMENTING IN THE DISTRICTS NEAR A RAILWAY STATION in contemporary times (XX-XXI centuries)	EXAMPLES OF CIVIL SITE DEVELOPMENT in contemporary times	COMPOSITIONAL AND PLANNING STRUCTURE OF A DISTRICT NEAR A RAILWAY STATION
historic apartment blocks	domestic buildings and hotels	apartment blocks of middle height	dispersed-nuclear
	sport objects	stadium "Philips"	
	science education objects	design academy, colleges	
electric light bulbs factory "Philips"	art and culture objects	workshops by Piet Hein Eek	
	trade enterprises	"La Piazza", "Bijenkorf Building"	
	managerial and business objects	bank departments	

CITY, COUNTRY — **Vienna, Austria**

scheme: district near a railway station in the structure of the city

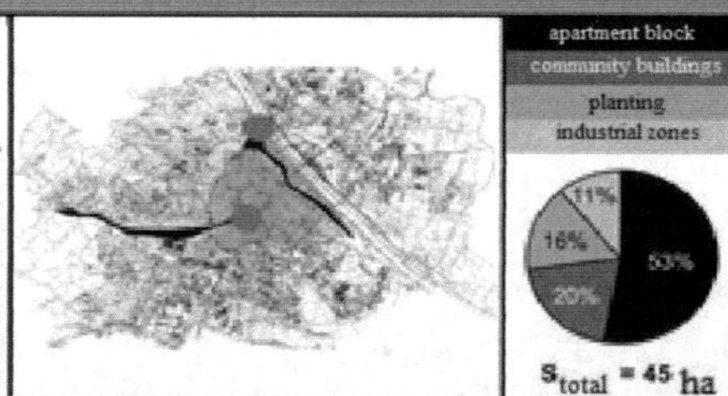

functional and planning scheme

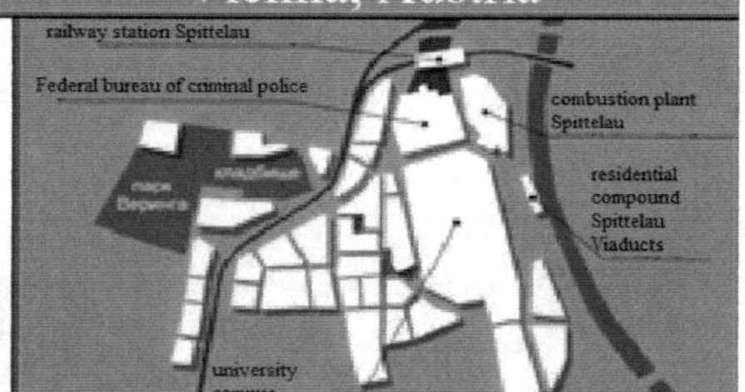

FUNCTIONS ESTABLISHED IN A DISTRICT NEAR A RAILWAY STATION historically (before XX century)	FUNCTIONS IMPLEMENTING IN THE DISTRICTS NEAR A RAILWAY STATION in contemporary times (XX-XXI centuries)	EXAMPLES OF CIVIL SITE DEVELOPMENT in contemporary times	COMPOSITIONAL AND PLANNING STRUCTURE OF A DISTRICT NEAR A RAILWAY STATION
historic apartment blocks	domestic buildings and hotels	residential compound Spittelau Viaducts	compact
	industrial buildings and housing	combustion plant	
Jewish cemetery	science education objects	university campus	
	managerial and business objects	criminal police bureau	

CITY, COUNTRY — **Oslo, Norway**

scheme: district near a railway station in the structure of the city

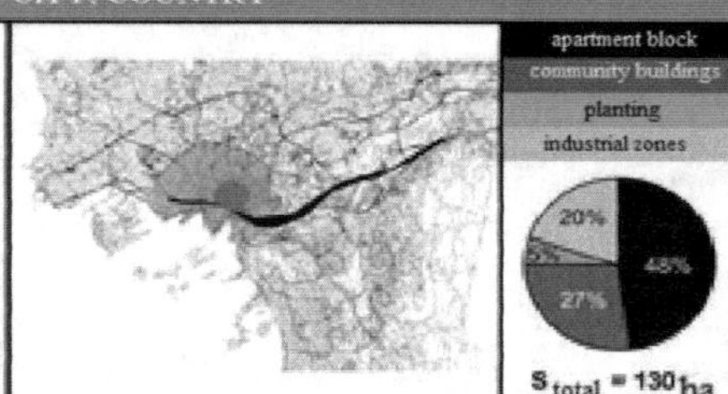

functional and planning scheme

FUNCTIONS ESTABLISHED IN A DISTRICT NEAR A RAILWAY STATION historically (before XX century)	FUNCTIONS IMPLEMENTING IN THE DISTRICTS NEAR A RAILWAY STATION in contemporary times (XX-XXI centuries)	EXAMPLES OF CIVIL SITE DEVELOPMENT in contemporary times	COMPOSITIONAL AND PLANNING STRUCTURE OF A DISTRICT NEAR A RAILWAY STATION
historic apartment blocks with the monuments of architecture and archeology	domestic buildings and hotels	Radisson Blu Plaza Hotel	compact, linear
St. Mary's Church, The Church of Saint Clement, St. Hallvard's Cathedral	sport objects	Oslo Spektrum	
	objects of cultureand art	opera theatre	
cemetery	managerial and business objects	Biskop Gunnerus Gate 14	

This theme continues to be topical a hundred years later at the beginning of the XXIst century. The researchers of city planning and architecture consider it in their scientific works. Mikhail Nikolaevich Kanunnikov reviews the problem of "site development and rational use of the territories adjacent to railway roads in large cities" [8] in his thesis for a candidate degree in architecture.

Moreover, many modern famous architect bureaus in their practical activity come across designing districts near a railway station. In this way in 2019 the project of architect bureau OMA on the reconstruction of the Belgian national railroad company's headquarters building was published [9]. The multistoried building on the very narrow plot along the railway road will include conference hall, classrooms, assembly hall and gym, restaurant, office and engineering equipment premises. In the end of 2019 Japanese architect bureau Nikken Sekkei was announced the winner in the competition in the designing of the architectural and city-planning concept of the site development on the territory of Riga goods yard in Moscow [10]. On the territory with the total area of more than 20 hectares the architects suggested to place large public complex accommodating the headquarters of the company "Russian Railways" and other office premises, sales premises, academic and medical organisations as well as housing accommodations open to the city. The complex is positioned by the authors as the key social and business object of this district which will allow developing further the adjacent city territories, attracting entrepreneurs, citizens and tourists there.

5. Conclusion

In such a way, districts near a railway station are important in the structure of the cities having appeared there on the cusp of the XXth and XXIst centuries and from that time having strategically important communicative functions.

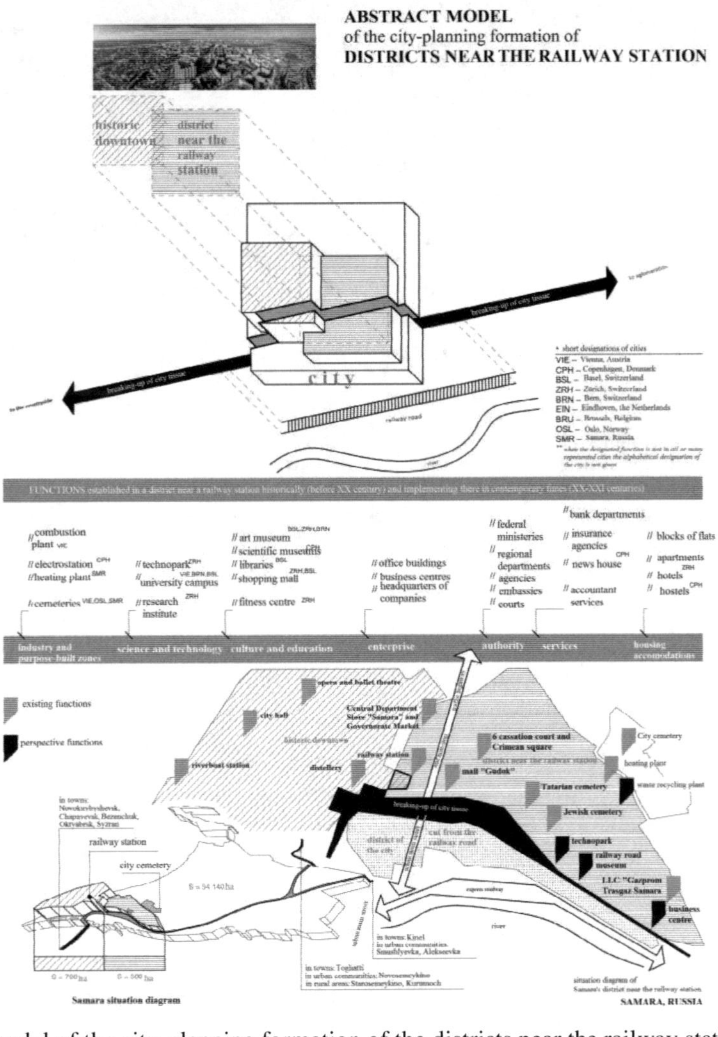

Figure 1. Abstract model of the city-planning formation of the districts near the railway station

IOP Conf. Series: Materials Science and Engineering **944** (2020) 012001 doi:10.1088/1757-899X/944/1/012001

The character of the site development of such districts has always had its peculiarities among which is chaotic site development without city development plans, the disbalance of functional programme and the lack of community buildings including the objects of social infrastructure, breaking-ups in the city tissue and the absence of transport links with the central districts of the city and others (Fig. 1). These and various other problems are being solved successfully in Russian cities including the ones with a population of 1 million or more such as Samara, the railway road in which appeared in 1875–1877 that cut the dwelling zone of the city from the river bank [11].

References

[1] Biktashev A I, Kolomina A I and Krasnobaev I V 2018 *News of the Kazan State University of Architecture and Engineering* **2(44)** pp 117–128
[2] Chaiko D S 2017 *Modern Construction and Architecture* **1(05)** pp 12–14
[3] Kublitskaya O V 2016 *The 72 students'science and technical conference* pp 32–34
[4] Federal law on January 10, 2003 #17-FZ *About the railway transport of the Russian Federation*
[5] SanRaN 2.2.1/2.1.1.1200-03 *Sanitary protective zones and sanitary classifiaction of enterprises, constructions and other objects*
[6] Lekareva N A 2015 City-planning Organization of Transit Territories Adjacent to the Railway Line (Publishing house Nauka obrazovaniya) pp 47–49
[7] Le Corbusie 2017 The New Spirit in Architecture (Moscow: Strelka Press)
[8] Kanunnikov M N 2002 *Multifunctional complexes in the territories adjacent to the railway in a contemporary city (as exemplified by Moscow)*
[9] OMA unveils "cliff-like" headquarters for SNCB in Brussels [Electronic resource] URL: https://www.dezeen.com/2020/02/25/oma-sncb-headquarters-architecture-offices-belgium (reference date: March 02, 2020)
[10] Project by Nikken Sekkei won in the competition on the project of Russian railway head-quarters in Riga railway station [Electronic resource] URL: https://archi.ru/news/85260/v-konkurse-na-proekt-shtab-kvartiry-rzhd-na-rizhskom-vokzale-pobedil-proekt-nikken-sekkei (reference date: January 21, 2020)
[11] Veretennikov D B 2015 *Vestnik SGASU. Town Planning and Architecture* **3(20)** pp 13–21

IOP Conf. Series: Materials Science and Engineering **944** (2020) 012002 doi:10.1088/1757-899X/944/1/012002

Ergonomic features and social demand for means of transportation for people with limited mobility

N V Bekk [1], M V Taube[1], I A Boychenko[1], M A Cherkasova[1]

[1] Kryachkov Novosibirsk State University of Architecture, Design and Arts, Novosibirsk, Russia

E-mail: 8dayofangel@mail.ru

Abstract. The article discusses the strategy and tasks of the development of means of transportation for people with limited mobility. The characteristics of people with limited mobility are reflected. The aims of the study: to reflect social issues for this group, to describe ergonomic requirements for the design of the facility, to give ergonomic diagrams and their rationale, to develop the concept of means of transportation. Possibilities of elements arrangement are shown.

1. Introduction

In recent years, the situation in our country has been developing in a positive direction for people who have any restrictions in moving through urban space, but the problem still exists. In 2019, according to the Federal state statistics service, the number of disabled people in Russia was 11 276 578 people, 183 417 of which live in the Novosibirsk region. These numbers may be higher since there are many persons who cannot pass through the morally and physically difficult process of obtaining or confirming disability.

Since 2011 the Accessible Environment program has been operating in Russia. The goal of this program is "... to create legal, economic and institutional conditions for the integration of persons with disabilities into society and to improve their quality of life" (from the Decree of the Government of the Russian Federation of March 29, 2019 No. 363 "On Approval of the State Program of the Russian Federation "Accessible Environment") [1]. In addition to the treatment and rehabilitation of a person, he/she needs adaptation in society for living a full life. The impossibility of contact with society due to physical ailments and misunderstanding, rejection on its part become the reasons for the isolation and frustration of a disabled person. Ensuring the mobility of a sedentary group of people among the disabled is the task underlying all these problems. Therefore, its solution is relevant and paramount [2].

One of the main factors in the design of a socially-oriented facility for people with disabilities is ergonomics.

2. Materials and methods

During the study, the following methods were used:

1. Empirical method (study of various sources of information, analysis of the data obtained, observation);
2. Qualitative method (focus group survey, expert interview, in-depth interview, focus group discussions);

3. Theoretical method (analysis, synthesis, analogy, classification).

The obtained from the research requirements for the designed object determined the initial ideas of the future shape of the object. The desired combination of a sporty look of a luxury car and the ergonomic requirements for the transportation of people with limited mobility, in particular wheelchair users, as the most complex spatial object, was determined.

3. Results

In order to take into account all the ergonomic requirements, a rectangle was taken as the basis for the shape. In the process of searching for the silhouette of a future car, those sections were cut off from the total volume that would not violate the special conditions of the designed object. For this reason, the main deformation occurs on the driver's side of the car.

After the first stage of sketching, the main options for the shape of the future car were presented. Each of them had their own characteristics, despite the fact that they were all made in the same style. The prototyping method revealed layout errors in several variants. If the maximum volume is saved from the optimum required, space appears that is not used in any way. When a strong lean at the rear of the car is added to the silhouette, there is not enough room to maneuver in the passenger area, and it also creates space in the very corner that is not used. Ergonomic errors are shown in Figure 1.

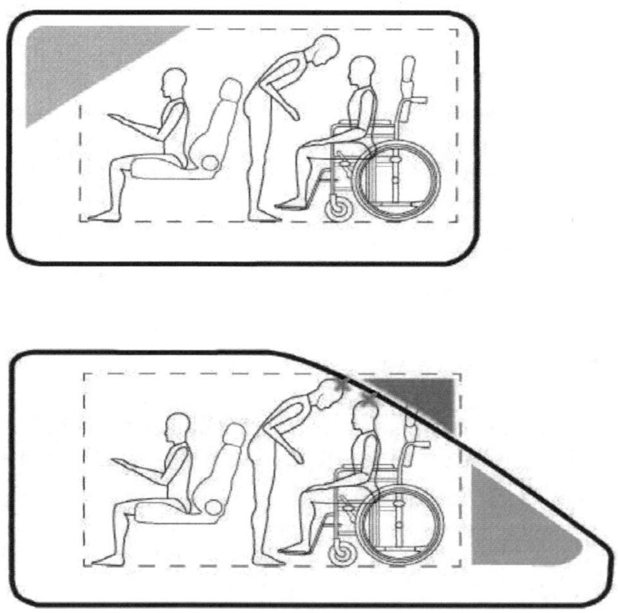

Figure 1. Ergonomic schemes. Examples of errors.

The result was an optimal ergonomic design: adding a slight slope at the front to make the most of the space and add streamlining to the car for aerodynamics as well as modern aesthetics.

The results obtained from the initial sketching, prototyping, modeling, and the errors identified made it possible to create the final 3D model of the social car, taking into account all the ergonomic features.

The car model was designed taking into account the results of the research work. In order to save money, the car has only one common entrance door, the dimensions correspond to the standard dimensions of any car of the "hatchback" type and the maximum used internal environment. The appearance is the result of a combination of the general theme of the project and the sporting nature. The last point is necessary in order to the topic of social problems to be interpreted by the population neutral or positively. Currently, the situation in Russia requires changing in minds of people by substituting concepts for the beloved and popular image, creating a positive image of the life of people with any restrictions, and ensuring their socialization.

IOP Conf. Series: Materials Science and Engineering **944** (2020) 012002 doi:10.1088/1757-899X/944/1/012002

The location of the internal components depends on the type of fuel used. It is assumed that the car is going to use electric power, but it is possible to use gas as fuel, all of these two options are much more environmentally friendly than gasoline. The selected engine type will not change the final shape. Figure 2 shows the layout of the power plant elements with electric traction.

The main internal components on which the ergonomic design depends are battery, inverter, differential, electric motor, front motor, air spring, damper, and electrically controlled rack and pinion steering.

The main dimensions of the transport were deliberately taken as minimal as possible since cost savings in this project are also important. But the data presented in the project can change during the production process, the concept is variable.

It is advisable to design from the salon to the outer body of the object. The silhouette of the car was made up of the simplest aerodynamic requirements (streamlining and tilt of the windshield) and the general composition of the object. During the design process, ergonomic diagrams and scenarios of interaction with the car were drawn up (Fig. 2).

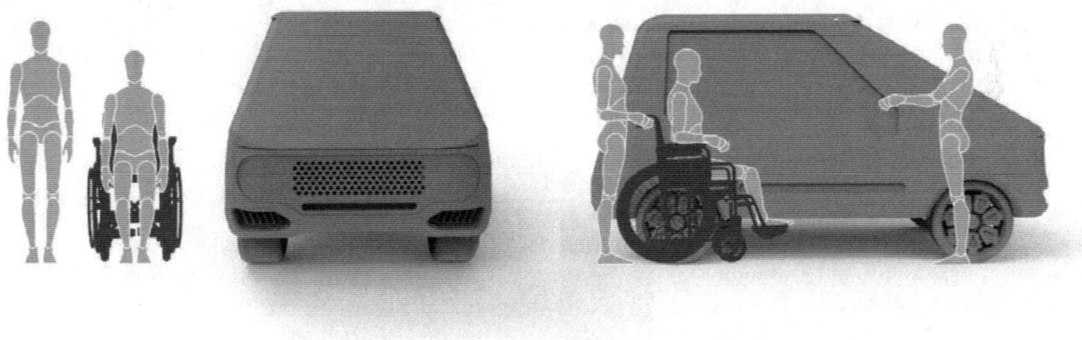

Figure 2. Ergonomic schemes. The ratio of the parameters of a person and a car.

The front door has been designed to be sliding on both sides to make it easy for passengers and the driver to enter the vehicle. A retractable lift has been designed for wheelchair users (Fig. 3).

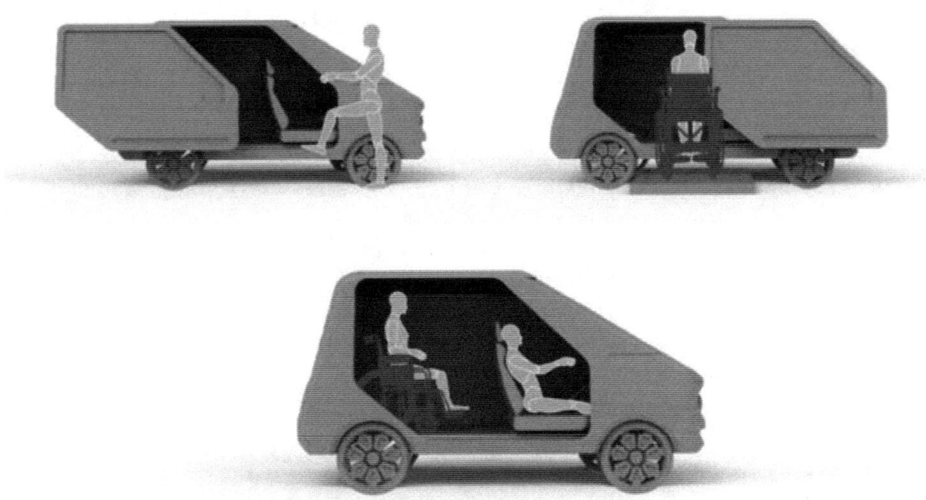

Figure 3. Ergonomic schemes: the location of the driver and passengers in the cabin.

The interior space is designed for one attendant, one wheelchair user, and a driver, or a driver and three passengers. This variability of placement within the vehicle allows the vehicle to be widely used as a "social taxi".

In the automotive industry, aluminum and steel are used for body and internal components. The interior uses leatherette materials or other wear-resistant fabrics. The rest of the inner and outer components of car bodies are made of plastic.

The car manufacturing process is divided into four main stages:

1. Pressing
2. Welding of automotive parts
3. Painting
4. Assembly

The manufacture of all components of the car depends on the material types and thickness, bends of the parts. In case the bends and/or the thickness of the part is small, pressing is used, the rest of the components are made using casting. In all cases, matrices are used. Matrix is a blank that forms the future shape of the product.

After the manufacture of the main parts, the welding stage follows. First of all, large and medium body parts are connected, then the interior space parts are welded and the gaps are filled. There are several types of welding, such as resistance welding, semi-automatic welding, arc welding, plasma welding, laser welding. Each type of welding has its own characteristics and is selected based on the technical goals, quality, and capabilities of the enterprise [3].

4. Discussion

Quality of life issues is constantly being considered in all spheres of society. The design of the human environment plays an important role in improving the quality of life. So far, the situation in our country is such that the design of the "accessible environment" remains on the sidelines [4]. And specialists working in the social sphere, as it were, "pull up", "adjust" the existing facilities to the needs of low-mobility groups of the population. However, the more correct, in our opinion, is the approach of full-fledged design of environmental objects and means of movement, taking into account modern aesthetics for people with limited mobility. The shared environment and the "accessible environment" must be one.

5. Conclusion

Recommended ergonomic schemes for vehicles for people with limited mobility and structural layout schemes were developed as a result of analytical research on the development of socialization of this group.

The analogs of all possible devices that increase the mobility of people with diseases of the musculoskeletal system were analyzed, a survey was conducted in groups to identify the needs in this area, expert opinion was taken into account. Both the main and indirect consumers of the product were identified. During the development, the technical, production, ergonomic and economic capabilities of the product were also identified.

An optimal ergonomic design will maximize the use of space and add streamlining to the vehicle for aerodynamics. The results obtained based on a 3D model make it possible to create cars taking into account all the ergonomic features. It was found that it is advisable to use only the bosom of the front door. The arrangement of the components is such that it is necessary to use gas and electric traction as fuel, depending on the possibilities.

In order to ensure a decent life for people with disabilities, it is necessary to develop concepts that make their life easier. One of the main factors in the design of a socially-oriented facility for people with disabilities is directly ergonomics.

IOP Conf. Series: Materials Science and Engineering **944** (2020) 012002 doi:10.1088/1757-899X/944/1/012002

The technical, production, ergonomic and economic capabilities of the product were identified during the development, as well.

Shortcomings which will inevitably be revealed for the first time of use are going to be corrected in the course of further implementation of the project. With an optimistic development option, it is planned to continue developing the product, but already with a larger capacity, and strategic decisions on the territorial distribution of the company will also be drawn up.

References

[1] Decree of the Government of the Russian Federation of March 29, 2019 No. 363 "On the approval of the state program of the Russian Federation" Accessible Environment " Passport of the State Program of the Russian Federation" Accessible Environment ". The purpose of the program. (in Russian)

[2] Stupakov N Malomobil'nyye gruppy naseleniya - eto kto? [People with limited mobility who are they?] [Electronic resource] URL: https://fb.ru/article/211299/malomobilnyie-gruppyi-naseleniya---eto-kto (in Russian)

[3] Lee J 2020 Apr 30. Frame Failures and Reframing Dialogues in the Public Sector Design Projects *International Journal of Design* [Online] 14:1 Available: http://www.ijdesign.org/index.php/IJDesign/article/view/3717

[4] Suib S, Van Engelen, J, and Crul M 2020 May 1. Enhancing Knowledge Exchange and Collaboration Between Craftspeople and Designers Using the Concept of Boundary Objects. *International Journal of Design* [Online] 14:1 Available: http://www.ijdesign.org/index.php/IJDesign/article/view/2876

IOP Conf. Series: Materials Science and Engineering **944** (2020) 012003 doi:10.1088/1757-899X/944/1/012003

Sustainability assessment of residential building architecture in terms of Nikos A. Salingaros's criteria

T Y Bystrova[1] **and O Y Rodnova**[1]

[1] Ural Federal University, 19, Mira ave., Yekaterinburg, 620002, Russia

E-mail: taby27@yandex.ru

Abstract. Modern 'sustainable' architecture requires clear criteria for the comprehensive assessment of its forms. An American theorist N. A. Salingaros deduced a number of laws governing the creation of a stable and adaptive architectural form that follows the structure of natural forms. These patterns serve as the basis for the authors of the article to derive criteria for evaluating architectural objects. The criteria are framed in the system, the testing of which was carried out in the analysis of residential complexes of the city of Yekaterinburg in 2010–2020.

1. Introduction

Globalization leads architecture to unification and arbitrary changeability of forms, loss of individuality and harmony [1]. Increasingly, users and critics forget about the status of architecture as an objective-spatial activity to create harmonious forms, reducing the building only to a set of engineering solutions. Attempts to strengthen sustainability are not always successful, since there are no clear-cut criteria for sustainable architecture. One of the leading ideologists of sustainable shaping is N. Salingaros (b. 1952), an apprentice of the prominent British architect Ch. Alexander [2, 3]. Their ideas for sustainable design are relevant and promising [4, 5, 6], but are not yet known enough [7]. They are not accepted by all architects and urbanists, who require even more specificity in the design [5]. The article defines and systematizes the criteria for sustainable architecture based on the approach of N. Salingaros, followed by a brief test in assessing modern residential architecture in Russia.

2. Materials and methods

The article develops the ideas of N. Salingaros presented in the book 'Algorithmic Sustainable Design: Twelve Lectures on Architecture' [8]. Here, in an appeal to the fundamental laws of the organization of matter and information, the principles of the formation of stable architecture close to the physical and mental arrangement of a man are substantiated. Not reducing scientific data, but presenting it to readers step by step, N. Salingaros offers mathematical algorithmic design solutions. He criticizes modernist and deconstructivist architecture, creating one-dimensional and inhumane forms [9]. The principles of creating sustainable adaptive architectural forms, including in relation to residential buildings, are given in the book rather briefly and were not developed by other authors. The structure of the book is quite complicated for direct design use. Thus, the first three parts are an intensive immersion in architectural mathematics and physics, which quite sharply turns into a general theoretical block, within which self-repetitions arise. Themes of universal scaling, references to the geometry of the living world, in particular, fractals, and a number of other fragments are uttered twice.

The most productive is the lack of unambiguous options for implementing the principles of organizing a stable form, the freedom of a specialist in choosing tools and techniques. Such a choice is dictated to a much greater extent by the features of the site and residents [8], turning architecture into a kind of 'membrane' between nature and civilization.

From the whole set of principles deduced by the architect, we will focus on the ideas of scaling and combinatorial complexity, for which it is possible to find a quantitative equivalent. Their clarity provides certain limitations in the design, helping the architect to get away from random or arbitrary ideas. By virtue of this, they can serve as criteria for assessing sustainable morphogenesis.

The empirical basis of the work is the material on 15 projects of residential complexes in Yekaterinburg (Russia), built in 2015–2020. Paradoxically, residential buildings are less often conceptualized from the standpoint of organic and sustainable architecture [10], but in the period after the pandemic we can expect an increase in interest in this approach in this particular segment.

3. Discussion

3.1. Ensuring the complexity of a form, according to N. Salingaros

According to N. Salingaros, we design by computing. Only calculations will bring architecture to sustainability. The first steps are mastering universal scaling, the Fibonacci series, the 'rule of 3' and other techniques that provide a holistic view of the projected object in the unity of all its elements and scales, from layout and facade to molding and window frames.

Why then are design algorithms needed? It is they that take a specialist away from the 'memory of a typology', the knowledge of which they received during their training. If an architect was taught 'modernist', then the 'memory of the typology' is not so easy to eradicate; a person makes certain design actions almost without accountability, without controlling themselves and, thereby, not allowing themselves to deviate from stereotypes. 'The algorithm allows us to be independent of memory, making us more creative.'

The living world and 'living' architecture have a complex multi-level device. N. Salingaros interprets their forms as information 'clots' arranged in a special way. For their existence and, especially, development, they need a certain complexity because the same units of information do not generate movement. This complexity assumes the existence of different levels of organization of the form – its different scales, symmetries, centers, self-similarity and other characteristics. A complex form is born step by step from simpler forms as a result of the action of various forces within it caused by these qualities. Based on this, in the first chapter of the work of N. Salingaros, five principles of the 'living' form, including the architectural one, are deduced (Table 1).

N. Salingaros offers design based on arithmetic sequences, for example, arithmetic recursion (Fibonacci series) to follow the scaling order. He explains the rule of increasing and decreasing scales, saying that the scale step is 1/3 or 30%. Another suggested sequence is exponential. When raising an exponent to some power, a series of numbers is formed that are used to create the form.

Universal scaling gives certain restrictions or rules for creating any form in architecture. The sizes of any element in architecture must be compared with the arithmetic sequence. Universal scaling involves a hierarchy of scales, where small and large scales create a single adaptive whole.

Biophilia as a general scientific term speaks of a person's innate desire to seek connections with wildlife [11]. Specifying it, N. Salingaros shows the connection at the genetic level of a man and the geometry of biological structures. He claims that human health depends on the surrounding shapes and geometries. The proximity of the architectural forms surrounding a man to biological structures makes it easy for them to process these forms, and therefore contributes to their well-being. It is important to understand that we are not talking about a primitive set of architectural elements, but about the concept of creating entire buildings and cities that obey the rules of biophilia. Ideas are already gaining followers [12, 13]. Combinatorial complexity is associated by N. Salingaros with the work of the nervous system and perception. He claims that the brain perceives information by combining it, combinations are perceived by a person more easily than monotonous repetition, which initiates comparison causing stress. Combinatorial complexity is determined by the number of identical parts, which are divided into groups that are convenient for perception.

Table 1. Organization principles providing scaling and combinatorial complexity in architecture (according to N. Salingaros).

№	*Principle*		*Description*
1	Recursion and Fibonacci sequence	a) Algorithmic design.	As a method of selecting the right decisions.
		b) The laws of morphogenesis.	Generating a form based on genetic information.
		c) Sequences (arithmetic and exponential).	Adapted Fibonacci sequence: 1, 3, 8, 21, 55, 144, 377, 987, 2584, … 1/3 or 30 % scale step.
2	Universal scaling	Consists in a hierarchy of scales, where small and large scales create a single form.	
3	Biophilia	Taking into account the relationship of man and the geometry of biological structures at the genetic level. Architecture directly affects the human condition.	
4	Scaling and division	Parallels the development of the embryo, which through the division of the cells that form the group, turns into a mature person.	
5	Combinatorial complexity	The brain perceives information, combining it, the combinations are perceived by a person easier than monotonous repetition, which causes stress.	

The principles given are not something new and unique; they have long been used by architects. It seems that today they are more relevant, both for external (urbanization, ecology, pandemic) and internal (lack of a ‘big style’, erosion of an aesthetic scale) reasons. An example of the implementation of the principle of universal scaling in modern conditions is the building at 1399 Park Avenue in New York, USA, from the architectural bureau Hill West Architects (Figure 1). The division of the facades and the difference in volumes create an alternative to standard rectangular high-rise buildings dating back to modernism. However, there is also a discrepancy with the principles of N. Salingaros, for example, monotony in the rhythm of windows and almost complete absence of small scales.

Figure 1. 1399 Park Avenue. New York, USA. 2020. Arch. Hill West Architects. Source: https://www.archdaily.com/.

Figure 2. ‘A House at Sea’. Bergeggi, Italy. 2019. Arch. Studio Daniele et al. Source: https://www.archdaily.com/.

IOP Conf. Series: Materials Science and Engineering **944** (2020) 012003 doi:10.1088/1757-899X/944/1/012003

'A House at Sea' (Italy) complies with the principles of sustainability in terms of scale and combinatorial complexity (Figure 2). The windows are designed so that they form groups by means of non-monotonous repetition. This building has a drawback – it lacks small-scale details.

3.2. Scaling and combinatorial complexity in the system of criteria for assessing architecture
We believe that the use of sequences is not always possible due to the design and technical requirements for architecture. Compliance with numerical sequences is difficult to bring into line with the existing standards. This is of great importance directly in the development of project documentation and in the construction process.

Biophilia and men's dependence on the geometry of biological structures are intuitively clear, but they cannot be translated into quantitative equivalents, which means that they cannot become parameters for assessing architectural objects according to the criteria for the stability of their shape. Such a 'translation' is possible for the following principles, deduced by N. Salingaros.

3.2.1. Scaling hierarchy (a system of a combination of elements of different scales). This explains the proportions of all elements of an object relative to each other. According to the laws of the physical world, a stable form requires a minimum of three levels of scale. N. Salingaros makes a projection this rule on architecture.

3.2.2. The formation of small forms by dividing the large ones using a universal scale step, approximately equal to 1/3. It can be used in dividing the total volume of the building, including when designing residential complexes. We presented the process of universal scaling in Table 2.

Table 2. The process of universal scaling.

№	*Description, according to N. Salingaros*	*Visualization*
1	Initial choice of form (search for the optimal template for further development): vertical / horizontal direction of construction	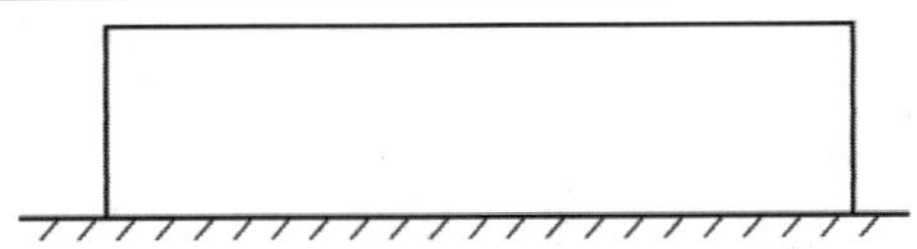
2	Conditional form division into modules	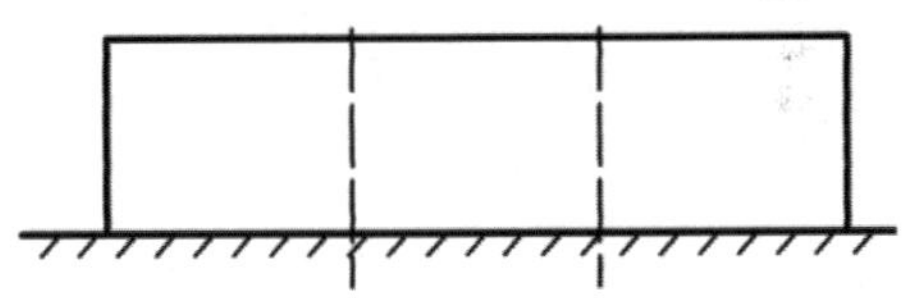
3	Selection of the defining element from which further scaling will be performed	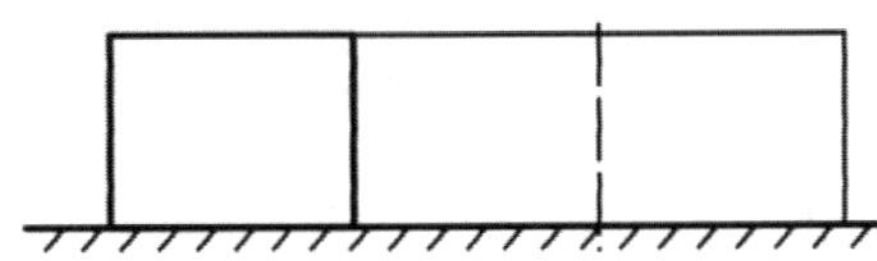
4	Correction of the selected form element: decrease / increase	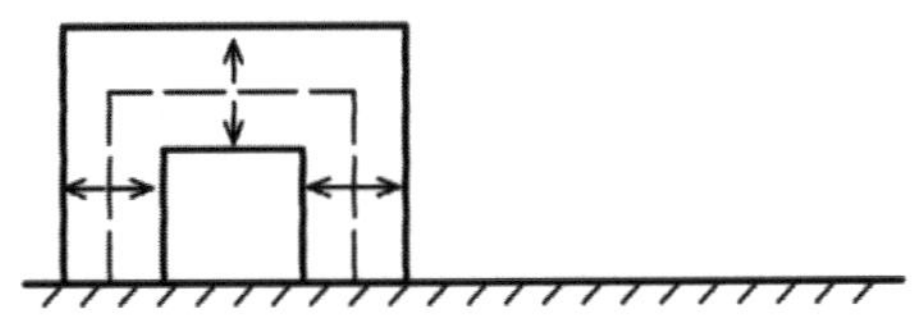
5	Selection of the defining element from which further scaling will be carried out	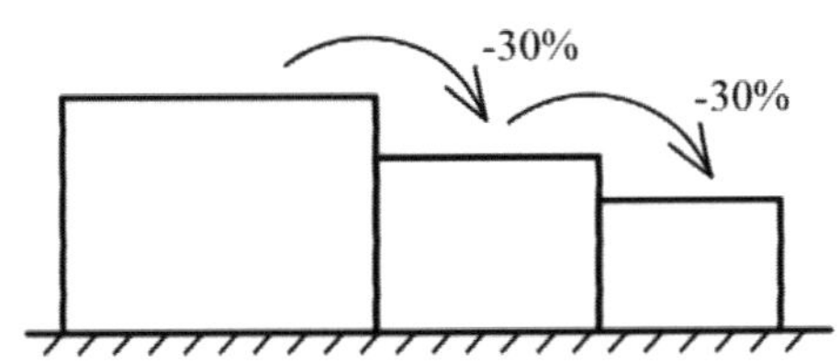

IOP Conf. Series: Materials Science and Engineering **944** (2020) 012003 doi:10.1088/1757-899X/944/1/012003

3.2.3. Small-scale grouping and medium-sized formation. Elements of the building are grouped at all levels of scale, thereby forming a larger scale. E.g.: windows are grouped vertically, then they are grouped horizontally. This is the movement from the window (small scale) to the whole building (largest scale) (Figure 3).

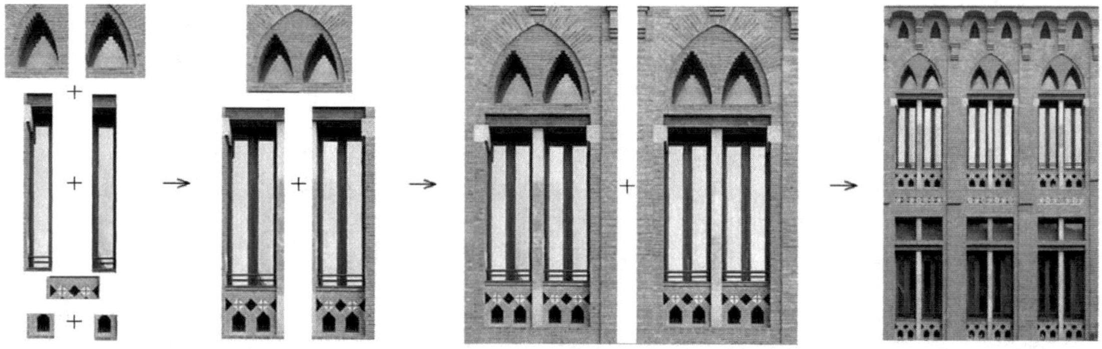

Figure 3. Maternity home. Arch. C. Oliveras, G. Guitart et al. Barcelona, Spain. 1883. Source: https://www.flickr.com/.

3.2.4. Variability on a small scale. Violation of monotony on a small scale can create textures, ornaments. The right choice of finishing materials significantly increases the architectural stability of the entire building. The more heterogeneous and textured is the lining, the more rigid is the object.

3.2.5. Refusal of monotonous repetition. On a medium scale, the rejection of monotony can be manifested through groupings, and on a small scale, it can be done by the variability of elements of the same type. This point is unifying for the previous two. It characterizes the synthesis of the two criteria.

4. Results

Based on the analysis of the principles for creating a sustainable architectural form, Table 3 is compiled. This is a system for assessing the sustainability of an architectural form in the aspect of scaling. It involves the criterion assessment of two points or four subpoints. The corresponding points are awarded for the compliance of the architectural object with a particular item.

15 residential complexes built in Yekaterinburg over the past five years have been examined (Figure 4). The analysis showed that only in 20% of residential complexes the forms correspond to the highest stability rating, another 20% have an average score, and the majority (60%) have a low rating. Figure 8 shows part of the evaluated buildings, which are divided into groups according to the assessment of their forms.

Table 4 presents the results of the evaluation of architectural forms in the projects of residential buildings in the city of Yekaterinburg, built in 2015-2020. In general, architects' attempts to get away from monotony are noticeable – if only because it is not remembered by accommodation buyers. Moreover, scaling more often has a superficial, decorative character and does not lead to a fundamental departure from modernist shaping. There is no dynamics over the years (the movement from less stable forms to more stable). There is a priority of economic considerations of developers and architects to architectural shaping.

Table 3. A system for assessing an object according to the criteria of a sustainable architectural form.

Group of principles	*Criterion*	*Compliance points*	*Note*
1 Scaling	1.1 Scaling hierarchy	2	Presence of **three** levels of scale: large, medium, small
		1	Presence of **two** levels of scale: large and medium or large and small
		0	Presence of **one** level of scale: large
	Total (max)	*2*	*Choosing one option*
	1.2 Small scale grouping	2	**Medium and small** scales are grouped
		1	Only **medium** scales are grouped
		1	Only **small** scales are grouped
		0	Scales are not grouped
	Total (max)	*2*	*Choosing one option*
2 Combinatorial complexity	2.1 Lack of monotony	1	Violation of monotonous repetition
		0	Presence of monotonous repetition
	Total (max)	*1*	*Choosing one option*
	2.2 Small scale variation	2	Small scale variation (**finishes and ornaments**)
		1	Small scale variation (**finishes**)
		1	Small scale variation (**ornaments**)
		0	Lack of variation
	Total (max)	*2*	*Choosing one option*
	Total (max)	***7***	

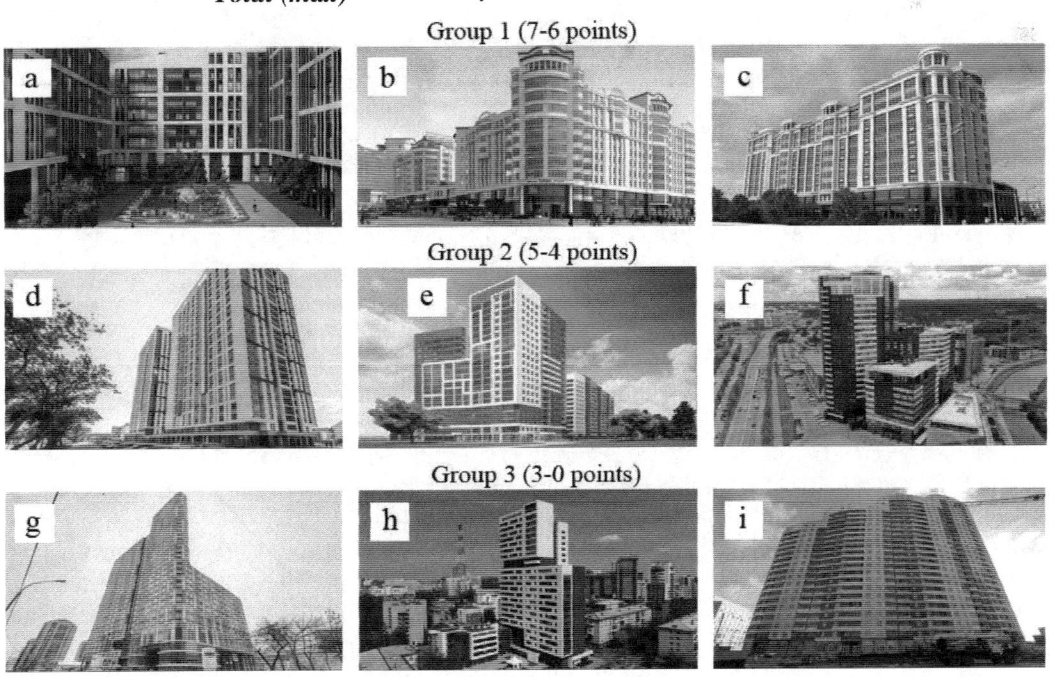

Figure 4. Residential complexes in Yekaterinburg. 2015–2020.

Table 4. A system for assessing an object according to the criteria of a sustainable architectural form.

Label	*Name of residential complex*	*Year of construction*	*Criteria*				*Total points*	*Group*
			1.1	*1.2*	*2.1*	*2.2*		
a	RC «Armada»	2019	2	2	1	2	7	1
b	RC «Club House Tikhvin»	2015	2	2	1	1	6	
c	RC «Club House Riviera»	2019	2	2	1	1	6	
d	RC «Moskovsky Kvartal»	2020	1	2	1	0	5	2
e	RC «Yantarnaya Dolina»	2020	2	1	1	0	4	
f	RC «Clever Park»	2019	2	2	0	0	4	
g	RC «Trinity»	2018	0	0	0	1	1	3
h	RC «Sedmoye nebo»	2017	1	1	1	0	3	
i	RC «Repin park»	2019	1	1	1	0	3	

5. Conclusion

The study confirmed that the principles deduced by N. Salingaros can be used as parameters of the stability of buildings and objects. Practical testing showed that these parameters can be systematized and applied to the scale of entire cities and countries. However, the derived criteria for scaling and combinatorial complexity relate only to a part of the characteristics of a sustainable architecture, so in the future it is necessary to refine the system for assessing the stability of architectural forms.

6. Acknowledgements

This work was carried out according to the FNI plan for 2020 with the support of the RAASC and the Ministry of Construction of Russia in accordance with the State Programme of the Russian Federation 'Development of Science and Technology' for 2013–2020, and the Programme of Fundamental Scientific Research of State Academies of Sciences for 2013–2020.

References

[1] Vlasov V G 2017 *Teoriya formoobrazovaniya v izobrazitel'nom iskusstve [Theory of shaping in the visual arts]* Uchebnik dlya vuzov (Saint-Petersburg: Publ. S-Peterb. un-ta) p 264
[2] Alexander Ch 1973 *Notes on the Synthesis of Form* (Harvard University Press) p 216
[3] Mehaffy M 2007 Notes on the genesis of wholes: Christopher Alexander and his continuing influence *Urban Des Int.* **12(1)** pp 41–49
[4] Roös P and Jones D 2017 Knowledge of Making Life: Design Pattern for Regenerative-Adaptive Design *KnE Engineering* **2(2)** p 203
[5] Jones B and Wong Y 2008 Virtual and physical: architect Christopher Alexander on living spaces (Melbourne, Australia) *Proceedings of VALA Libraries, Technology and the Future. 2008 Conf.*
[6] Lea D 1994 Christopher Alexander: An Introduction for Object-Oriented Designers *ACM SIGSOFT Software Engineering Notes* **19(1)** pp 39–46.
[7] Salama A M 2007 Book Review: Nikos A. Salingaros: A New Vitruvius for 21st Century Architecture and Urbanism *Archnet-IJAR-International Journal of Architectural Research* **1(2)** 114–131
[8] Salingaros N 2019 *Algoritmy ustojchivogo proektirovaniya. Dvenadcat' lekcij ob arhitekture [Algorithmic Sustainable Design: Twelve lectures on architecture]* ed T Bystrova (Moscow–Ekaterinburg: Kabinetnyj uchenyj) p 272

[9] Salingaros N 2017 *Anti-arhitektura i dekonstrukciya. Triumf nigilizma [Anti-architecture and Deconstruction: The Triumph of Nihilism]* ed T Bystrova, E Duylovskaya and A Makarova (Moscow–Ekaterinburg: Kabinetnyj uchenyj) p 296
[10] Bystrova T 2019 Concept of Organic Architecture in the Second Half of the XXth Century in the Context of Sustainable Development *IOP Conference Series: Materials Science and Engineering* **481** p 012020
[11] Fromm E 2010 *Dusha cheloveka, eyo sposobnost' k dobru i zlu [The Heart of Man: its genius for good and evil]* ed V A Zaksa (Moscow: AST, Astrel') p 256
[12] Downton P, Jones D and Zeunert J 2016 Biophilia in Urban Design: Patterns and principles for smart Australian cities *International Urban Design Conference* (Canberra) pp 168–183
[13] Amirova N and Stasinopoulos T N 2017 Faces of Biophilia in Contemporary Turkish Architecture *Living and Sustainability: An Environmental Critique of Design and Building Practices, Locally and Globally. AMPS Proceedings* **9** pp 115–127

IOP Conf. Series: Materials Science and Engineering **944** (2020) 012004 doi:10.1088/1757-899X/944/1/012004

Pictorial techniques of Russia and China in environmental design

Chang Xiaogeng[1]

[1]College of Humanities and Social Sciences, Beijing Institute of Petrochemical Technology, 19 Qingyuan North Road, Daxing District, Beijing, 102617, People's Republic of China; Department of Art History, Saint Petersburg Stieglitz State Academy of Art and Design, 13 Solyanoy Lane, Saint Petersburg, 191028, Russian Federation

E-mail: changxiaogeng@bipt.edu.cn

Abstract. Environmental design in Russia and China in academic terms refers to the field of education in the areas of art and culture, in methodological terms it is based on the methods and principles of fine art. The article presents a creative project of the photo exhibition "Dance of Leaves", which was created by the author using techniques of Chinese traditional painting and Russian abstract art, in particular the ideas of V. V. Kandinsky. The project applied the method of integration of disciplines, for example, color science, cultural studies, engineering etc. A special feature of the photo exhibition was the presence of various lighting technologies, as well as the game of colors and lines with symbolic cultural connotation.
Keywords. Environmental design, photo exhibition, lighting technology, color symbolism, abstract art, Chinese painting.

1. Introduction

Environmental design as an academic discipline from the perspective of classification and methodology has a close relationship with fine arts. According to the scheme on the Figure 1, based on the materials of the Russian Classification on Academic Specialties (OKSO) [1] and a similar Catalog and Content of Undergraduate Educational Programs in the universities of China [2], the specialty "Environmental Design" refers to the "Art and culture" field of education, and to the enlarged groups of "Fine and applied arts" in Russia and "Design" in China.

The study program of the specialty "Environmental Design" is focused on the formation of the "imagery" of thinking and the "artistic" aesthetic view of students, since the key disciplines, in addition to the design itself, are drawing, painting, composition. This characteristic fundamentally distinguishes "Environmental Design" from the related disciplines such as "Design of the architectural environment" and "Landscape architecture".

In methodological terms, environmental design is based on fine art. The content of environmental design consists in the study of the "image". The concept of "image" arose within the framework of fine art. The objective of fine art and environmental design is to coordinate relations. The designer of the environment, like the artist, pays great attention to the process of creativity, which indicates the "processuality" of the objective of coordinating relations between the environment and a person [3].

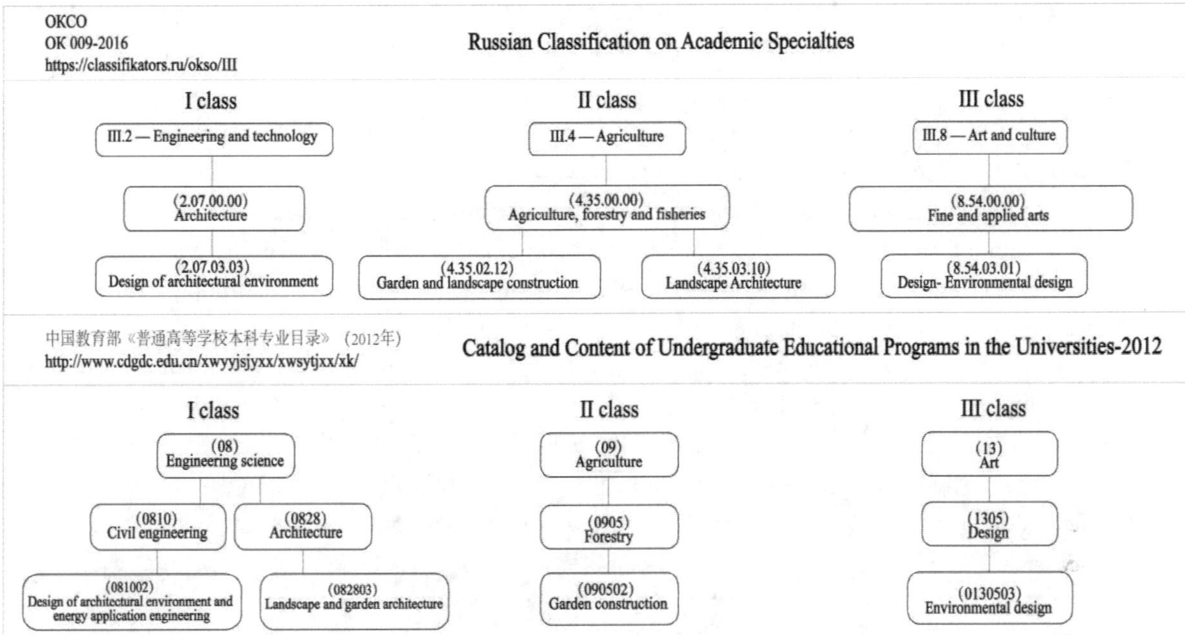

Figure 1. Classification of the specialty "Environmental Design" in Russia and China.

2. The project "Dance of Leaves": materials, methods, technologies

During the creation of this project, the author uses the method of integration of the following disciplines: cultural studies, color science, engineering, ergonomics, and sociology.

The project "Dance of Leaves" is an exhibition of a series of photographs called "Food" taken by a Chinese photographer Zhang Da (张达). The natural theme was a key focus of the photo exhibition: exhibition structures were made in the shape of trees, branches, leaves and tree burrs (Figure 2) [4].

The entire exhibition space was divided into several parts, according to the four seasons of the year. Each zone was dominated by the characteristic colors of structural elements and photos: thus, "Spring" zone was colored with white and cool green; "Summer" - warm tones of orange and green; "Autumn" - dark red; "Winter" - black and white. At the same time, the main color of the walls of the exhibition space was white. Black and white colors were used when creating the frames of structures.

Most of the exhibition stands were made of an electrically conductive composite material based on carbon fiber. It is light in weight, easy to use, and flexible enough to create the shape of exhibition structures that follow the curves of wood.

The author used several forms of stands in the project. The first type was the floor stands with "growing up" branches of "trees". The second type was stands consisted of hanging "falling" branches of "trees". Photos and colored rag items in the shape of leaves were placed on these stands. Each "leaf" was attached to the "branch" with a metal chain; if desired, it could be removed from the "tree". The third type of stands is called "burrs" in compliance with the wood theme. Zhang Da's film photos were inserted to the "burrs", and could be viewed through the viewfinder with "zoom in" function. All stands met the requirements of ergonomics, were made in accordance with the average parameters of human height, which provided convenience in viewing photographs and made the exhibition areas quite passable.

One of the features of the exhibition space was working with light, which provided the necessary effect of communication [5] between the designer and visitors. The interactivity of lighting technologies was important for the author to create a sense of the "natural" environment of the exhibition. The author used the following sets for working with light:

- light bulbs with motion sensors: each of the stands was equipped with them;
- the system of reed switches: they were placed inside the "branches" of hanging and floor stands in the shape of trees; when removing the "leaf", the lights located on the stands lit up;
- the system of image projectors (two HD high-resolution projectors, 1920 × 1080P) on the ceiling structure of the exhibition space: the beam of the projector was directed to the floor and showed photographs of the series called "Food";
- infrared synchronizers used for flash: they were located in the "Autumn" zone; this lighting technology was needed to demonstrate the culmination of the "natural" cycle;
- viewfinder system with triple magnification: it was necessary for viewing film photos of Zhang Da.

Figure 2. Elements of the "Dance of Leaves" project.

3. The influence of the ideas of Russian abstract art and traditional Chinese painting on the creation of the project "Dance of Leaves".

Since the 20th century, Western art has been deviating from realism, as an example, that was the time when abstract art appeared. The concepts of a "point", "line" and "plane" became symbols of painting in the European region and were closely associated with the name of V. V. Kandinsky, whose ideas made a huge contribution to modern painting and design.

The materials used in Chinese traditional painting, differed from those in the West, and early creative ideas were strongly influenced by the concepts of Confucianism, Buddhism and Taoism that have shaped the traditional style of *wenrenhua* ("文人画", "literati painting"), which was expressed in the phrase: "Convey the spirit with images" ("以形写神", yi xing xie shen). Undoubtedly, the use of

IOP Conf. Series: Materials Science and Engineering **944** (2020) 012004 doi:10.1088/1757-899X/944/1/012004

points and lines still has an indispensable place in Chinese painting art.

The author selected the results of studying the symbolism of points and lines in Russian and Chinese painting art, which became a kind of theoretical guide when creating the exhibition "Dance of Leaves", in particular, forms of exhibition modules in the shape of trees and leaves. The use of points and lines is a kind of technique for abstract artistic expression, and the most direct and striking example of it is written symbols. In both Russian and Chinese languages, during the writing process there should be pauses, change of signs' form and thickness to display the signs' symbolism, their content and hidden context. In "the hyper-industrial society" [6] due to changes in writing tools and standardization of fonts, people gradually stop paying attention to the thickness of characters and their "rhythm". The author would like to convey this characteristic feature of today's society, so he did not concentrate on the thickness of the exhibition modules, but sought to show the direction of the growth of "trees" so that they looked like real ones.

According to V. V. Kandinsky, "the geometric point is the ultimate and most singular union of silence and speech" [7]. In the project "Dance of Leaves", the "point" represents the embodiment of the "pause" on the branch line. These "pauses" became interactive positions where, for example, a reed switch was placed and film photographs were presented. "Pause" is what attracts attention.

"A form which, when on the otherwise empty basic plane, may still be considered to be a point, must be termed a plane when, for example, a very thin line appears with it upon the basic plane" [8]. In the exhibition modules of the project "Dance of Leaves", "pause" is a "point" that hides a surprise for visitors when bulbs are lighting, as well as a plane on which the "branches" with photos are placed.

Using these "pauses", the artist and designer can control the behavior and visual direction of the "readers" of the artwork (in this particular case, they are visitors of the exhibition), and contribute to the appearance of some thoughts of the "readers". In this way, the artist builds relationships between the artwork and a "reader".

According to V. V. Kandinsky, "the concept, element, can be understood in two different ways: as an external, and, as an inner concept. Externally, each individual graphic or pictorial form is an element. Inwardly, it is not this form itself but, rather, the tension within it, which constitutes the element. In fact, no materializing of external forms expresses the content of a work of painting but, rather, the forces - tensions which are alive within it" [9]. The creative photo exhibition "Dance of Leaves" is accompanied by the "elements" of lighting. The purpose of these "elements" was to "awake" the minds of visitors to provoke their certain associations. According to the author, associations are manifestations of the internal tension of art forms. In other words, the exhibition "Dance of Leaves" is an opportunity to influence the mood and thoughts of visitors through interactive methods of working with the light. It is also a kind of experiment that explores the degree of tension in a design work created with use of artistic language.

As V. V. Kandinsky said, the artist should take into account the relationship between the point and the plane, as well as the relationship between the point and other forms on the plane, and as the designer of the exhibition "Dance of Leaves", the author should pay particular attention to the relation of lines and color forms in space. It is well-known that V. V. Kandinsky formulated the theory of color, according to which color affects the consciousness, it is able to generate certain feelings that are not related to visual perception [10]. As for color, in the exhibition "Dance of Leaves" the author applied

IOP Conf. Series: Materials Science and Engineering **944** (2020) 012004 doi:10.1088/1757-899X/944/1/012004

the method of dividing the exhibition areas according to the four seasons, and each area was prevailed by the specific colors. That method made it possible to influence the feelings, emotions and consciousness of visitors.

In addition to the specific use of color in the exhibition "Dance of Leaves", the author sought to combine some art techniques of Russian and Chinese painting when creating the project. From the standpoint of research on the artistic depiction of lines, the author believes that Russian abstract art mainly focuses on the direction and tension of lines in drawing, while the basis of Chinese painting is the frequency and intersection of lines. These differences do not contradict, but can complement each other.

In Chinese painting, it is believed that "thread is bone, ink is flesh" ("线为骨，墨为肉" , xian wei gu, mo wei rou). Therefore, there should be used a soft and long brush to write or draw the lines that have internal power. It is not only a test of the painter's art, but also a kind of embodiment of the truths of Confucianism that tell us the following phrase: "A nobleman should be gentle in his treatment, but firm in his decisions" ("君子外柔内刚", junzi wai rou nei gang), and Taoism that can be demonstrated in the next passage: "Yin and Yang exist inside everything" ("万物阴阳相济" , wanwu yin yang xiangji). Therefore, each line of the structures of the exhibition "Dance of Leaves" had to combine the hardness with the softness of the embodiment. This is one of the important reasons why a composite material based on carbon fiber was chosen to create structures: it provides beauty and power of visual perception.

Since ancient times, the art of traditional Chinese painting has been rooted in the principle that says: before creating an artwork, it is necessary to work out its concept in detail. Therefore, the special state of mind called "enlightenment in silence" ("静思顿悟", jingsi dunwu) is so important in traditional Chinese painting, and this idea has been preserved until nowadays. Thus, due to the fact that the cultural environment in the development of Russian and Chinese painting traditions is not the same, there are differences in types of painting, where the use of lines is also not exceptional. The basis of Chinese painting is the pursuit of true knowledge, which is most often realized through the black and white lines. The key of Russian painting is the search for the expression of human emotions, which is usually embodied in a combination of color and lines.

Moreover, the technique of "scattered perspective" is mainly used since the very beginning of Chinese painting tradition [11] (in Chinese painting, there are no rules of perspective, similar to the Western one, in which perspective ensures that the entire space is filled with a drawing). This aesthetic taste was influenced by Taoist ideas about "emptiness of material world" ("虚空", xukong). In the 40th Chapter of the Lao Tzu's work "*Tao Te Ching*" (道德经，Daode jing) it is said: "All things in the cosmos arise from being. Being arises from non-being" ("天下下生生有，有生生") [12]. Only by making the mind and body empty, one can approach the enlightenment. In other words, the focus of Chinese painting tradition is mainly laid on the emptiness. The presence of empty spaces helps and allows the "reader" of the artwork to further imagine the picture and think of its story by him/herself. Combining the ideas mentioned above, the author chose black and white lines as the main colors prevailing in the exhibition space in accordance with Chinese style, which ensured the existence of empty spaces in the composition.

At the same time, the author while creating the tree structures, used the painting method called

IOP Conf. Series: Materials Science and Engineering **944** (2020) 012004 doi:10.1088/1757-899X/944/1/012004

"crab claw" ("蟹爪画法", xiezhua huafa), which is based on the image of lines in their frequency or cross lines (Figure 3). The origins of this method can be traced in the paintings of the album "Mustard Seed Garden Manual of Painting" ("芥子园画撰", Jieziyuan huapu). This album is very famous in China; it contains pictures of various dynasties. Until nowadays artists still use the painting methods that were previously used in the pictures of this album.

Figure 3. Painting method "crab claw".

A reference to the ideas of abstract art can be the fact that the author used a certain length of tree branches depending on the conditions of space, thereby forming the orientation of the shape of exhibition structures. By combining the colors of photos, the shape of exhibition structures, and color classification, the author sought to create the internal tension of exhibition objects.

The direction of the "branches" in the structures was different: in hanging stands, there was the effect of "falling" branches, in floor stands there was the effect of "growing up" branches. The "falling" branches had no trunks, and it seemed as if they were suspended in the emptiness. The predominance of "falling" branches in the "Spring" and "Winter" zones reflects a state of calm in the emptiness. In addition, in the "Winter" zone, photos in black and white were attached to hanging stands, which reflected the idea of "falling asleep" of nature. In the "Summer" and "Autumn" zones, on the contrary, there was a predominance of the floor stands in the shape of trees with growing up "branches", which symbolized the "splendor" of nature. Through the use of color photographs and small items in the shape of green and red leaves, the author sought to reflect the internal tension of the exhibition works in the "Summer" and "Autumn" zones.

4. Conclusion

Light technologies, a game with a color palette, - all this created a sense of interactivity of the exhibition "Dance of Leaves" and attracted the attention of visitors. Each exhibition structure met the requirements of space ergonomics. The natural theme and symbolism of color and lines, which are typical for Chinese and Russian painting traditions made the project aesthetically attractive.

The author believes that there is no fundamental difference between Russian and Chinese painting. The study and application of points and lines in the art of Russia and China have their own characteristics. The cultures of both countries have given the world deep and memorable masterpieces. They have become the creative basis for the modern development of environmental design, which in methodological and academic terms reveals deep relations with art.

IOP Conf. Series: Materials Science and Engineering **944** (2020) 012004 doi:10.1088/1757-899X/944/1/012004

The author intended to ensure that the project of the photo exhibition "Dance of Leaves" from theoretical and practical angles met the current interdisciplinary approach, had an experimental character, combined development in the art of the West and the East, and at the same time preserved cultural continuity.

References

[1] *Russian Classification on Academic Specialties 8.54.02.01 – Design (divided into groups)* Available from: https://classifikators.ru/okso/8.54.02.01

[2] *Catalog and Content of Undergraduate Educational Programs in the Universities* China Academic Degrees & Graduate Education Information Available from: http://www.cdgdc.edu.cn/xwyyjsjyxx/xwsytjxx/xk/xkzyml/276559.shtml

[3] Chang X 2019 Theoretical Boundaries of Environmental Design. *Phil. and Cult.* **11** 10

[4] Chang X 2019 Project «Dance of leaves» - an example of organizing an exhibition space within the framework of environmental design. *Man and Cult.* **5** 131–45

[5] Lasswell H D 1948 The structure and function of communication in society *The Communication of Ideas* ed L Bryson (New York: Harper and Row) pp 37–51

[6] Veltz P 2018 *Hyper-industrial Society, the New Productive Capitalism* (Taiwan: Lianpu Shufang) p 176

[7] Kandinsky W W 1947 *Point and Line to Plane* (Michigan: Cranbrook press) p 25

[8] Kandinsky W W 1947 *Point and Line to Plane* (Michigan: Cranbrook press) p 29

[9] Kandinsky W W 1947 *Point and Line to Plane* (Michigan: Cranbrook press) p 33

[10] Kandinsky W W 1946 *Concerning the Spiritual in Art* (New York: Solomon Guggenheim Foundation) pp 44-78

[11] Kovalevsky Y E 2008 Space in Chinese landscape painting. *World of Sci, Cult, Edu.* **5(12)** 138–40

[12] *Daode Jing* Translated by Muller A C Available from: http://www.acmuller.net/con-dao/daodejing.html

IOP Conf. Series: Materials Science and Engineering **944** (2020) 012005 doi:10.1088/1757-899X/944/1/012005

Reconstruction of cultural and leisure buildings of the Soviet period

I N Chechel [1], **I D Mikhailova**[1], **I P Chechel**[1]

[1] Belgorod State Technological University named after V.G. Shoukhov, 46, Kostyukov str., Belgorod, 308012, Russia

E-mail: Golden-Line7@yandex.ru, irinakitaeza@gmail.com

Abstract. In Russia there is an impressive fund of Soviet heritage objects of cultural and leisure activities of population. The most massive construction of club establishments was in the 1960s-70s in the USSR. Today the cultural houses in Russia are undergoing dramatic changes. Not all existing buildings are of architectural value, while they are often structurally safe, proved their value in practical operation and continue to maintain a working condition, but they are outmoded and in need of reconstruction in different ways. It is found that when carrying out measures for the reconstruction of cultural houses, it is necessary to provide an expanded functional composition of premises for the organization of the leisure of population in accordance with modern requirements. Moreover the reconstruction of existing buildings requires taking into account many factors associated with the increase in loads on foundations, increase in the intensity of use of adjacent territories, work in the space-limited conditions of a construction site and need to modernize engineering communications. The most important are engineering and technical measures related to the examination of the technical condition of the supporting structures, identification of the actual design scheme of the inspected structural element, actual loads and effects to carry out measures to strengthen structures. However, the modernization of the audience halls as the main compositional and space-planning elements of cultural, leisure and entertainment buildings seems is especially important.

1. Introduction

Culture is an integral part of a civilized society. The functional implementation of culture in society can be performed out only with a developed material and technical base, ensuring and maintaining continuous interaction of the population with public buildings and structures serving them. A large place among them is occupied by structures, buildings and premises for cultural and leisure activities of population. This huge group includes entertainment, club and leisure facilities. These include concert halls - philharmonic halls for symphonic and gala concerts, organ concerts, pop, concert and theater, universal concert, entertainment and leisure, Houses and Palaces of Culture, clubs. The group of club and leisure and entertainment establishments is represented both as independent objects and integrated ones with cultural, entertainment and leisure establishments as an addition to the premises of the audience complex. It is also included in the main functional and planning element, cultural, leisure and social centers.

During the operation, the architectural and design environment of such buildings and their open areas must adapt to changing conditions, technologies and the way of life of society, based on the expansion of cultural, leisure, communication and recreational zones [11].

IOP Conf. Series: Materials Science and Engineering **944** (2020) 012005 doi:10.1088/1757-899X/944/1/012005

2. Problem statement

Today in Russia there is an impressive fund of Soviet heritage sites of cultural and leisure activities of population (Figure 1).

a) b) c)

Figure 1. a) cultural center in Shebekino, Belgorod region, 2020; b) the building of Belgorod state philharmonic hall (former Cultural center "Zheleznodorozhnik") in Belgorod, 2009; c) Palace of Culture "Yubileiny" in Belgorod, 2020. Source: Internet sources of free access

The mass construction of club buildings dates back to 1960-1970 (Table 1).

Table 1. Growth of the network of clubs in the USSR

Year	Total clubs, thousand	Including	
		in urban areas	in rural areas
1913	0,2	0,1	0,1
1922	12,2	3,6	8,6
1940	118,0	10,0	108,0
1945	94,4	6,5	87,9
1950	125,4	9,3	116,1
1960	128,6	14,1	114,5
1975*	135,1	20,2	114,9

Source: http://stroy-spravka.ru/article/ctanovlenie-sovetskogo-kluba

One of the main directions of club activity is the development of folk art, the expansion of types and genres of amateur performances, increase in the number of its participants, increase in the ideological and artistic level of repertoires and performing skills [1]. However, after 1991, the role of club activity declined sharply. Many of the clubs or cultural houses, especially in rural areas, were closed due to lack of funding and / or began to be used not in accordance with their original purpose [2].

Nowadays, this type of institution, like a cultural center, is undergoing dramatic changes. According to the report of the Accounts Chamber of the Russian Federation, the number of cultural houses decreased by one thousand in 2014. This tendency is especially typical for rural settlements. For example, in the Belgorod and Kaliningrad regions, the number of rural clubs decreased by 75% and 31%, respectively. In addition, in a number of constituent entities of the Russian Federation, the number of clubs simultaneously decreased. The most significant reductions were in the Belgorod, Sakhalin and Amur regions. [3].

The city-planning science of the USSR assigned cultural, entertainment and leisure buildings the important place in the planning structure of cities and towns. They were placed in public centers of residential areas (neighborhoods, settlements). They were dominant in the architectural and spatial composition of residential formations. They were centers of attraction for the population. As a rule, these objects were placed on sites convenient for the free access of population. The area for cultural events was in front of the main facades.

IOP Conf. Series: Materials Science and Engineering **944** (2020) 012005 doi:10.1088/1757-899X/944/1/012005

Today, in the conditions of a shortage of free space for the construction of similar objects (which is especially important for the central regions), there is a need to include in the work of the obsolete, but working objects of the Soviet heritage from the point of view of constructive safety[1].

During the reconstruction it is necessary to expand the functional content of objects including the introduction of an expanded training group of premises in order to solve the increasing problem of illiteracy (including computer) of the population of different age groups [10]. The development of social infrastructure in this direction will improve the urban environment in accordance with the needs of citizens [13] and contribute to the development of the regional settlement system as a whole within the framework of the implementation of regional programs and strategies for socio-economic development [14]. The most important element in the formation of a modern network of entertainment, club and leisure and entertainment establishments is the reconstruction of existing buildings of the Soviet period with the modernization of their main constituent elements. The number of facilities in the Russian Federation, the technical condition of which has been brought in compliance with modern regulatory requirements through reconstruction and modernization should be significantly increased [8, 9].

As an example of the modernization of an outmoded building and its inclusion in the work of a modern network of cultural and entertainment buildings, we can cite the reconstruction of the building of Belgorod State Philharmonic Hall.

Belgorod State Philharmonic Hall is the main concert venue of the city. It is located in the building 56A on the Belgorodsky Polk Street since 2004

Until that time, the building was called Zheleznodorozhnik cultural center. It was used as a city club and was a 2-storey rectangular building with an audience hall for 600 people, a cafe for 50 people and a small number of rehearsal rooms (Figure 1b, Figure 2b) [8,9].

a)

b)

Figure 2. Belgorod State Philharmonic Hall, 2010.
a) the building after reconstruction, b) the building before reconstruction
Source: Internet sources of free access

After the reconstruction, the total area of the building was increased from 7777 to 15125 meters^2. New extension on the main facade and a superstructure over a part of the existing building completely absorbed and modified the old building. The object was included in the Shortlist of the Best Building competition "House of the Year" of 12 best buildings built in the Russian Federation in 2010. According to the organizers "... the building of Belgorod State Philharmonic Hall, which the architect Irina Chechel designed on the site of the former Cultural House "Zheleznodorozhnik", is perhaps one of the most unexpected buildings of 2010. During the reconstruction, the cultural center was transformed due to the original extension and additional upper floor. It is absolutely impossible to

[1] Note: the monuments of architecture and cultural heritage are not included in this list

recognize the Soviet Cultural Palace as it is in the deconstructivist volume, decorated with two inverted glass cones" (Figure 2a) [15].

Today the cultural building has three concert halls of various types and capacities. The large classical music hall can accommodate up to 580 people. It consists of a parterre, two boxes, an amphitheater and a balcony. It is located on the first floor of the building. It hosts concerts of classical music, plays, festivals, instrumental arrangements sound, plastic performances and New Year's musicals. A small, multifunctional hall occupies one of the parts of the extension and has 300 seats. It is used for dance evenings, concerts, round tables, buffets, and performances of ensembles of folk instruments.

The organ music hall for 180 listeners (Figure 3) is situated in the attached part.

Figure 3. Organ Hall of the Belgorod State Philharmonic Hall, 2011 [1b]

The hall of classical music Belgorod State Philharmonic Hall was created by reconstruction of the existing hall of the former cultural center "Zheleznodorozhnik". According to its space-planning solution, the original hall was created in the form of a traditional theater hall with an extensive stage, a pronounced portal opening and an orchestra pit. The dimensions of the stage box in plan are 18x10 m, with a height of 7.5 m up to the grate. The hall in the plan also close resembles rectangle with dimensions of 21x24 m, with an average height of the amphitheater of spectator seats of 7.5 m. The total air volume of the audience section of the hall is 3700 cubic m^3.

The architecture of the hall was changed by the introduction of a second tier - a balcony, which accommodates 100 additional seats for audience and side VIP boxes. The proportions of the room unsuccessful for the music hall were corrected by a number of engineering and technical measures, the competent selection of finishing materials and the architectural solution of the interior details, justified by the calculations of architectural acoustics of the hall (Figure 4).

a) b) v)

Figure 4. Belgorod State Philharmonic Hall, 2010:

a) a hall before reconstruction, b) a hall of classical music after reconstruction, c) a balcony of a hall of classical music (https://archi.ru/russia/35290/dom-goda-na-finishnoi-pryamoi)

The main task of the reconstruction of cultural, entertainment and leisure facilities is the modernization of the largest element of this group of buildings – the audience hall, concert hall or

IOP Conf. Series: Materials Science and Engineering **944** (2020) 012005 doi:10.1088/1757-899X/944/1/012005

cinema. The purpose of the hall determines its architecture and, determining the volumetric-spatial structure, lay the basis for the architectural concept of the building.

The technological solution of the hall is especially important. High requirements are imposed on the technological filling of the halls, in search of the optimal correspondence to which there is a certain range of hall categories: stationary, semi-stationary and fully transformable, suitable for certain genres of performances.

A stationary concert hall is a completely immovable audience technological space: the seats, stage and floor of which can not be mechanically moved in space. Thus, a concert hall of this category is not capable of undergoing any global spatial changes during the operational period.

The examples of such systems are most concert halls of past centuries, as well as modern highly specialized performance spaces, the contents and functions of which meet the requirements of a specific, unchanging situation. These include a modern concert hall in Copenhagen, Denmark with a capacity of 1800 people, with a specific musical purpose of the hall. The next example is the Second Stage of the Mariinsky Theater in St. Petersburg, Russia for 1,830 seats with a stage box device, which allows the use of various methods of show organization. In its turn, the stage box for such a hall is determined according to its direct purpose, capacity and other number of factors, the provision of which allows creating an ideal concert space that meets the specific criteria (Figure 5).

a) b)

Figure 5. a) Copenhagen Concert Hall (without stage box): Hall interior. b) The second stage of the Mariinsky Theater in St. Petersburg (with a stage box): the interior of the hall [4]

The transformable concert hall is a dynamic space for audience and artists, the technological mechanisms of which are designed to serve as a single multifunctional venue for different types of performances. With this design of the hall, the criteria for the reliability and mobility of the transformable elements and the multifunctionality of the audio-visual equipment are imposed on the equipment [5].

A similar transformable element can be presented by the partitions dividing a large space into smaller parts, both on the scale of a small conference hall and the whole cultural house. The example of such a hall is the Rusakov House of Culture, designed by architect K.S. Melnikov. Its 1500-seat concert hall is equipped with three 250-seat balconies. It is equipped with a partitioning system; the balconies can also serve as isolated spaces for other uses. The corresponding constructive solution of the hall determines its unique architectural appearance (Figure 6).

IOP Conf. Series: Materials Science and Engineering **944** (2020) 012005 doi:10.1088/1757-899X/944/1/012005

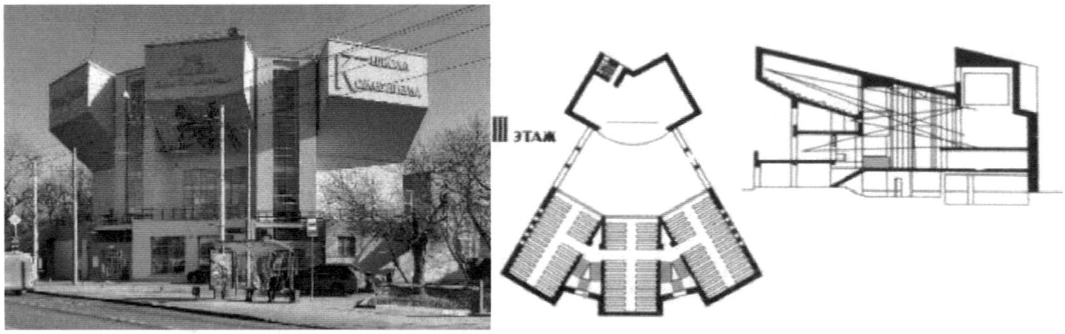

a) b)

Figure 6. a) Facade of the Rusakov House of Culture. b) Plan and section of the Rusakov House of Culture. Source: Internet sources of free access

The system of double-sided turntables with rows of seats on one side of the platform and a covering similar to the floor covering on the other side can also transform a concert space. Each row is a part of a full-fledged mechanism, the platform of which, on one side, is equipped with seats, on the other - a covering similar to the floor (sometimes stage) of the hall. With automatic action, the entire hall, in whole or in part, is turned over with the help of special mechanisms with the right side of the platform up. However, due to their uniqueness and novelty, such technologies are not widely used in the design of concert halls. Nevertheless, with small volumes the platforms transform the hall on one level into a multi-level audience hall as in the River Rock Casino Resort, using the GALA hall rotation system (Figure 7).

Figure 7. Hall transformation process of River Rock Casino Resort [6].

The transformable floor is also a tool to change the space of the hall. It allows creating a flat surface or amphitheater. The common floor of the hall and the stage is equipped at the level of the stage table and at the floor level of the first row of the parterre [7]. The Dee and Charles Wyly Theater Auditorium in the USA is equipped with similar technology, which can be an exhibition space and a single or multi-level auditorium. Its hall can be both a space for exhibitions and a single-level or multi-level audience hall. In this particular case, the transformations are mostly not automatically: the seats must first be stored in a special area of the stage, which, using electrical mechanisms, will transfer them to the storage under the stage, and then separately manually fix all the differences in the floor of the hall (Fig. 8).

Figure 8. Dee and Charles Wyly Theater Auditorium transformation process [6].

Another popular transformation tool is "bleachers" - movable stands with folding seats. Most often, such systems can be found in the design of sports facilities. However, the ease of their use allows using them in the design of concert halls. The disadvantage of this design may be the design of the stands, which can not be hidden from the viewer's eyes with a certain device in a folding form. However, today there are opportunities for solving this problem by moving the bleachers under the stage space. This technology in comparison with the others listed above, has a lower cost and is often installed locally in large crowded places (for example, shopping centers) in order to serve small concerts of modern performers. Such a system provides high mobility of audience space. However, such transformations allow using the hall for other purposes (by extending and sliding the required number of rows of stands).

A transformable type of a concert hall can also be equipped with other transformation tools. However, today the above mentioned types are the most popular. They can be found in one form or another in the semi-stationary category of halls. Such spaces are equipped with both absolutely static and transformable elements. Thus, the stalls of the Opera House (the space of the balconies remains static) in Vienna is often used for theatrical masquerade balls and the stalls of the Crocus City Hall concert hall in Moscow turns into a dance stall (Figure 9).

a) b) c)

Figure 9. Belgorod State Philharmonic Hall, 2010
a) Opera House in Vienna, b) Concert Hall of Crocus City Hall, Moscow, c) Dance stalls of Concert Hall Crocus City Hall, Moscow
(https://www.tourister.ru/world/europe/russia/city/belgorod/theatres/34364).

Conclusion

1. During the research it was found that in the middle of the XX century in the USSR (1960-1970) there was a massive construction of buildings for cultural and leisure activities of population – cultural houses. All conditions were created for the wide development of mass cultural, educational and leisure

institutions. As a rule, they were built according to standard designs for the organization of leisure and met the ideological, sanitary and hygienic, architectural and construction requirements of their time.

2. With the change of ideological attitudes in the country after 1991, the role of club activity sharply decreased. Many of the clubs or cultural houses were closed due to lack of funding and began to be used for other purposes. Today in Russia, the cultural houses as a type of institutions for cultural and leisure activities of population are undergoing dramatic changes. Their number is decreasing every year. They lost their cultural, educational and educational role. At the same time, the buildings themselves proved their value in practical operation and continue to maintain a working condition, although they are outdated and in need of reconstruction.

3. During the reconstruction of buildings of cultural centers, it is necessary to provide an expanded functional composition of premises for the organization of leisure of population in accordance with modern requirements.

4. It is especially important to modernize the audience halls as the main compositional and space-planning elements of buildings. This will expand the functional purpose, attract the attention of citizens and become an investment attractive object.

References

[1] Gavrilenko A Ya Big Soviet Encyclopedia *Cultural institutions of the USSR S-229-264* [Electronic resource] URL: http://stroy-spravka.ru/article/ctanovlenie-sovetskogo-kluba

[2] Latysheva M Cultural house as a propaganda tool in the USSR [3] URL: https://tass.ru/kultura/2590530 (Date of access: 20.07.2020). [4] Ideas for Life Magazine: Amazing Concert Halls of the World [Electronic resource]. URL: https://novate.ru/blogs/220315/30520/

[5] Group of companies "Antanor". Complex equipment of the transformer hall [Electronic resource]. URL: https://atanorgroup.ru/model-decisions/zal-transformer/

[6] GALA SYSTEMS INC. Solution: Multipurpose hall systems [Electronic resource]. URL: https: //www.galasystems.com/en/solutions/multipurpose-hall-systems/

[7] Official site of Theatrical Technical Corporation: production of theatrical mechanics. [Electronic resource]. URL: http://ttcstage.ru/sistemy-transformatsii/

[8] Chechel I P and Chechel I N 2019 Analytical review of the space-planning solutions of the building of the Belgorod State Philharmonic Society *"Youth and Science, the Way to Success": a collection of scientific articles of the III All-Russian Scientific Conference of Promising Developments of Young Scientists* (March 21-22, 2019) **4** ML-34 p 119–126

[9] Chechel I P and Chechel I N Analytical review of constructive solutions for the building of the Belgorod State Philharmonic Society *"Future of Science -2019": collection of scientific articles of the VII All-Russian Scientific Conference of promising developments of young scientists* (April 25-26, 2019) Vol **5** ML-36. p 138–142

[10] Chechel I P and Chechel I N New trends in the design and design of schools in different countries *Technical aesthetics and design research* **1 (3)** p 19–29

[11] Chechel I P and Chechel I N 2019 New tendencies in a school designing in different countries *Technical Aesthetics and Design Research* **1 (3)** p. 19–29

[12] Ilvitskaya S V, Perkova M V, Petrova L V and Shulginova O A 2019 Design security adaptation of communication spaces for the modern educational forms' implementation in the educational buildings of universities *CATPID-2019. IOP Conf. Series: Materials Science and Engineering* **698** (2019) 033024, doi: 10.1088 / 1757-899X / 698/3 / 033024IOP

[13] Perkova M 2018 The Methodology for identifying and resolving urban planning conflicts and contradictions in development on the example of small towns in the Belgorod region *Bulletin of BSTU named after V.G. Shukhov* **9** p 74–84

[14] Perkova M V, Vaitens A G and Lebedeva Yu D 2018 Method of social adaptation of adjoining territories of apartment buildings *Bulletin of BSTU named after V.G. Shukhov* **10** p 90–102

[15] Perkova M V 2016 Regional settlement system *International Journal of Pharmacy & Technology* **8 (4)** p 26621–26634

IOP Conf. Series: Materials Science and Engineering **944** (2020) 012006 doi:10.1088/1757-899X/944/1/012006

Design of rehabilitative and developmental board games for people with impaired motor and cognitive functions

O B Chepurova[1], **T A Putintseva**[1], **O P Tarasova**[1] **and D M Tuysina**[1]

Orenburg State University, avenue Pobedy, 13. Orenburg, 460018, Russia

E-mail: garry_ic@rambler.ru

Abstract. The article presents the results of the development of rehabilitative and developmental board games of varying complexity for people of all age groups with a violation of the musculoskeletal system and cognitive functions by the author team of the Orenburg State University design department. Games are designed to restore and develop motor skills, visual perception, cognitive abilities, logical and spatial thinking. Particular attention is paid to the variability of games and parameters of game chips configuration, with the aim of gradual and continuous improvement of patients' skills in the rehabilitation process.

1. Introduction

In modern society, there is an increasing social need for the development of rehabilitative and developmental games for people with musculoskeletal disorders that restore both motility and brain activity; intended for all age groups; available for use both in medical institutions and at home.

Throughout the world, the socialization of society part that, due to physiological, psychological and cognitive characteristics, experiences certain difficulties in realizing its labor capabilities, in organizing its life, and in interacting with society, remains relevant. Many of them cannot overcome these barriers due to impaired function of the musculoskeletal system or cognitive impairment acquired in connection with the disease or injuries. Only in our country at present, 12.1 million is granted the status of a disabled person. Many of them have a diagnosis associated with disabilities of different severity.

For example, 70-80% of survivors of a stroke remain disabled and only 20-30% of them irretrievably lose their ability to independently perform elementary functions. Currently, the age of stroke is getting younger and becoming one of the main socio-rehabilitation and medical problems.

The most common violations that limit the ability to perform certain vital functions, communication – impaired memory, speech, motor abilities. The first stage of lost abilities restoration occurs under the direct supervision of medical personnel. However, the recovery process can last more than one year and in the periods between medical recovery procedures, the patient is forced to overcome difficulties at home independently or with the help of loved ones.

The market of medical rehabilitative and development equipment is rapidly developing in the direction of robotization, computerization, and others progressive, but expensive technologies, designed for specialized medical or rehabilitation centers. Domestic and foreign manufacturers are developing personal equipment to continue rehabilitation exercises on their own. Basically, it is also based on computer technology aimed at restoring and developing one or more lost functions in the post-stroke or post-traumatic period and also belongs to the price category above the average.

Almost all categories of patients need special organization of rehabilitation conditions in the home environment. This environment and its content should be accessible to all segments of the population, both wealthy and low-income. In addition, there are practically no proposals for introducing into the home life the simplest to learn and inexpensive, but methodically developed rehabilitation and development technologies that provide a comprehensive restoration of various human functions. For example, the simultaneous restoration of motor functions, memory, etc.

To solve the problem of organizing a patient's continuous rehabilitation and development process, an integrated approach is needed, combining rehabilitologists, occupational therapists, speech therapists, psychologists, teachers, design engineers and other specialized specialists. Designers are also actively involved in scientific and practical developments. The game techniques used by them in the development of rehabilitation and training kits, first of all, remove the psychological barrier of inferiority in patients and motivate for further rehabilitation actions. The influence of gaming technologies on the effectiveness of rehabilitation and developmental procedures has been confirmed by many world studies and does not require confirmation [1].

Integrated development of rehabilitative and developmental board games for people with impaired musculoskeletal and cognitive functions, performed by the author team of the Orenburg State University design department (Chepurova O.B. – head of the design department, PhD in Art criticism, associate professor; Putintseva T.A. – PhD in Art criticism, associate professor; Shleiuk S.G. – PhD in Art criticism, associate professor; Tuysina D.M. – PhD in Art criticism, associate professor; Tarasova O.P. – PhD in Pedagogic Sciences, associate professor; Chepurov I.V. – senior teacher, Ryabov S.V. – assistant, with the participation of design department students), contribute to the successful solution of this problem.

2. Materials and methods

Rehabilitative and developmental board games are designed to restore and develop motor skills, visual perception, cognitive abilities, logical and spatial thinking of people with impaired musculoskeletal and cognitive functions for all age groups.

Their use is envisaged in various fields: at home, in specialized institutions, including children's (medical institutions, rehabilitation centers, schools, kindergartens).

It is known that gaming technologies are effective as a method of rehabilitation and training. Thesis research on the subject of games and gaming technologies was carried out by Associate Professor of the OSU Design Department Tuysina (Popova) D.M. [2]. The use of gaming technology in the design of design objects provides a high degree of focus on cognitive activity, includes logic, motility, memory, attention, emotional experiences; have a positive effect on the adaptation of a person in everyday life.

However, it is worth noting that gaming technology is actively used for rehabilitation and set various tasks. During the game, a person not only shows his interest in interaction with game elements, but also fixes the cognitive process in a light form of perception. Therefore, many gaming technology focused on rehabilitation, use this effect for the task. The game as a process of patient interaction with game elements allows not only to maintain concentration, but also to fix the acquired skills. During the design, scenario modeling of interaction with game elements was presented: the game process was studied from the point of view of the logical task in the game and the tasks set, ergonomic parameters of all elements and clear graphic design. The task of restraining the interest in game technologies in patients contributed to the decision to introduce a system of replaceable game elements to expand the game variability. Thus, the emotional state and behavior of the patient in the gameplay were predicted [3]. It was the variation that allowed us to include the phasing of the game process during rehabilitation from simple to complex, depending on the function and task of the game.

When creating rehabilitation and restorative means, a number of methods were used that made it possible to facilitate the interaction of the user contingent with the game elements, taking into account the adaptive capabilities of the person and the disease severity. Among them, the anatomical and anthropometric methods, the scenario modeling method, the percentile method, the method of

IOP Conf. Series: Materials Science and Engineering **944** (2020) 012006 doi:10.1088/1757-899X/944/1/012006

compiling a checklist of requirements for the design object are significant. So, data on the structure of the human body and its functional capabilities were required when calculating the parameters of the chip and the playing field (overall dimensions, dimensions of reach of the hands while sitting or standing, muscle strength), taking into account the specifics of the game activity and the health status of the player. In this regard, we studied static, dynamic, and functional features that allowed us to determine the design parameters and scale relationships of the parts and the whole game:

- the height of the working surface (playing field), taking into account the height of the elbow point of the seated player;
- the size and layout of the game elements on the playing field, taking into account the reach of the patient's hands from different positions and poses that he may occupy due to illness, as well as the nature of the movements that need to be performed during the course of the game;
- weight and size of the chip, taking into account the hand size, patient's muscle strength and the hands joints mobility;
- determination of the optimal size of the workspace, etc.

In addition, We took into account the specific features of human interaction with the game elements: the need for visual control over the action execution, the need to perform a number of operations with the right or left hand, etc. This approach to the design of game elements allows to optimize the game process due to the ability of the user to take a relatively comfortable pose, because of which his movements will not cause difficulties, as well as increase the duration of the game process, which will contribute to the desired therapeutic effect.

Particular attention in the process of project activities was given to the search for chip parameters. According to the game scenario, when it moves across the field, the player uses the end grip. This method is not always applicable (for example, it cannot be used by patients with some visual agnosia, since with this grip the image located on the upper part of the chip is closed), but it is most optimal with the given restrictions (board game). The design parameters of the chip were selected empirically taking into account the wrist size, the characteristics of the product retention and the patient's ability to control the process of transferring the product along the playing field. At this design stage, the scenario modeling method and the percentile method were used. It was important to identify product parameters that are equally convenient for both men and women, since the contingent of users consists of people with significantly different anthropometric signs of the wrist (transgression within 50-60%) [4] and age characteristics. It is known that with the indicated diseases in age-related patients, the adaptive capabilities of the body are significantly reduced – with age, mobility in the joints decreases, coordination of movements is lost, muscle strength decreases. Therefore, the task of the designers was to determine the weight of the chip and its size, based on optimal efforts and energy costs of all categories of users.

In the course of scenography, a general idea about the proportions and sizes of the projected element was formed, an object structure was formed, based on the principle of a tumbler, a color solution, materials were selected that provided the required shape and function. When using the checklist compilation method, the requirements were formulated for the game elements - its quality characteristics, parameters, graphic solution, functions, etc. User satisfaction with the chip, in particular, suggested that it corresponded to the anthropometric characteristics and wrist functionality, finger mobility, using a positive color scheme and patient-friendly graphics. However, the big difference between anthropometric parameters of the game users that determine the boundaries of its capabilities, as well as their features, required, for example, to unify the object and add allowances for adaptive movements to players with very different body sizes. In turn, the comparative method for compliance and the analysis method were used to search for the degree of stylization of images at the stage of creating thumbnails of graphic game elements, taking into account the presence of a player with a post-stroke state of visual agnosia.

The application of eco-design methods and 3D printer technology made it possible to produce test samples of the game kit basic elements from environmentally friendly material. Polylactide, which includes waste from corn and sugar cane, was selected as the raw material for the manufacture. This is

IOP Conf. Series: Materials Science and Engineering **944** (2020) 012006 doi:10.1088/1757-899X/944/1/012006

a relatively soft plastic that is used in industry for packaging food products, medicine containers, and surgical sutures. Due to the natural composition of the material components, the disposal of products in natural conditions is possible.

3. Results and Discussion

The rehabilitative and developmental games presented in the article relate to indoor games on the boards with small moving playing objects. Many previous developments have been made in this direction.

One of the constructive analogues is the toy "Vanka-vstanka" containing a body with a hollow spherical base and a weight made in the form of a permanent magnet mounted in the body with the ability to move. The magnet is suspended by means of a flexible thread, it has a magnetic interaction with the ferromagnetic material placed in the closed cavity of the base (SU 1519728 A1, 11/7/1989, Bull. No. 41). The disadvantages of this device can be noted its instability and a long period of swinging.

Also known is a magnetic chess device containing a magnetized chessboard, a set of flat metal figures, a leading playing chip with a built-in magnet (RU 2092210, 10.10.1997). But this device does not have playing elements with the necessary anthropometric parameters for people with a violation of the musculoskeletal system and cognitive functions.

One of the modern designs offered in the market of rehabilitative and developmental games is the RehabMedical sound panel "Palletto", which has a high-quality design of the external shape and color. It uses the Braille technique for the development of associative thinking, sounds are recorded and reproduced with a certain set of pictures. However, the high cost of this device allows its using only for a wealthy category of people.

Since 2003, teachers and students of the OSU Design Department have been working on the research and design of rehabilitative and training sets of games for children with musculoskeletal disorders in collaboration with the All-Russian Research Institute of Technical Aesthetics as part of the Design and Life Quality of the Population program. The exhibition "Art Furniture 2004" exhibited and certified the result of joint work on the design of specialized gaming complexes.

In 2008, in the book of PhD in Art criticism, Professor Grashin Alexander Alexandrovich, Head of the Department of Theory and Methodology of Industrial Design and Ergonomics of the Research Institute of Technical Aesthetics of MIREA “Design of a Children's Developing Subject Medium”, the results of a design solution for rehabilitation and developing subject-spatial environment, as a base matrix for additions to any rehabilitation and recovery complex in the system of material and educational development facilities for patients with musculoskeletal disorders [5].

The first project proposals of the design department presented in this book demonstrate rehabilitative and developmental games, in their anthropometric parameters, designed only for children. They include playing chips that do not contain replaceable cards, which deprives the game system of variability; the base of the playing chips is made in the form of a hemisphere, which leads to instability and prolonged swaying of the playing chip during the game; the side surfaces of the playing chips are even and smooth, which makes it difficult to fix them in the player’s hand. Testing of the developed products made it possible to identify a number of significant shortcomings and continue further research, taking into account their formative qualities, methodological content, technological design and the expansion of functional use for all age categories.

Thus, the main tasks for us in developing rehabilitative and developing board games for people with impaired musculoskeletal and cognitive functions were:

- giving a set of properties to the playing chips which are necessary for people with musculoskeletal disorders of all age groups: the necessary anthropometric parameters, improving the quality of fixing the playing chip in the player’s hand, increasing the stability of the playing chip in combination with the ability to take a vertical position from an inclined position;
- increasing the degree of variability of the game.

IOP Conf. Series: Materials Science and Engineering **944** (2020) 012006 doi:10.1088/1757-899X/944/1/012006

The rehabilitative and development board games developed by us for people with disorders of the musculoskeletal system are equipped with a game field, game chips, exchangeable cards for the playing field and exchangeable cards for game chips, packaging. An example of the game is shown in Figure 1. Consider its structure and principle of operation.

The game chip consists of a base and a top with a lid, as can be seen in Figures 2 and 3. These elements are fastened to each other by inserting on the elastic properties of the material. As mentioned above, the game chip is made of lightweight impact-resistant eco-friendly plastic using 3D printing technology.

The base of the game chip has a hollow hemisphere design in which a metal load and filling material are located.

The top consists of a cylindrical part with a cover.

The cylindrical part of the pommel is made in the form of a sleeve with a ribbed side surface, for maximum ease of holding it in the hand of a participant in the game. On the lid of the pommel there is a magnetic layer for holding replaceable cards of game chips.

Dimensional and structural characteristics of the game chip correspond to the anthropometric parameters of the hands of the players, anatomical plastic; power, speed, biomechanical and other capabilities of the players.

Figure 1. An example of solving rehabilitation and developing board games for people with impaired musculoskeletal and cognitive functions. The graphic part was performed by Victoria Trubitsyna, a student (D. Tuysina, teacher).

The playing field has several layers: a magnetic layer is placed on the basis of the playing field (for holding game chips of exchangeable playing field cards by the attraction method). Above is the layer of the graphic solution for the playing field. The game field contains sections for the location of removable cards.

Replaceable playing field cards and replaceable playing chip cards have a magnetic layer for attracting, respectively, to the field and to the chips. The media layer with a graphic image is the top layer of removable cards.

Graphic images located on removable cards correspond to the scenario and semantic concept of the current game and may change with the start of a new game.

All game elements are compactly placed in a package.

Using the principle of shifted center of gravity in the organization of the shape of game chips ensures their stability. To do this, in a specially prepared cavity at the base of the game chip is placed a metal load with a large mass. This is the main innovative constructive and functional solution to the game.

The functional value of this principle lies in the comfort of the game process: a game chip, at any deviation from the vertical, restores its original position on the playing field.

The stability of the game chips is enhanced by the magnetic attraction of the metal load to the magnetic layer of the playing field.

The logical concept of games is as follows. During the game, the participant performs exercises to restore and develop hand motor skills (interaction with game chips), as well as to restore and develop memory and logic (remembering and repeating basic visual images and logical sequence of actions).

Games are aimed at expanding the zone of immediate opportunity. Due to the varying complexity of the games, it is possible for people of the combined contingent of different age groups to participate; simultaneous participation of one, two or more people, participation of an assistant.

Figure 2. Game chip; game chip with a removable card placed on it. Chips were developed and designed by: Chepurova O.B., Putintseva T.A., student Nikitenko T., Ryabov S.V.

Figure 3. The base of the game chip; the cylindrical part of the game chip top; game chip cover.

The principle of the game: in the sectors of the playing field are replaceable cards of the playing field of the chosen topic (by magnetic attraction), on the playing chips are replaced by cards of playing chips of the corresponding theme (by magnetic attraction), the player takes the playing chips by hand and puts them on the replaceable cards of the playing fields corresponding in meaning.

The semantic concept of games for compliance consists in the associative recall by the game participants (using semantic associative pairs) of the correspondences, situations, phenomena, images that are most often encountered in life and necessary for a full perception of the world. It is always necessary to restore the logical connections of everyday objects in everyday human activities. It can also be interesting and informative for the patient to develop and restore knowledge about natural laws, etc. In this case, the participant puts a game chip with a replaceable card on the playing field on the corresponding pair of replaceable cards.

A more complex level of games is the work with letters and words. Letters are placed on replaceable cards of game chips; words are composed of letters.

IOP Conf. Series: Materials Science and Engineering **944** (2020) 012006 doi:10.1088/1757-899X/944/1/012006

The options for the semantic concept of games are infinitely diverse - tic-tac-toe, checkers, chess and many other favorite games.

Thus, games possess high variability, visibility for players. The graphics of game chips and sectors of the game field can be changed depending on the level of complexity and plot of the game, replacing replaceable cards of game chips and replaceable cards of the game field. Thus, if necessary, you can easily change the theme of the game to another.

Graphic design is an important factor in the degree of complexity of the gameplay. In the course of work on illustrative material using the method of comparative analysis, graphics sketches of game elements were developed. It is worth noting that the graphic component of the game allows to include different degrees of difficulty in perception, thereby being variative and multi-level. At the same time, interchangeable graphics for the playing field and chips allow setting certain development goals. The chosen technique of illustrative material execution allows to quickly and clearly read the silhouettes of objects depicted on the chips and the playing field. The silhouettes are pastiche; proportional relationships between themselves in the images are preserved. Each game contains its own script depending on the tasks and topics of the illustrative material.

Figure 4 presents one of the concepts of graphic design, which already has the second level of visual perception complexity, in contrast to simple geometric shapes. It is intended to study the color ratios of images on chips to achromatic images on the playing field. The color layout was selected taking into account visual perception and tonal correlation of shades among themselves [6]. For greater contrast, the background on the playing field and chips remains white. Each color and tone in a person is associated with emotions and feelings. Thanks to the correct layout of the color the attention and memory of the patient can be improved. For reliability, the color gamut of each image was selected from the original image, so as not to mislead the person. Isolation of the desired silhouette from similar color and tonal ratios increases the degree of complexity. The silhouette outline of each image is crisp and original.

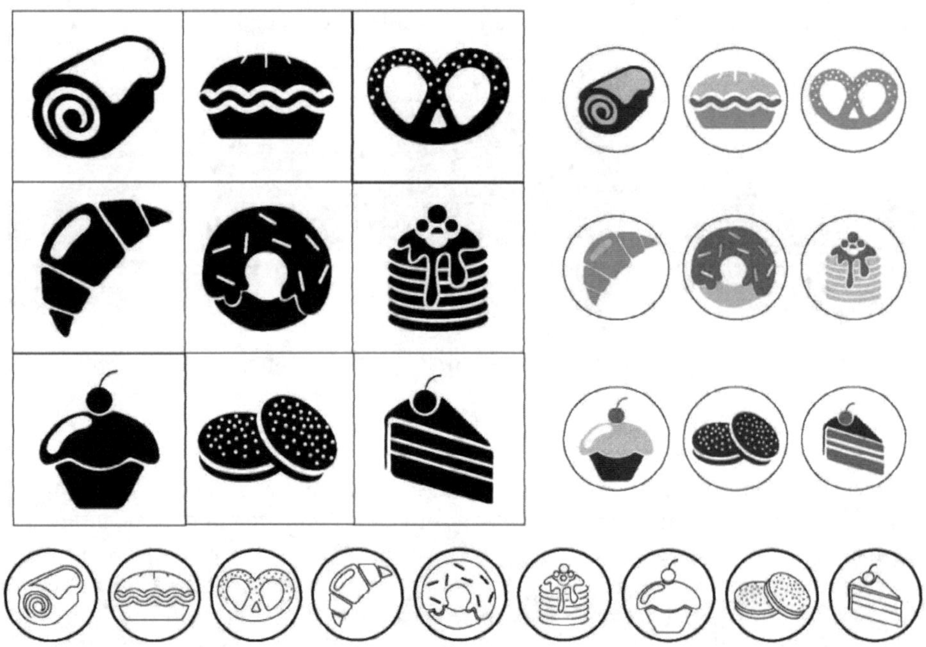

Figure 4. An example of the graphical solution concept to rehabilitative and developmental board games for people with musculoskeletal disorders. The graphic part was carried out by student Drozdova Tatyana (Tuysina D.M., teacher).

IOP Conf. Series: Materials Science and Engineering **944** (2020) 012006 doi:10.1088/1757-899X/944/1/012006

The second version of the game process includes searching not color dice, but a silhouette solution of images and their contour execution. This solution has a more complex degree of visual perception. Monochrome design makes it easier to find the silhouette of the necessary chips with the image on the playing field. Graphic stylization of a three-dimensional object in a flat form simplifies the perception of the patient and increases the speed of visual recognition of the image. Such a solution maintains the patient's interest in game elements as a solution to the problem and encourages further gameplay. The presented concept of the graphic design of the game is designed for corrective action on the visual apparatus of the patient. And game therapy tasks are aimed at accumulating visual information from the player.

The requirements for the designed objects are objectively reflected in the formulation of criteria and indicators for assessing their quality. Criteria, as signs, on the basis of which the design object is evaluated, contribute to obtaining the most complete picture of its quality condition during operation. The indicators, in turn, describe the characteristics and features of the created product. When evaluating the criteria and indicators of the rehabilitation and developmental game, the following list of criteria and indicators was used:

- aesthetic, which determines the external attractiveness and artistic expressiveness of the game, due to which there is interest and sympathy for the object of the user. This fact is very important, as in many patients the existing way of life is destroyed as a result of the disease, which often causes psychological trauma and refusal of rehabilitation therapy;
- ergonomic criterion assumes that the design object matches the anthropometric, psychological, psychophysiological and other features of the target audience, taking into account the specifics of the gaming activity tasks in order to restore lost functions and the conditions for rehabilitation and developmental games in a medical institution or in everyday life;
- functional criterion assumes that the product meets its intended purpose in terms of design, levels and variety of practical applications, the intended rehabilitation and restoration effect;
- environmental criterion is designed to assess the degree of psychological and physical safety in the use and safety of the product;
- operational criterion requires ensuring easy maintenance and storage of product elements, strength, wear and shape stability of the product, allowing its active use to maintain consumer qualities and attractive appearance;
- economic criterion suggests the possibility of acquiring a set of games for self-sustaining continuation of rehabilitation therapy to a large group of patients with impaired functions.

4. Conclusion

The presented rehabilitative and developmental board games for people with disorders of the musculoskeletal system and cognitive functions contribute to the successful solution of the problem of restoring lost skills and adapting them in everyday life.

Unlike well-known technical solutions, these games contain a playing field with a magnetic layer, replaceable cards of the playing field with a magnetic layer, replaceable cards of playing chips with a magnetic layer; playing chips have anthropometric characteristics for people with disorders of the musculoskeletal system of all age groups, a ribbed (textured) side surface to improve the quality of fixing a playing chip in a player's hand, a magnetic layer for fixing replaceable playing card cards, a metal weight in the base for the ability of a chip to accept vertical position from inclined; the base of the chip has a complex shape - the surface contour of the base of the playing chip is presented in the form of conjugations of circles.

In developing the design of the games, ergonomic criteria came in first place - the ability to easily capture and hold the game chip, the correspondence of its shape to the anthropometric parameters of the players' hands, anatomical plastic; compliance with power, speed, biomechanical, etc. As a result, the design of the chip provides ease of capture by the player, a quick return from any position to the original vertical.

All elements of the games are made from environmentally friendly materials, safe and harmless to use, simple and convenient during the game. Original authoring graphics and a thematic approach to creating games form vivid imaginative solutions of a positive orientation.

The author team of the design department obtained a patent for the invention of a rehabilitative and developmental board game for people with disabilities, designed to restore hand motor skills, visual memory, cognitive abilities and logical thinking of people with impaired musculoskeletal and cognitive functions of varying complexity [7]. The rehabilitation and developing board games developed by us are an element in the rehabilitation and restoration of the subject-spatial environment.

Further development of design in this direction involves the sets of games development with high variability of configuration and parameters of game chips, variability of the graphic solution in order to diversify rehabilitation and recovery games. This will make it possible to gradually and continuously improve the skills of patients of all age groups of any severity of the disease in the process of rehabilitation.

5. Acknowledgments

The opportunity to implement the rehabilitative and developmental set of games developed by the OSU department staff and to test it in medical and rehabilitation institutions of the Orenburg region appeared with the financial support of the Elena and Gennady Timchenko Charitable Fund in the framework of the All-Russian grant "Active Generation-2017". Due to this, the set of games made in the material was tested at the Orenburg Regional Clinical Hospital, the state budgetary health care institution, the Regional Center for Medical Rehabilitation, the state budgetary health care institution, and the Dubovaya Roscha rehabilitation and rehabilitation sanatorium.

According to the conclusions of rehabilitologists, the result of testing made it possible to recognize the set of games with methodological material of varying complexity as effective, which allows combining recovery procedures for speech, cognitive and motor functions in a single process. Game technologies have become a motivator for patients to continue to interact independently with the kit both in specialized institutions and at home, which has helped to organize a continuous rehabilitation process and improve recovery processes.

References

[1] Ya-Xuan Hung, Pei-Chen Huang, Kuan-Ta Chen and Woei-Chyn Chu 2016 What Do Stroke Patients Look for in Game-Based Rehabilitation. Retrieved from https: https://www.ncbi.nlm.nih.gov/pubmed/26986120

[2] Tuysina (Popova) D M 2013 Children's toy book as a developing design form *Extended abstract of Candidate's thesis* **24**

[3] Podprugina V V and Blinnikova I V 2012 Study of ideas about emotions through picturesque tests *Psychology Issues* **3** 134–142.

[4] 2009 *Ergoddesign of industrial products and the subject-spatial environment: textbook. Allowance for students. Universities enrolled in special. "Design", "Ergonomics"* ed. V I Kulaykin, L D Chaynovaya (Moscow: Humanities. ed. VLADOS Center) p 311.

[5] Grashin A A 2008 *Design of a children's developing subject environment* (Moscow: Architecture-S) p 296 ISBN 978-5-9647-0146-0

[6] Johannes Itten 2004 *Fundamentals of Color* (M-D.: Aronov) p 53.

[7] Chepurova O B, Novikov A I, Putintseva T A, Shleuk S G, Chepurov I V, Tuysina D M, Tarasova O P and Ryabov S V 2018 *Rehabilitation and development board game for people with musculoskeletal disorders*. Patent for invention RU 2670410 C1, 10.22.2018 Application No. 2018108279 dated 06.03.2018

IOP Conf. Series: Materials Science and Engineering **944** (2020) 012007 doi:10.1088/1757-899X/944/1/012007

Modular system-based urban equipment design

N D Dembich[1], A V Streltsov[1], M A Zyrina[1], A A Marchenko[1]

[1]State University of Russia named after A. N. Kosygin (Tehnology.Design.Art), Moscow, Russia

E-mail: dembich@yahoo.com; mr.Streltsov@gmail.com; maria.zyrina@gmail.com; dargo-96@bk.ru

Abstract. The article examines the problem of abundant squares of the administrative centers of town, typical for the Soviet period, and its solution by using the design of modular systems for creating urban equipment. The organization of a comfortable space under constantly changing conditions, the ability to quickly adapt to the needs of the consumer makes the development of modular architectural structures increasingly acute. The article considers basic principles of designing modular systems for organizing urban spaces for various public events. The main principles for optimizing the design of such modular systems have been formulated, such as multifunctionality and versatility, flexibility and adaptability, reusability, transportability and mobility. The main design features, units, parts and materials of the modular system are described. Based on the study of trends in the design of such systems, a conceptual project of modular open structures and universal pavilions for the city of Rubtsovsk, Altai Territory, is proposed. In addition to the module, a planning solution for the central square was proposed. The proposed modular system allows to improve the ergonomic and aesthetic properties of urban spaces, taking into account the climatic conditions of the region. The system has good adaptive characteristics, as it easily adapts to the needs of consumers. The cost-effective, flexible and growing structure of the modular system allows to quickly create a new image of the center of the settlement and contribute to its natural development. Design tools have become key elements in organizing the squares. The offered modular objects can form a single environment complex and perfectly decorate the shopping area. The created space can be used for theatrical events and festivals, since a stage, seats for spectators and recreational areas can be built from the elements of the modules.

1. Introduction
Number of activities like fairs, festivals, open air forums continually grows in a modern city. The prosperity of public life is always associated with the emergence of new public spaces, within which people interact not only as spectators, consumers or "assets", but also as active actors. This environment forms the modern consciousness of the townspeople. New forms of social interaction contribute to the development of collective participation and unpredictable cooperation, which are becoming increasingly important in social life today. At the same time most of former Soviet small and medium-sized towns still have unused spaces of central parts of settlements – one of the best reserved squares for citizen activity [1,2,3]. Organization of comfortable space in a continuously changing environment, ability to quickly adapt to consumer request leads to urgent development of modular architectural constructions. After studying current trends, a conceptual project of modular open construction and pavilions of universal purpose is proposed for city of Rubtsovsk, Altay region.

2. Materials and methods
Modular conception assumes that parts of an object are independent and could be used autonomously, as well as interconnect in different combinations [4, 5, 6]. This approach allows to change a way of interaction between different parts of the product, as requirements change individual parts can change as well without affecting the basic design. If individual elements are damaged, they can be easily replaced. When reused, the exterior design may be refreshed while leaving the most expensive parts

unchanged. This allows physical changes to be minimized, but still to achieve the desired functional modifications. The presence of unified units and parts, which are used in different combinations, allows to transform the structure in different versions. The use of modular combinatorial design is considered the highest form of standardization. Standards identify and empower the most effective design methods and techniques. This contributes to the unification of the structural elements of the product, which leads to high efficiency of their production and operation.

At the initial design stage, analogues of modular sectional objects were considered. Preference was given to art objects (art installations), furniture and kiosks. The design takes into account works of such popular designers as Alik Levy, Philippe Negro, Manuel Sanz, Irene Brea and others. The analysis of world trends in the design of street modular systems has shown that modular structures are focused on the practical distribution and implementation of new technologies and materials in various areas of life, trade, information exchange, leisure centers, art objects, etc. Pre-project research of the scope of modular constructions allowed to formulate the principles for streamlining methods of functional and aesthetic development of modular systems for environmental space organization:

- System multifunctionality and universality;
- Flexibility and adaptability of system to situational changes (variability and combinatorics);
- Reusability (collapsible constructions);
- Transportability and mobility;
- Suitability of modular system for persons with disabilities;
- Taking into account the peculiarities of visual perception of environment design objects;
- Light and color spatial-layout study of design objects;
- Harmonious and integral appearance of modular system in open urban space conditions of post-soviet towns; and
- Economy and efficiency.

3. Results

Developed module is a construction based on a system of repeatable elements, easily assembled for a necessary configuration for any situation (Fig. 1).

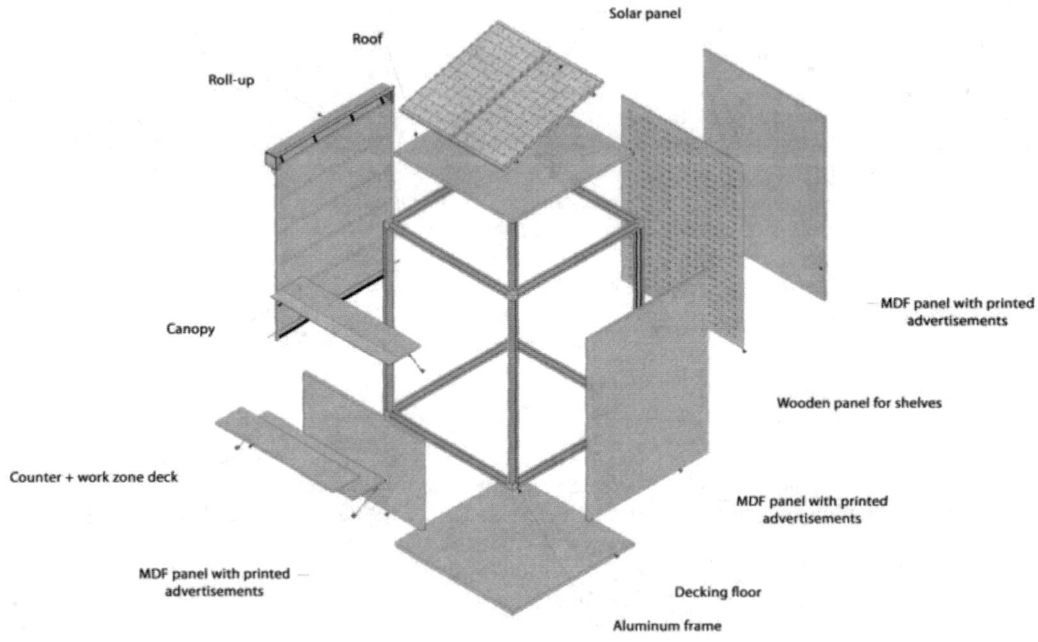

Figure 1. A module design.

IOP Conf. Series: Materials Science and Engineering **944** (2020) 012007 doi:10.1088/1757-899X/944/1/012007

Laconic module space allows for reduction of price and production time, allows for easy and free assembly of several modules into a single system. The framework consists of aluminum sections with grooves that allow interconnection of elements without self-detachment concerns. For extra durability the system can be secured with bolts. Aside from interconnecting framework elements and finished constructions, grooves can be used for locking module walls, flooring and roofing. One of important components are holed wall panels that are secured to the framework. Holes are used to secure elements for holding shelves. Shelf length varies according to customer request, but width (300mm) and thickness (18mm) are fixed. Framing walls are supposed to be produced from MDF, also winter option is provided with heat insulation and a sliding window. External panel side could be used for printed advertisements. Roof allows for solar battery placement. Interior has lighting, electrical outlets, small heaters if needed. Doors — roll-ups. Aside from enumerated elements, modules are equipped with folding sunscreens, work areas for serving orders, production, packaging. Built-in furniture is placed inside module, all other elements are unified for quick production and assembly. Module dimensions are 2000x2000x2600mm. Within the available space we are able to accommodate all the necessary elements for production and sale of drinks, snacks and small souvenirs with respect to ergonomic requirements. Based on recommendations of design theoreticists [7,8,9], certain varieties of module connections are proposed (Fig. 2, Fig. 3).

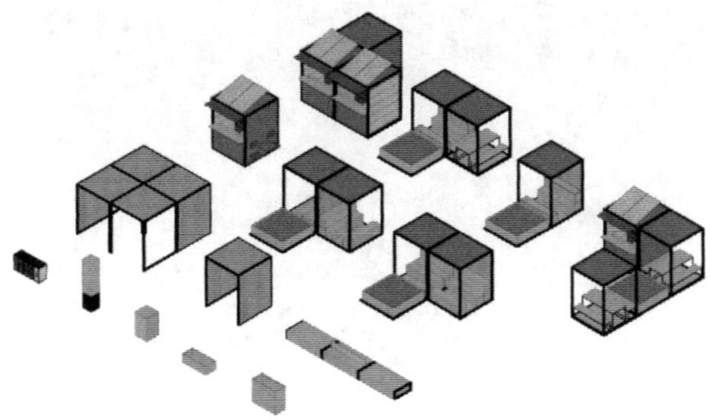

Figure 2. Module connections in clusters.

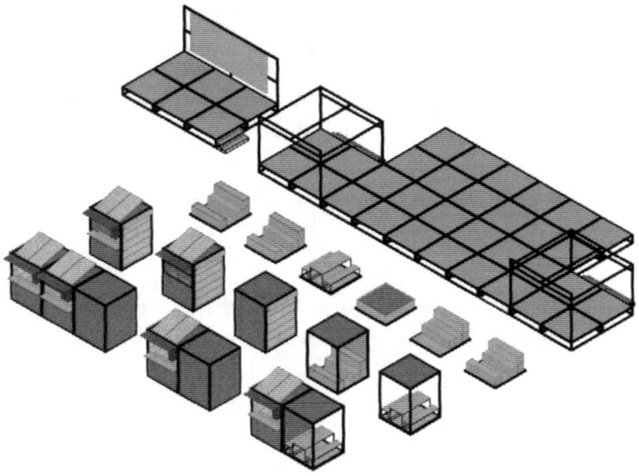

Figure 3. Module connections in rows.

Besides module, there is a proposition of planning solution of central square in the city of Rubtsovsk, Altay region (Fig. 4, Fig. 5).

Figure 4. Planning solution visualization.

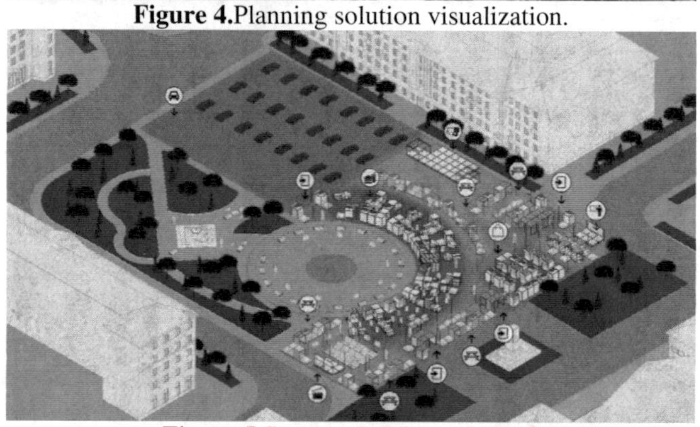

Figure 5. Square planning solution.

The center of the composition of the territory is a fountain framed by a compass pattern. Zones and pavilions are arranged in a radial way. Smooth radial movements of visitors will create a pleasant atmosphere. Based on the analysis of the potential visitors' needs, equipment has been developed in detail for the following socially significant zones - information entrance, public catering, cinema zone, concert zone, souvenir sales, lecture hall, quiet recreation area, sanitary zone, parking. Zones of cinema, concerts, lectures and master classes are separated from each other and are located at different ends of the square. The food zone is located along the ring of the fountain and forms the boundary between the festive zone and a passage to the park. Due to the radial arrangement, trading modules can be equipped with many seating places. Each food point is isolated and equipped with trash containers and disinfectants. Recreation areas are scattered in small clusters throughout the territory. The sanitary zones are placed throughout the festival area and are represented by mobile dry closets and bins with garbage sorting function. Illuminated navigation modules are located throughout the territory. Since the modules are mobile, compact and do not require much space, the festival will take up only ½ of the space, 1/3 will be occupied by parking, which is located along the main roadway. There are four entrances to the territory, equipped with information boards, ticket offices and security frames.

4. Discussion

Today, when the significance of the central squares of former socialist cities has been lost, these real-life spaces as geographical and symbolic centers where people familiar and unfamiliar with each other meet can acquire new authentic forms of existence thanks to universal modular systems [10].

Hosting active city events with a thought-through design has a positive impact on urban growth. The symbiosis of images and products, the volume and coverage of the image market at the national or even global level and the role of the symbolic economy form the image of the city as a whole, contributing to its development [10].

Natural products at the farmers' market, all kinds of entertainment, cultural events, holidays, festivals, fairs, etc., all these allow an area or a town to be placed on the "visual pleasures map" and to become a place of gentrification. Consumers of culture are involved in the interconnected production of symbols of space. Modular systems are the form and vehicle for any such event, they provide redundant city squares and become places of visual attraction, desirable social "oases".

Social culture is constructed at the micro level and is created from the variety of episodes that make up everyday life in urban spaces. As citizens are joining social life, its structure constantly changes.

The process of forming the culture of public spaces will proceed organically, since the design of the equipment of the modular system has the ability to be modified according to the new desires of consumers, to adapt its form in accordance with the ever-changing content.

The proposed modular system allows improvement ergonomic and aesthetic properties of urban space. Takes into account the climate conditions for current region. This system has good adaptive properties, because it easily adapts to customer requirements. Implementation of such project with regards to the interests of the residents, tourists, state institutions and businesses could help transform spatial ensemble for city square into an area of active city life, uniting everyone. Design means became the main key elements in square areas organization. The usage of geometrically unsophisticated elements as a constructive basis is due to the necessity to fit the objects into the old Soviet urban environment and the economic capabilities of the region. Proposed modular objects can put together a uniform environment complex and exceptionally decorate the retail space. Created space could be used for on-stage holidays and festivals, because module elements could be used to construct stage, grandstand and recreational areas.

5. Conclusion

The analysis of trends in designing modular systems suited for organizing urban space allows to conclude that large empty spaces in administrative squares of Soviet period in small and medium-sized towns are best adapted for active use by residents and tourists with help of modular equipment. Such equipment could be used in any city or region depending on the territory size and configuration, weather conditions, residents' requests', and local infrastructure features. Worth mentioning that profitable, flexible and growing structure of modular system allows to create a new image of settlement and assist its natural development as soon as possible. The specificity of the environmental objects, for which the modular system has been developed, presupposes a large capacity and the possibility of simultaneous occupancy of the entire territory with a mass of people. Therefore, it is necessary to equip the modular structure with electronic media, digital means and communication networks. This will be the main cost for the project. Such a multifunctional urban infrastructure should include not only electronic means to improve the quality and efficiency of information, navigation, advertising, entertainment, education, training, but also identification devices, motion sensors, cameras, etc. These devices will make the modular system electronically transparent and more democratic, which will open up new perspectives for a "flexible environment" [11].

6. Acknowledgments

We would like to thank the manager of master's program "Theory and practice of environmental objects design" professor Volkodaeva Irina Borisovna for her guidance and holder of master's degree in Environmental design Burova Yulia Aleskandrovna for allowing to use her images in the publication.

References

[1] Obednina S V 2013 Modular principle for form creation in design *Academical journal of UralNIIproekt RAASN* **1** 85–90

[2] Zyrina M A, Dembich N D, Gryazeva I V, Kazakova N Yu 2020 Visual perception of the city environment in the context of the modern costume *Herald of Slavic cultures* **1** 287–302

[3] Kuznetsova E I 2016 Social technologies as a basis of designer's creativity *Herald of the Institute of Economics RAS* **4(88)** 100-12 Accessed on Jun. 09, 2020. [Online]. Available:cyberleninka.ru/article/n/sotsialnye-tehnologii-osnova-dizaynerskogo-tvorchestva/viewer

[4] Ulrich K, Eppinger S 2007 *Product design and development. Industrial design: creation and manufacture of a product.* (Moscow: Vershinina)

[5] Maier M, Rechtin E 2000 *The Art of Systems Architecting, second edition* (Boca Raton: CRC Press)

[6] Shimko V T 2006 *Architectural-artistical design of urban environment* (Moscow: Architektor-S)

[7] Shubnikov M V 2006 *Structural patterns of architectural morphogenesis.* (Moscow: Architecture-S)

IOP Conf. Series: Materials Science and Engineering **944** (2020) 012007 doi:10.1088/1757-899X/944/1/012007

[8] Konstantinovich A M 2009 *Key prenciples of variability and transformation of constructions.* (Minsk: MIU Publisher)
[9] Saprykina N A 2004 *Architectural Form. Statics and dynamics.* (Moscow: Architecture-S)
[10] Zykin Sh 2018 *City cultures* (Moscow: Novoe literaturnoe obozreniye)
[11] McQuire S 2014 *The Media City: Media, Architecture and Urban Space* (Moscow: Strelka Press)

IOP Conf. Series: Materials Science and Engineering **944** (2020) 012008 doi:10.1088/1757-899X/944/1/012008

Architectural and planning principles of STEAM-space organization

O V Morozova[1], E S Dukhanina[1]

[1] Kryachkov Novosibirsk State University of Architecture, Design and Arts, Novosibirsk, Russia

E-mail: o.morozova@nsuada.ru

Abstract. The article discusses the issue of organizing the educational space for the implementation of STEAM technology. The purpose of the article is to reveal the features of STEAM learning and to formulate the principles of organizing the learning space for this type of learning. On the topic of the research, a brief overview of the implemented architectural projects of educational spaces using STEAM technology was carried out. As a result, the authors identified architectural and planning features and principles of organizing spaces for STEAM training and created a model of an art and technology lyceum using the obtained principles for organizing a block of informatics and 3D modelling.

1. Introduction

STEM technology (science, technology, engineering, mathematics) is one of the most promising educational technologies, which confidently takes its place in the global educational trends. In 2015, a large-scale study of the European Schoolnet was carried out by an international organization consisting of the ministries of education of 31 European countries. [6] According to the results of the European Schoolnet study, 80% of countries noted STEM education as their priority and are developing several educational reforms that involve the use of STEM technologies.

STEM (science, technology, engineering, mathematics) approach is a wide range of actions, practices and techniques that are focused on the application of knowledge and skills from different scientific fields in one project. [4] For arousing of students' interest in technical subjects and give the finished product an aesthetic and philosophical component, it was decided to add an art component to STEM technology, resulting in STEAM technology.

For the successful implementation of this technique in educational practice, a particular educational space is required. This article analyzes the features of the implementation of STEAM technology and identifies the principles of organizing the learning space for STEAM technology. The authors have created a model of an art and technology lyceum using the principles obtained for organizing the block of informatics and 3D modelling.

2. Materials and Methods

The main theoretical research methods are the analysis and generalization of scientific literature and electronic publications on the topic, analysis and systematization of the essential characteristics of the space for organizing training in STEAM technology. For creating a model of the lyceum, the project method, the method of "getting used for the role" and the modelling method were applied.

3. Principles of organizing educational space for the implementation of STEAM technology

3.1. Necessary conditions and features of the organization of the STEAM learning space

The basis of STEAM training is formed by interdisciplinary projects; it is a kind of problematic situations that are set by the teacher. The goal of STEAM technology is to develop creative thinking, skills in using an engineering approach to solving real problems, understanding the importance of design, and understanding the role of technology in solving them. [1]

For being successful in implementing this educational technology, it is necessary to simulate an appropriate learning environment that is different from the standard classroom in the school. [3] At the stage of developing spaces that will correspond to the project-based and interdisciplinary nature of STEAM-learning, it is necessary to take into account many conditions and features of organizing the space for STEAM-learning:

1. The project activity is the main one, i.e. most of the classroom space should be dedicated to creativity and project development. [5] The geometry of the room can also be not standard rectangular, but square or hexagonal, which may allow for better positioning of furniture.

It is worth considering that when organizing project activities, active training, and a constant change of activity is a need. With this form of training, traditional lectures are rarely used, preference is given to discussions and creative practices.

2. Availability and active use of technology. The educational space should be filled with the necessary presentation equipment that allows teachers and students presenting their ideas (mobile magnetic whiteboards, plasma panel, multiple displays and a projector). Students have quick access to laptops or tablets with the necessary software, microphones and headphones. Uninterrupted Wi-Fi and power supplies are of great importance, which students can use not only during but also after classes, in the course of independent work on a project.

3. Continuous learning, which involves working on a project not only during the lesson. After classes, students should be able to engage in discussion and revision of the project in special informal spaces – coworking spaces, laboratories and other types of activities. Access to such premises should always be open.

4. An interdisciplinary set of educational programs means a specialised curriculum and a convenient location of classrooms for related disciplines and joint projects. This circumstance is in line with the trend to use an integrated curriculum rather than teach related subjects in isolation.

3.2. The principles of organising the educational space for the implementation of STEAM technology

Today STEAM-training has established itself as a progressive educational technology. Over the past ten years, attractive educational spaces have appeared in the world that reflect not only the ideology of STEAM learning but also modern aesthetics and design. The following examples reveal the architectural and planning features and principles of organising spaces for STEAM training.

ARC / Architectural Resources Cambridge, 2013, UK.

In the ARC project, the educational organisation needed to provide space for the implementation of an integrated curriculum. For this, the architects have completely united two different "wings" of the building – "Science" and "Mathematics". In the new project, science and math classrooms create a single space, excluding separate study laboratories and classrooms. Courses such as computer science, robotics, and high-level DNA research are also being incorporated into the new centre, which in turn will strengthen the technological and engineering components of the STEM curriculum. [2]

Laboratories, classrooms and ancillary facilities are designed to accommodate a variety of activities. Students and teachers of presentations, group discussions and projects, individual research work (Fig. 1), can use space for demonstration. The flexibility and mobility of the classroom allow varied the learning process, and the presence of transformable furniture allows changing the environment for individual or group exercises.

IOP Conf. Series: Materials Science and Engineering **944** (2020) 012008 doi:10.1088/1757-899X/944/1/012008

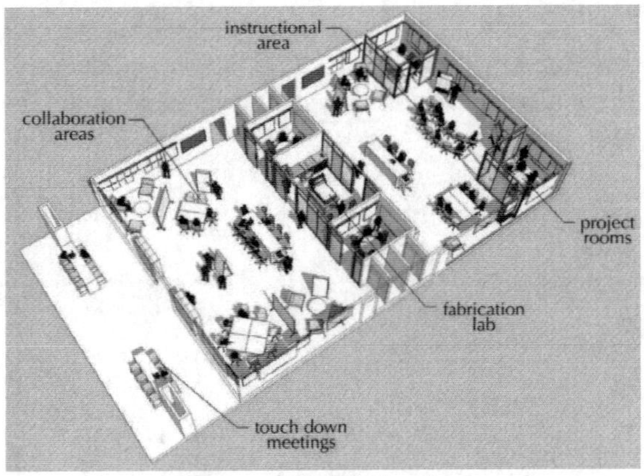

Figure 1. Mobile class
Flexible classrooms
Source: https://arcusa.com/node/292

3.3. HMC Architects, Fremont Academy of Engineering and Design in Pomona, Calif., 2015, USA
At the Fremont Academy of Engineering and Design in Pomona, California, architects from HMC have solved a variety of planning challenges. The architects designed the space to deliver a program for grades 7-12 based on the Project Lead the Way curriculum. This integrated curriculum assumes the presence of both special engineering classes and classes for English and mathematics classes that are related to an engineering focus.

The new laboratory is part of a newly built building located on the existing Fremont campus. The building is very spacious with high ceilings and large windows making it very bright. Some of the walls were deliberately left white and blank for displaying material through a projector. The engineering laboratory has mobile furniture that can be easily moved around the study space and is also adjustable for the convenience of students. The laboratory equipment includes 3D printers, a drilling machine, a panel saw and a CNC milling machine. This room is suitable both for individual work at the machines and for group project discussions. [7]

In the centre is a room for storing projects and materials, which works for both rooms and connects them together.

The design lab is on the other side of the warehouse and has perimeter storage and movable furniture for group work, presentations and various computer and design activities (Fig. 2).

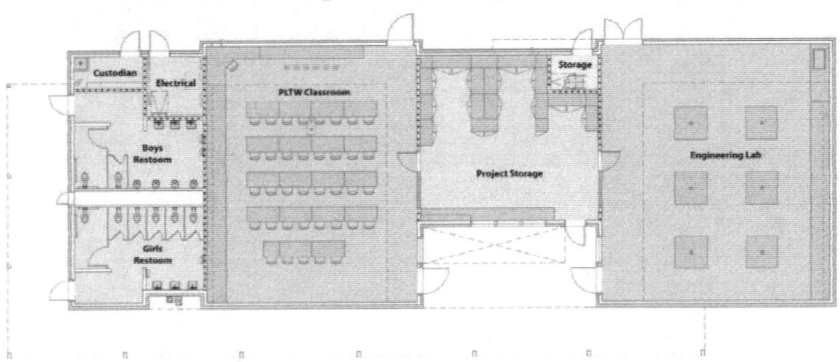

Figure 2. Plan of the Academy of Engineering and Design Fremont
Source: https://hmcarchitects.com/news/planning-a-stem-facility-despite-the-big-unknowns/

Based on their pedagogical and technological features of STEAM-teaching and the considered examples of implemented educational spaces, it is possible to formulate the principles of organizing the educational space of STEAM-learning.

1. Mobility of spaces with free planning and transforming classes, where the teacher and students have the opportunity to organize the educational process independently. All laboratories and classes should be connected, or have a shared space – a hall for constant meetings and discussions.

2. High manufacturability of materials and active use of technologies – the use of innovative and low-cost energy-saving engineering solutions and thoughtful design.

3. Organization of an informal space for students, where they can independently work on an assignment or project outside the classroom.

All of the above principles of space organization work only with an integrated curriculum and an interdisciplinary approach.

4. Results

Educational spaces for the implementation of STEAM learning have the following characteristics:

• these spaces are intended for project activity, which is the main one in STEAM training;

• ensure the availability and active use of technology;

• implement lifelong learning, which implies working on a project outside of school hours in specialized and open spaces;

• designed for an interdisciplinary set of programs.

During the analysis of the implemented projects of educational spaces for STEAM training, the following principles of organizing the educational space for STEAM training were identified:

1. mobility of spaces with free planning and transforming classes,
2. high manufacturability of materials and active use of technologies,
3. organization of an informal space for students.

The principles obtained were applied in the project of the Art and Technology Lyceum in Novosibirsk to organize a STEAM space located in the architectural and art block (Fig. 3). At the request of the customer, STEAM-space is implemented for the module of information technology disciplines - computer science, 3D-modeling, prototyping and design activities. In the structure of the Lyceum, STEAM-space occupies part of the 3rd floor and is a single space divided into zones – study rooms for individual subjects and a central common hall with mobile furniture for project discussions (Fig. 4).

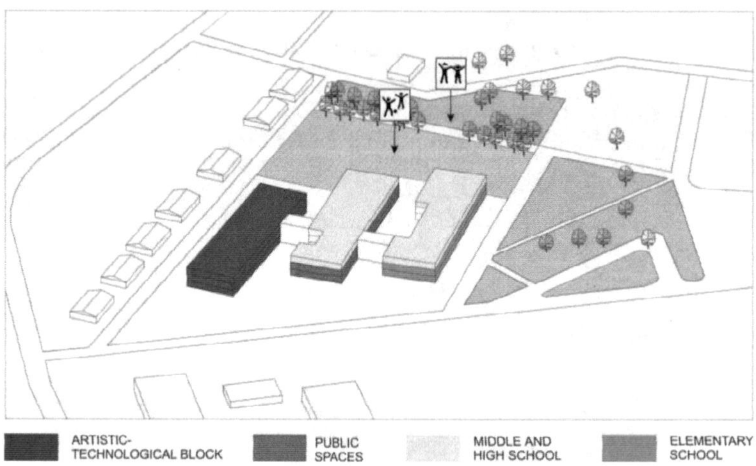

Figure 3. Functional blocks of the building of the Art and Technology Lyceum in Novosibirsk
Source: authors archive.

IOP Conf. Series: Materials Science and Engineering **944** (2020) 012008 doi:10.1088/1757-899X/944/1/012008

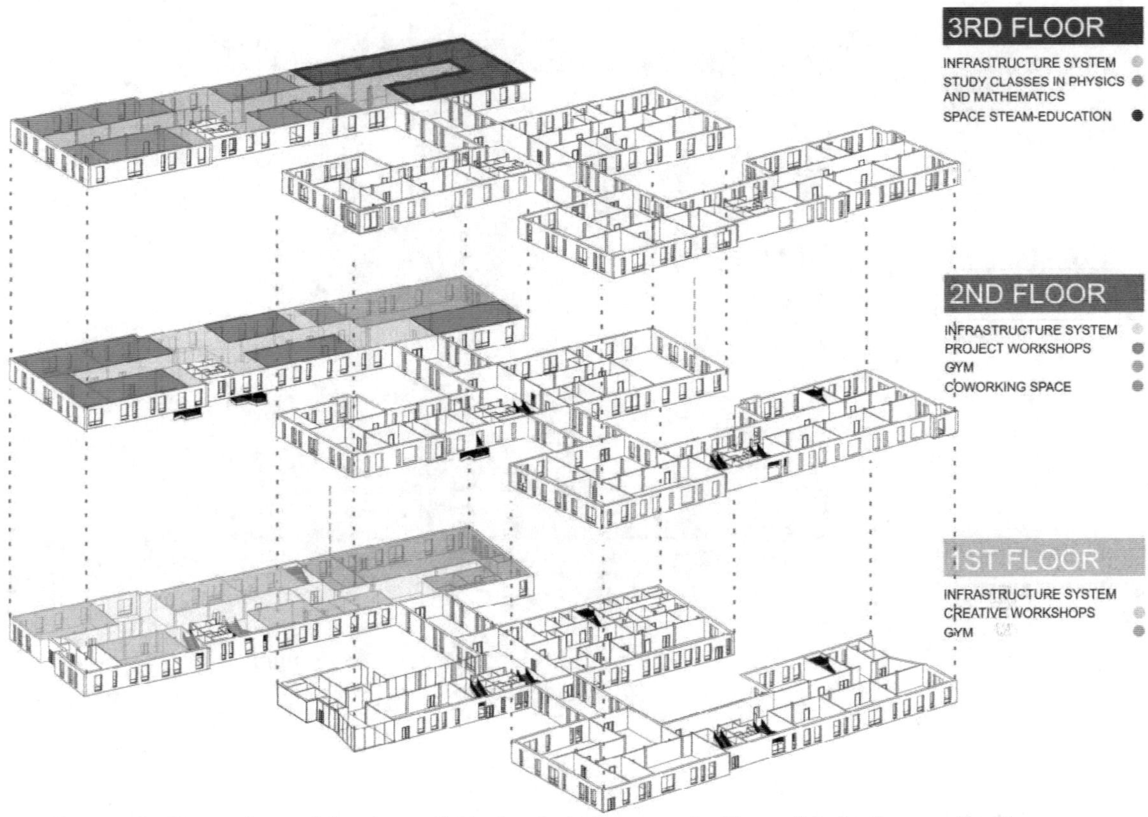

Figure 4. The project of the Art and Technology Lyceum in Novosibirsk. Space-planning scheme of the lyceum, STEAM-space for the module of information technology disciplines
Source: authors archive

The common hall, if necessary, can become an area of communication or teamwork, it can be subject rooms. The mobility of spaces is achieved by a sliding partition between the computer science rooms, which can be used as two separate classrooms or as one ample space for holding a competition or developing a team project. Also, all the cabinets of the block are interconnected by glass partitions and doors (Fig. 5).

In the block of informatics and 3D modelling, there are computer classes with projectors and mobile TV panels, a design workshop adjacent to a class in 3D modelling and 3D printers, a laboratory assistant is equipped with workplaces for teachers and a place to relax. The interior design uses ecological materials, light walls with cork boards for the presentation of ideas or the process of discussion.

For independent work in the lyceum, there are coworking and recreation spaces with mobile partitions, which, if necessary, create a working area or a presentation area. Also, the library provides space for independent work and a small area for projects.

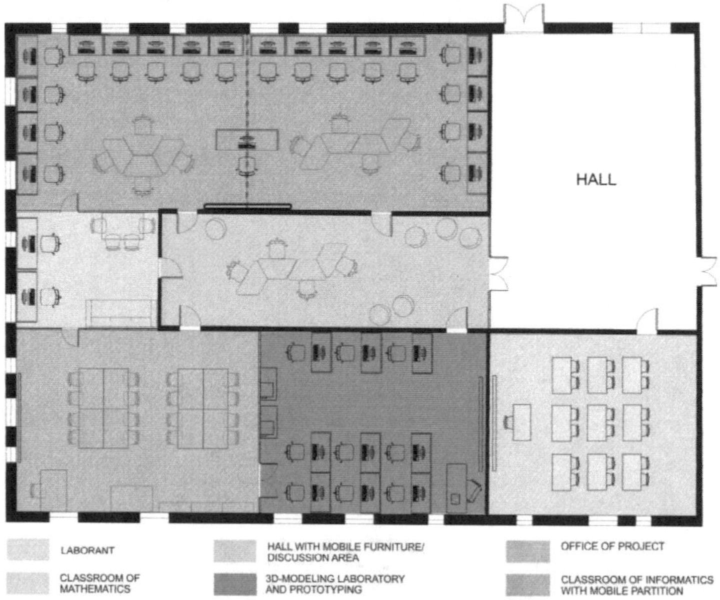

Figure 5. STEAM education space plan
Source: authors archive

5. Conclusion

Modern learning technologies, such as STEAM learning, have a direct impact on the architectural and planning organization of the learning space. Successfully implement STEAM technology needs to create a unique learning environment that meets the conditions and features of this technology.

Architecture and pedagogical technology are interdependent. When designing any educational space, the set of pedagogical technologies and the form of teaching should be initially determined. STEAM training is based on the project method, active learning and the use of technology. Therefore, in the project of the Art and Technology Lyceum in Novosibirsk, special attention was paid to creating a mobile space with available zoning of the educational space, taking into account the building codes and regulations of the Russian Federation. Also, a large number of informal spaces were created at the Lyceum for students to work independently during off-hours.

It should be emphasized once again that a simulated space for STEAM learning only works with interdisciplinary projects and an integrated curriculum. Based on the formulated principles and the author's model of the Art and Technology Lyceum in Novosibirsk, it is planned to develop variable layouts of the main functional blocks of premises required for the implementation of STEAM technologies. Architects when designing new school buildings or educational grounds using STEAM technology can use this set of diagrams and particular recommendations.

References

[1] Barrows H S 1980 Problem-based Learning: An approach to medical education *Springer series on Medical Education* (NewYork)

[2] Cottone N, AIA, LEED AP. STEM Turning the Buzz into Functional Design, 2013. URL: https://arcusa.com/node/292

[3] Design Trends Influencing STEM Education, 2019. URL: https://spaces4learning.com/Articles/2020/02/19/Design-Trends-Influencing-STEM-Education.aspx?Page=2

[4] Grand-Clement, Sarah, Axelle Devaux, Julie Belanger, and Catriona Manville, Digital learning: Education and skills in the digital age. RAND Corporation and Corsham Institute 2017 URL: https://www.rand.org/pubs/conf_proceedings/CF369.html.

[5] Shafaee M and Madani R (2010) Design principles of educational facilities for children based on the child's creativity *Journal of Technology of Education* **4(3)** pp 215–222.

[6] STEM-podkhod v obrazovanii idei metody praktika perspektivy Minsk 2018 URL: http://edu4future.by/storage/app/media/camp/stem-podkhod-v-obrazovaniiprint.pdf

[7] Wilkeson K Planning a STEM Facility Despite the Big Unknowns 2015 https://hmcarchitects.com/news/planning-a-stem-facility-despite-the-big-unknowns/

IOP Conf. Series: Materials Science and Engineering **944** (2020) 012009 doi:10.1088/1757-899X/944/1/012009

A new approach to the design of street furniture in Krasnoyarsk

P P Zvonareva[1], I V Filipskaya[1], I V Yangulova[1]

[1] Institute of architecture and design, Siberian Federal University, Krasnoyarsk, Russia
E-mail: zvono4ek2008@yandex.ru, ifilipskaya@gmail.com, yangirina@yandex.ru

Abstract. The article describes the trends in the field of small architectural forms and outdoor furniture in the public environment of Krasnoyarsk. The objects of study are outdoor furniture located in central and popular public places that shape the image of the urban environment. The article contains a study of street improvement in the city of Krasnoyarsk, which made it possible to assess the current state of furniture used in the city. The research methodology consists in conducting an Internet survey of respondents in the social network "VKontakte". The survey showed the advantages and disadvantages of existing furniture objects according to ergonomic, artistic, aesthetic, environmental and economic criteria. The collected material made it possible to determine changes in the approach to the design of outdoor furniture for public urban spaces, made it clear the feasibility of an integrated solution for environmental objects. The article demonstrates the design concepts of products with progressive design and improved qualities. The authors-developers propose a modern approach to the organization of public spaces and the design of products, taking into account the anthropometric and physiological characteristics of the population. They explain the use of modern materials in terms of ergonomics and ecology, such as wood-polymer composites, plywood, modified wood. The recommendations of the expert in the person of the chief architect of the city, in the design and implementation of such objects, were taken into account. The results of the analysis have both theoretical and practical significance and will be used in the improvement of urban spaces within the framework of the “Comfortable Urban Environment” program in Krasnoyarsk.

1. Introduction
Recently, the city is experiencing a new wave of renovation – rethinking public and business spaces. In Krasnoyarsk, as in any industrial metropolis, the urgent problem is the total problem of the improvement of public space and places for recreation of citizens. This problem is caused by the choice of non-ergonomic outdoor furniture, ineffective arrangement of objects, the absence of a number of functional elements, such as ballot boxes, lighting masts, bicycle parking. It is worth remembering that public spaces: parks, squares, squares are the best place for the implementation of interesting youth projects. Such projects give the city a new look, gradually introducing into it a modern, progressive, innovative solutions and ideas [3].

2. Materials and Methods
The objects of study are places for recreation, architectural forms, outdoor furniture located in squares in significant areas of Krasnoyarsk. The authors analyzed the needs of Krasnoyarsk residents, the

latest trends in the design and arrangement of public spaces in the city. For articles were selected results of the survey the most popular city places for recreation.

The research is based on the method of analyzing the results of public opinion polls, expert interviews, presenting conceptual questions and conclusions in professional communities at scientific conferences. Also, a technical examination of the quality of furniture and materials was carried out.

In Krasnoyarsk, attempts are made to improve the appearance of the city, it can be clearly seen how the shape of the central streets - Prospect Mira (Fig. 1). However, the location of urban objects and their layout calls into question their relevance among citizens. For example, benches are placed in large numbers in small areas, which is why half of them are not used. Obviously, people can rest "face to face" with such an arrangement of seating positions, which is contrary to ergonomics and a comfortable psychological atmosphere.

Fig. 1. City center, Prospect Mira, Krasnoyarsk.

The materials for furniture are often coniferous woods growing in the Krasnoyarsk Territory, but in the summer at elevated temperatures, such products secrete resin, which leads to damage to clothing and even injuries. Sculpture “Valentin and Valentina” - the territory of the former fountain is one of the well-known projects. The place has no public attractiveness to the extent that is possible, because the reconstruction of the layout is based on multiple duplication of similar forms of benches on the perimeter of the square, and some benches are located facing the road. (Fig. 2).

Fig. 2. Sculpture “Valentine and Valentine”, Krasnoyarsk.

3. Results

Objects in the examples are the result of an unsuccessful approach to the design. These examples did not take into account such factors as ergonomics, wind protection, lighting, a sufficient number of trash bins, rational use of materials.

To achieve a good result in the creation and improvement of outdoor furniture, one should take into account the principles of functional diversity, a comfortable environment for communication, harmony with nature in terms of ensuring the diversity of the visual appearance of the territory, the use of environmentally friendly and practical materials such as wood-polymer composite, birch plywood, modified wood [1].

The Internet survey has become an operational method for researching and identifying furniture disadantages. The survey included questionnaires, data processing, and obtaining a graphic representation of the results. This method is an automated process, does not require material costs and saves time. The "VKontakte" social network with a target audience of about 100 people aged 19 to 40 was selected for the survey. The questionnaire was attached to the publication about the improvement in the city of Krasnoyarsk and consisted of four questions with the proposed answer options, and the interviewed audience could express their opinion (Fig. 3). This method is very popular on social networks and is highly effective.

Thus, it was possible to identify and understand what details of the arrangement of the existing park zones of the city of Krasnoyarsk cause dissatisfaction, as well as what primarily attracts people in places of street space. First of all, the respondents expressed a lot of complaints about the design and appearance of street furniture, city residents would like to see the objects modern and attractive. Secondly, there were remarks about ergonomic characteristics - sharp angles on the benches, angles of inclination of the backs that do not correspond to a comfortable rest, splinters and resin remain from poor-quality materials. Third, based on the existing need for waste sorting, the interviewed audience spoke about the need to install additional waste bins for plastic. Some of the respondents were dissatisfied with the irrational organization of space, the arrangement of benches, the lack of visors, and bicycle parking. Structural projects were also noted, if any parts break, repair and replacement are impossible, in such cases the dismantling of the entire structure and the installation of a new product are required, and sometimes due to lack of funding, the objects of street furniture are broken for several months.

Students of the Design Department of the Institute of Architecture and Design of the Siberian Federal University have developed projects to modernize and fill these public spaces with furniture, which creates a favorable psychophysiological atmosphere among the townspeople.

In the "Delta" project, attention is paid not only to filling the territory with furniture, but also to the development of the structure, layout of the area in the urban environment (Fig. 4). Since the square is a corner element of urban planning, this means that in the process of compositional planning, it is necessary to consider the factor of visitor movement. This problem reveals the main task - to captivate the visitor by providing him with an unobtrusive choice – "to go further or to give in to the enchanting image of the atmosphere and hide for a moment from the noise of the big city".

The furniture in this project is a parametric design. The elements are modeled using the Rhinoceros program [4].

The seating position in the unfolded position is held together by a bar, which allows the vertical modules to take a stable position. When the elements are folded, the fold goes up, and the moving part takes a horizontal position. Then it is fixed with the same bar, already passed through the entire folded structure through the provided holes. Thus, the panel changes its functional role in the decorative on the condition of minimum costs of dismantling, in the project term. This solution makes it possible to make a few benches from one due to the transformation of the elements (Fig. 5).

IOP Conf. Series: Materials Science and Engineering **944** (2020) 012009 doi:10.1088/1757-899X/944/1/012009

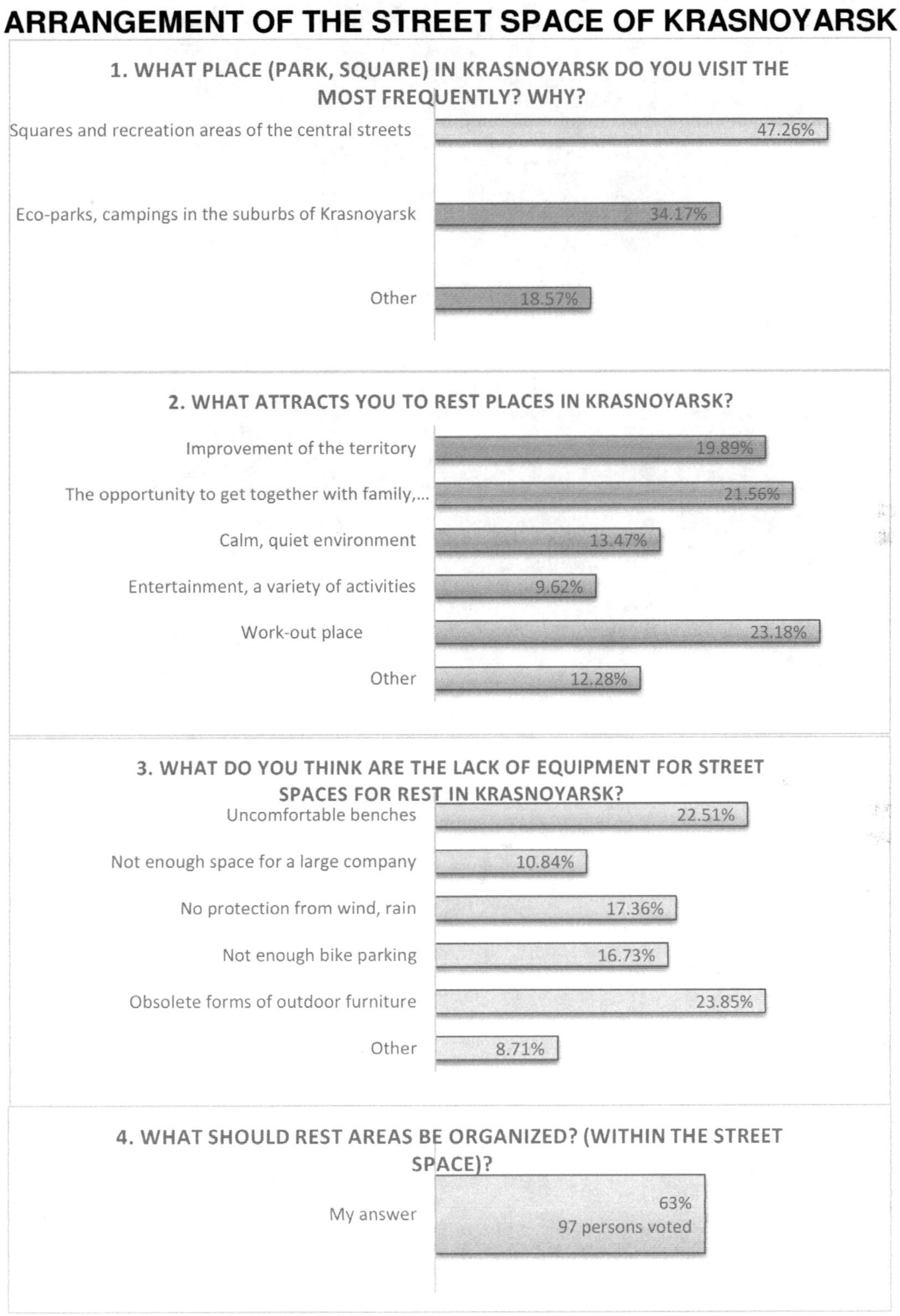

Fig. 3. The results of the public survey

Fig. 4. The "Delta» project.

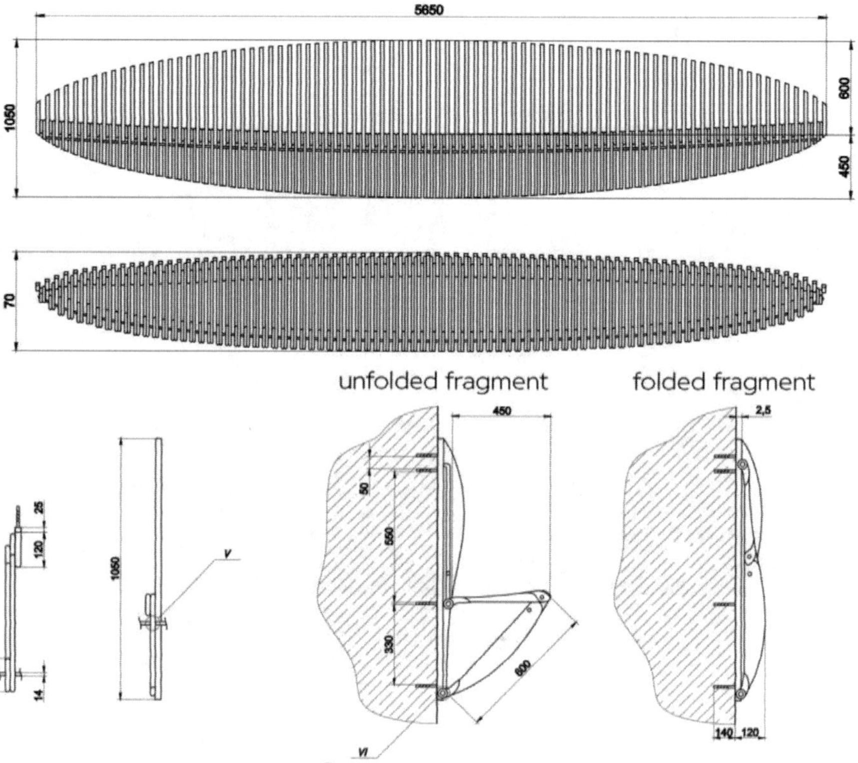

Fig. 5. The scheme of transformation and assembly of the bench.

The aesthetic qualities of objects are also very important. Benches can be composed in various compositions using a single line, which will help to create a unique image of the area, and the lights can be adjusted n height and inclination, which also provides additional configuration options. It is necessary to use environmentally friendly, affordable and durable materials - concrete, metal and wood as the main construction materials. (Fig. 6).

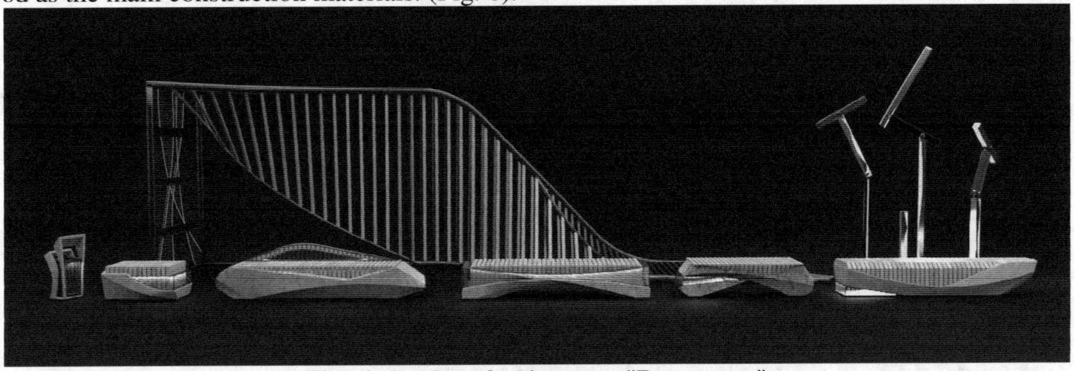

Fig. 6. Outdoor furniture set "Resonance".

A distinctive feature may be the presence of a swing complex (Fig. 7). For the manufacture of such products, it is planned to use traditional and modern materials, such as metal, plywood and wood-polymer composite, which is growing in popularity as a structural material for outdoor facilities, it does not swell and does not require a protective finish [7].

Fig. 7. Outdoor furniture for SFU public space.

Outdoor furniture design should take into account the ergonomic and functional needs of the townspeople. This style of modern performance of small architectural forms is already conquering the parks and squares in the developing cities of Russia [5].

4. Discussion

In general, today the city of Krasnoyarsk is ready for transformation, for interesting design solutions that should satisfy the needs of the townspeople. According to the results of the survey, among respondents aged 19-40 years, the following problems of organizing public urban spaces were noted, such as a shortage or irrational arrangement of waste bins, outdated forms of objects, uncomfortable benches, an insufficient number of bicycle parking, irrational compositional solution of seating places for a large company, an overabundance of asphalt pavements, cheap furniture materials and a low aesthetic component. It is worth remembering that public spaces such as parks and squares are an amazing place to implement interesting youth projects.

The results of this study will be used by the Department of Municipal Services in the implementation of measures within the framework of the "Comfortable Urban Environment" program operating in Krasnoyarsk, which provides improvement of public areas. Krasnoyarsk is preparing to celebrate its 400th anniversary, therefore, grants for the improvement of squares and parks for the installation of art objects and outdoor furniture are regularly held [6].

5. Conclusion

The problem of a competent approach to the design of outdoor furniture objects is increasing its relevance. Town planning departments are interested in the selection of high-quality samples of outdoor furniture, designed as a result of an integrated approach, which takes into account the color-texture solutions of materials, laws of rational composition and technology. They choose furniture that provides conditions for a healthy, comfortable, convenient life, both for an individual and for all residents of the city of Krasnoyarsk.

Recognizable outdoor furniture will become a hallmark. The city will acquire a recognizable appearance and will attract guests from other cities and countries.

6. Acknowledgements

The authors are grateful to A.V. Kuklina, N.V. Beloshedova, T.M. Chizhova for the research carried out in the process of graduation design, proposed developments and projects of street furniture.
Special thanks to Yu.A. Solovarova to the chief architect of the city for an informative conversation and advice on the improvement program.

References

1. Runge V F and Manusevich Yu P 2005 *Ergonomics in the design of the environment: a tutorial* (Moscow: Architecture-S)
2. Kurochkin V A 2018 Humanization of urbanized spaces by environmental design means [Electronic resource] *Architecton: news of universities* **1(61)** URL: http://archvuz.ru/2018_1/13
3. Lyaptsev N N and *Naumova* S V 2018 Urban public spaces: the evolution of the problems (the contemporary design context) [Electronic resource] *Architecton: news of universities* **3(63)** URL: http://archvuz.ru/2018_3/16 (date of the application: 29.01.2020)
4. Designing parametric structures in Grasshopper [Electronic resource] URL: http: // https://kd3d.ru/3d-modelirovanie (date of the application: 29.07.2020)
5. March, technology and marketing for museums [Electronic resource] URL: http://mart-museum.ru/2014/05/28/kinetic/ (date of the application: 14.03.2020)
6. Electronic fund of legal and normative-technical documentation: On the approval of the municipal program "Increasing the efficiency of the activities of city government in the formation of a modern urban environment" for 2018-2024 [Electronic resource] URL: http://docs.cntd.ru/document/446636060 (date of the application: 20.06.2020)
7. Kielė A, Vaičiukynienė D, Tamošaitis G, Pupeikis D and Bistrickaitė R 2020 Wood shavings and alkali-activated slag bio-composite *Holz als Roh und Werkstoff* **3(78)** pp 513–522

IOP Conf. Series: Materials Science and Engineering **944** (2020) 012010 doi:10.1088/1757-899X/944/1/012010

Microalgae in architecture as an energy source

S V Ilvitskaya[1], A G Chistyakova[2]

[1] Federal State Budgetary Educational Establishment of Higher Education “State University of Land Management” (GUZ), Department of Architecture, 105064, Kazakova Street, 15, Moscow, Russia
[1] Federal State Budgetary Educational Establishment of Higher Education “National Research Moscow State University of Civil Engineering” (NRU MGSU), Department of Architecture, 129337, Yaroslavskoye Shosse, 26, Moscow, Russia
[2]Federal State Budgetary Educational Establishment of Higher Education “State University of Land Management” (GUZ), Department of Architecture, 105064, Kazakova Street, 15, Moscow, Russia

E-mail: ilvitskaya@mail.ru

Abstract. Microalgae possess unique properties: in the course of their life, microalgae, through the absorption of carbon dioxide, sunlight and nutrients, carry out the process of photosynthesis and turn into biomass. Biomass is further processed and many types of products are produced from it, such as oils, nutraceuticals, livestock feed, fertilizers and other substances. Some of them can be used as a renewable energy source, for example, for the production of biofuels. The production of biofuels using biomass as a renewable energy source, which is an inexhaustible source for solving the energy problem, is becoming increasingly important. The use of algae in architecture has brought many benefits, such as energy saving, reducing carbon dioxide emissions, oxygen generation, biofuel production, wastewater treatment at the micro level using building facades and at the macro level by integration into urban space. This article analyzes the problems associated with these aspects and proposes their solutions, which must be taken into account when further integrating photobioreactor systems into architecture.

1. Introduction

Today it seems impossible to live in a world without energy, its constant use has become an integral part of our lives. Most of the energy comes from fossil fuel sources, the burning of which causes environmental pollution and climate change. At the same time, a large amount of carbon dioxide is released into the atmosphere [1]. Fossil fuels, especially oil, coal and natural gas, provide about 85% of our energy needs worldwide. The biggest problem with fossil fuels is that they are the ultimate source, and they are in danger of disappearing in the near future. Today, traditional sources of energy, such as oil and coal, etc., are used to a greater extent than renewable sources of energy. To achieve a reduction in carbon dioxide emissions, the use of renewable energy sources such as nuclear energy, solar energy, geothermal energy, wind energy and microalgae biomass energy is inevitable [2]. Unlike fossil and nuclear fuels, alternative energy is provided by eternal natural sources (wind, sunlight, geothermal energy and microalgae biomass). Using these resources to meet our energy needs supports sustainable development by reducing greenhouse gas emissions. The development and use of renewable energy

IOP Conf. Series: Materials Science and Engineering **944** (2020) 012010 doi:10.1088/1757-899X/944/1/012010

sources will bring significant benefits to all countries of the world, including energy production, environmental protection, pollution reduction and the development of modern technologies.

With the growing energy consumption in the world and the depletion of fossil fuel reserves, we must be sure that in the near future we will be ensured energy security when traditional sources of energy are exhausted. Currently, the energy we use comes from fossil fuel sources, and when it is produced, greenhouse gas CO2 is released, and then it is transported over long distances. In this regard, energy needs local production from a sustainable source. Buildings contribute 30% of the total energy consumption, and therefore, if buildings can generate their own energy, this will have a big impact on its total consumption. Currently, wind generators and photovoltaic panels are popular energy-saving installations. These systems can sustainably generate electricity, but they cannot consume carbon dioxide to lower their level. Microalgae can simultaneously reduce CO2 emissions and produce energy [3].

2. Materials and methods

Numerous parameters affect algae cultivation and the lipid content of algae affects the biomass harvested. Photosynthesis is a reaction that provides the first conversion of sunlight into energy. Consequently, all the components involved in photosynthesis contribute to the growth of algae. CO2, water consumption, supply and concentration of nutrients (nitrogen, phosphorus), light, O2 removal, temperature, pH and salinity. In addition to environmental parameters that can photosynthesis of microalgae to efficiently collect microalgae, operating parameters are also very important. [4] These operational parameters relate to the conditions for growing microalgae in the photobioreactor system. Lighting, mass transfer, mixing, dilution coefficient, collection frequency, scaling, process control and cleaning are important operational parameters to create optimal conditions for growing microalgae.

The fuel component is converted into biodiesel, and the remaining biomass can be used for fertilizers, animal feed and other auxiliary products. The nutrients in the system are removed from the water and processed to produce biomass. [5] A portion of the biomass used is used to produce electricity that is consumed during biomass production or for biogas production by anaerobic digestion. Excess electricity can be stored in batteries or sold by the city network. Photovoltaic cells can also be used to protect the electricity generated by the microalgae system. Anaerobic digestion waste can be used as a nutrient-rich fertilizer. Hydrogen gas is also released from the microalgae biomass. But as a source of hydrogen, microalgae are still under research and development.

Oil is obtained by drying microalgae. After oil extraction, anaerobic fermentation of algae sludge with a high content of organic substances is carried out. Thus, biogas energy is obtained due to the high content of organic substances in microalgae used for the production of biodiesel [6]. Converting biogas to biogas produces electricity that can be reused for growing microalgae. Thus, biodiesel, biogas, animal feed and electricity are generated from microlites at the same time.

3. Results and discussions

According to studies, biomass energy is considered the main source of renewable energy, as it has significant advantages over others. For example, it is available all over the world, it is relatively simple to process, without requiring expensive equipment, in addition, it can be stored for a long time and reused. To produce biomass energy of the first generation, which is considered an alternative to traditional fossil fuels, by producing biodiesel from oils extracted from various land plants, such as soybean, sunflower, palm, corn, coconut and others, large agricultural areas are needed. [7] Thus, the use of agricultural land with a limited area capacity and a limited capacity of fuel obtained from agricultural products, which form the basis of the fuel product, makes algae more productive among other types of biomass [8]. To reduce the use of agricultural land, as well as to reduce the cost of transporting fuel, it is proposed to use buildings and urban space for the implementation of these energy sources. New types of photobioreactors make it possible to convert even more natural resources into energy. By incorporating this system into architecture, we get buildings that are independent of the city

IOP Conf. Series: Materials Science and Engineering **944** (2020) 012010 doi:10.1088/1757-899X/944/1/012010

network. Autonomy will create safe conditions for medical facilities in case of interruptions or natural disasters, as well as more comfortable conditions for living in an urban environment [9]. Despite the fact that photovoltaic solar panels and wind turbines as integrated renewable energy sources are very common in architecture, algae biomass as the most efficient source has limited use in this area.

Microalgae-based photobioreactor systems are typically used in architecture in three different ways: to create integrated photobioreactors as a second facade, to form holistic urban spaces, and to create individual small architectural forms of microalgae [10].

The use of a photobioreactor integrated in building facades.

1) B.I.Q House, Hamburg, Germany (Figure 1).
2) Photo.Synth.Etica, Dublin, Ireland.
3) Process Zero Concept Building, Los Angeles, California, USA.

Figure 1. B.I.Q House, Hamburg, Germany.

The formation of integral urban spaces.

1) Carbon T.A.P. (Tunnel Algae Park), Philadelphia, USA (Figure 2).
2) Culture Urbaine, Geneva, Switzerland.
3) AlgaEnergetyCity, Istanbul, Turkey.

Figure 2. Carbon T.A.P. (Tunnel Algae Park), Philadelphia, USA.

Separate small architectural forms based on microalgae.

1) Urban Algae Canopy, EcoLogicStudio (Figure 3).
2) Street lamp, Pierre Callech.
3) Living beings, Jacob Dunias and Ethan Frier.

Figure 3. Urban Algae Canopy, EcoLogicStudio.

4. Conclusions

Throughout the evolution of the architectural environment of the city, gardening has largely determined the quality of urban areas under the conditions of modern super-urbanization of the urban environment, the resource for the formation of parks and city squares has been practically exhausted [10]. The greening of the buildings themselves often also cannot completely neutralize the negative impact of anthropogenic factors, and offset the reduction in the productive surface of the Earth. Under these conditions, one of the promising technologies is the integration of microalgae cultures into urban spaces. According to a number of indicators, this technology is four times more effective than conventional landscaping methods, which allows almost completely preserving the productive potential of territories alienated from nature, devoting 75% of the site area for construction.

Thus, using photobioreactive systems as a second facade in buildings and targeted implementation of them in architecture at micro and macro levels, it was noted that this integration has many advantages, such as energy saving, oxygen supply to the air, and reduction of carbon dioxide emissions into the atmosphere, biofuel production and wastewater recycling. This situation demonstrates a holistic approach to sustainable development, which can cover the whole city with the integration of photobioreactors developed using microalgae. Their unmatched benefits through a combination of technical, biological, and chemical cycles in architecture begin an innovative approach to architecture using renewable energy sources by integrating the values of ecological architectural design and shed light on future research.

Referenses

[1] Ilvitskaya S V and Polyakov I A 2014 *J. Natural and technical sciences* **78** 443–444
[2] Saprykina N A 2017 The formation of an eco-sustainable living environment of the future *Theory. Practice. Perspectives (Saarbrücken: Palmarium Academic Publishing 2017)* 232
[3] Lukyanov V A, Stifeev A I and Gorbunova S Yu 2013 *J. Bulletin of the Kursk State Agricultural Academy.* **9** 55–57
[4] Dah-Wei Tsai D, Honglay Chen P, Meng-jung Chou C, Feng Hsu C and Ramaraj R 2015 *J. Ecological Engineering.* **84** 386–389
[5] Voerman S E, Glasby T M, Gladstone W and Gribben P E 2017 *J. Marine Environmental Research.* **131** 205–214
[6] Mathimani T and Pugazhendhi A 2019 *J. Biocatalysis and Agricultural Biotechnology* **17** 326–330

[7] Alvarez J, Maider A, Lopez G, Olazar M and Bilbao J 2018 *J. Bioenergy and Biofuels* **1** 173–202
[8] Frederiksen P and Bogers M 2013 *J. Biomass and Bioenergy* **55** 1–2
[9] Usov Y Yu 2013 *The formation of the architectural and planning structure of bioclimatic residential buildings* (Moscow: MARCHI 2013) 30
[10] Ilvitskaya S V and Lobkova T V 2018 *J. Architecture and construction of Russia* **1** 108–114

IOP Conf. Series: Materials Science and Engineering **944** (2020) 012011 doi:10.1088/1757-899X/944/1/012011

Concept of a hybrid contextual model for designing environmental objects

K V Ishmukhametova[1], D R Kudasheva[1], S H Baymuratova[1] and A N Tuktarova[1]

[1] Ufa State Petroleum Technological University, 30, Kosmonavtov Street., Ufa, 450064, Russia

E-mail: ikv23@mail.ru

Abstract. This article is devoted to the hybrid model as a new paradigm for contextual design of an environmental object that has dualistic features within its genius loci. Features of the concept of contextual hybrid are revealed in the article on the basis of experimental design material devoted to problematic environmental objects of cities with unclear differentiation of contextual aspects. The strategy of building a hybrid contextual design model is described in the article on the example of the territory of "Cherkalikhina Gora" in Ufa. The authors introduce the concept of "urban memory" as an ideological platform for the project model of the territory. Dualistic features of contextual aspects of the environment through the rod of urban memory are impaled on the physical platform of a contextual hybrid model such as "nature", thereby assembling a single hybrid contextual model of environmental object design.

1. Introduction

Aspects of the urban environment are complementary. But they need to be restructured in order for the designer to have reference points for the project model of a new environmental object [1,2].

The methodology of designing environmental objects is based on the concept of binary opposition of compositional and contextual methods of shaping [4].

The compositional design method is considered as a process of borrowing and author's interpretation of cultural forms and prototypes [4].

Cultural prototypes of some particular places may resonate with current trends in environmental shaping. So nomadic ethnicities have no analogues of distinctive settled structures, and they are borrowed prototype forms from European culture. This approach leads to the unification and depersonalization of local environmental objects, giving them the features of an average "quality" architecture [6].

The second element of the considered binary opposition is the contextual approach of environment design [1,4].

The urban organism is the result of the vital activity, cultural and political will of previous generations. Therefore, it does not exclude the products of vital aberrations and, as a result, the appearance of antagonistic features in the deterministic aspects of the environmental context of specific urban locations [3].

Differentiation within the context of individual urban areas can be difficult by the absence of distinct characteristic parts of it. At the same time, the inseparability and dual essence of the context of these locations are elements of the self-identity of the entire city.

IOP Conf. Series: Materials Science and Engineering **944** (2020) 012011 doi:10.1088/1757-899X/944/1/012011

The structure of the environment object context can be conditionally divided into two mutually complementary "horizontal" and "vertical" levels [1].

The horizontal level is humanistic. It defines the process-event nature of the environment, set by the functional and behavioral logic of movement and interaction of people within a specific location.

The vertical level implies a change in the position of a person in the city as in space as in time. The behavioral logic of a person here goes to the mental, phenomenological level. The city becomes a field of meanings and spatial images for people.

2. Materials and methods

Using the horizontal and vertical levels of context, the authors propose the concept of "Contextual hybrid" as a way to form a contextual project model obtained by "interbreeding" various antagonistic environmental aspects [3].

The antagonism of environmental aspects is shown in the following. Figurative emanations of the functionality of neighboring objects are antipodes to each other. The historical building is mentally indistinguishable from the name of its owner who has no public sympathy. There may also be contradictions in the need for changes to the urban place related to the possibility of exploitation for people and at the same time in the preservation of an authentic silhouette outline of a natural or urban landscape of the same place.

The dualism of oppositional pairs reveals features that meet the value criteria in the field of spiritual and cultural, positive mental and imaginative, and functionally justified for the interested parties of the projected space [5,7].

Each aspect does not carry the potential to implement a strategy based on it for creating a memorable image of the space [5]. However, strengthened by the mutual influence of the hybrid context, they create a new contextual reality, which already claims to build on its basis the identity of the entire projected space.

3. Application of the concept of a hybrid contextual design model.

The territory of Cherkalikhina Gora in Ufa became an experimental platform for testing the concept of contextual hybrid. This area is densely built up and has a polar functional composition.

The territory also has a "hidden" historical and cultural potential. Historical buildings in this location are not rated as architectural monuments. The local population does not perceive the semiotics of these buildings as historical codes of past epochs [7]. The public marks them as traces of life's perturbations. These objects have a fixed mental image of the "old", "not functioning", which leads the citizen to a spiritual dilemma – "leave as a memory" - "remove as evidence of decrepitude and backwardness".

The Muslim cemetery of this territory has ambivalent features of a mentally shaped component. It is both a spiritually intimate territory and socially significant as a burial place for iconic personalities of Ufa. There is both a religious and sacred site centered in the mosque "Gafran" and everyday life processes of the local population. It has the imagery of mourning mysticism and optimism of spiritual revival as a place of Sunday school at the mosque and a summer children's camp.

The idea of "urban memory" became the basis for a positive perception of ambivalent images of cemetery and historical buildings with weakly expressed features as cultural objects and historical monuments [7]. Perceiving an object on a mental level, a person understands that it represents both the past, present and future of the territory. History objects become "assets" of the context, carrying the potential as starting semantic interpretations for new objects in the environment [2].

Thus, the water tower of the early twentieth century and candy and gingerbread factory of D. P. Brenstein, built in the late nineteenth century, become "pillars of memory". They mark benchmarks and reference points for growth and development on the "body of urban organism" [7].

The idea platform "urban memory" for the model of contextual hydride for the projected new environmental object begins to generate positive features of the surrounding dualistic context. The new object is an environmental connective tissue between the existing objects of the territory. So the new environmental object is a single historical and cultural complex that absorbs all the objects of the existing environment, equalizing and not contrasting them with each other, but alternately demonstrating and activating as a person moves within the new space.

The unique natural landscape of the place with viewpoints on the Belaya river on the cliff of Cherkalikhina mountain, a special pedigreed variety of plants with characteristic endemics of this place – all this is a special natural context of the territory. Nature becomes a contextual platform for assembling the physical aspect of a new hybrid context. As common places of all contextual aspects of the territory, the chosen contextual platform connects the urbanized territories of the center of Ufa (built according to the plan of V. Geste's quarter grid) with nature with its naturalistic geoplastics, biodiversity and panoramic views.

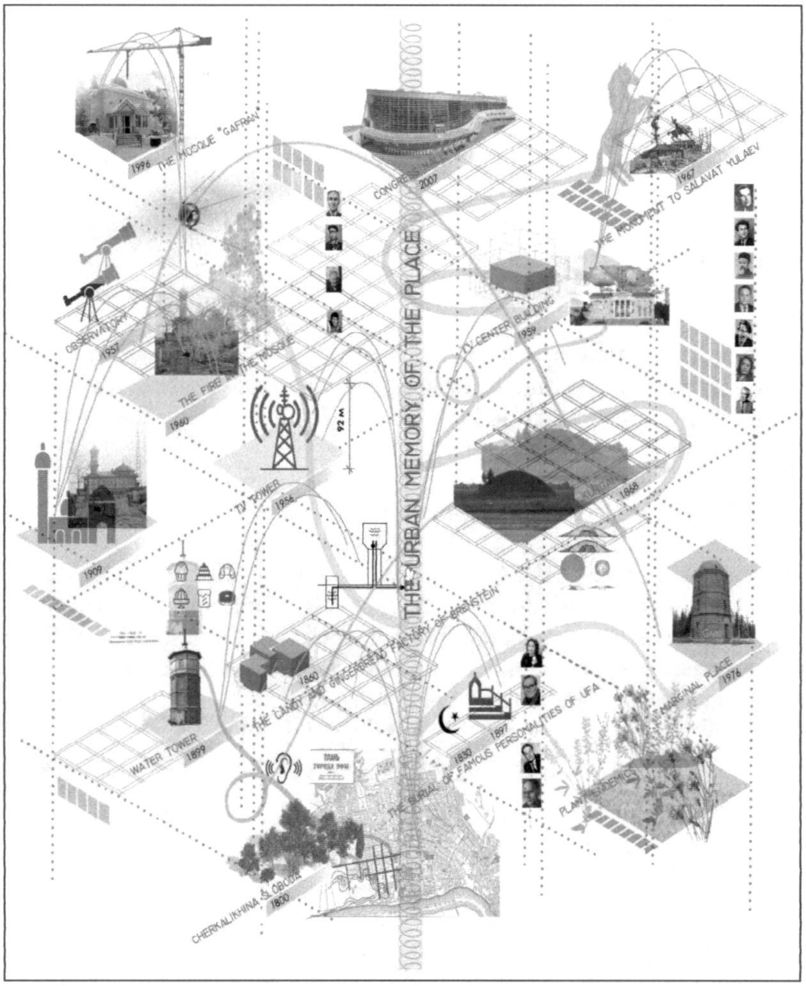

Figure 1. Hybrid contextual model of "Cherkalikhina Gora".

In this way, the metaphysical aspects of the environment through the ideological platform of urban memory are immersed in the physical aspect of the hybrid model. The physical platform of the hybrid model is the method of merging the natural component of the environment with the orthogonal grid of the existing development of the city center. The city merges with nature. Straight streets are extended in one plane and acquire the softness and outline of natural terrain in another plane. Breaking off a mountain "Cherkalikha " the streets become viewing platforms. Moving along the streets from the city center to the sites that open magnificent panorama of nature and distant objects of the urban environment, a person understands the metaphysics of the urban organism growing from the depths of centuries and looking to the future [5].

4. Conclusion

All the different aspects of the completely perceived context of the environment are united by the condition of temporary formation and relate to the results of the past and history.

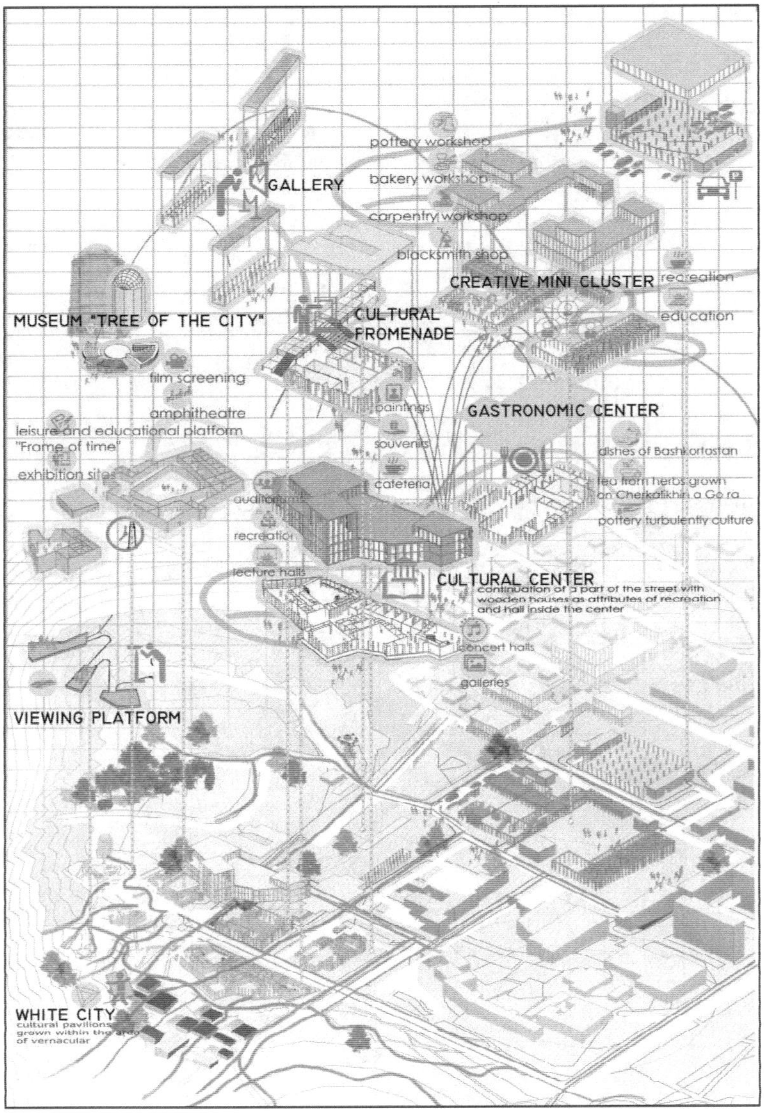

Figure 2. Design model of the historical and cultural complex "Cherkalikhina Gora".

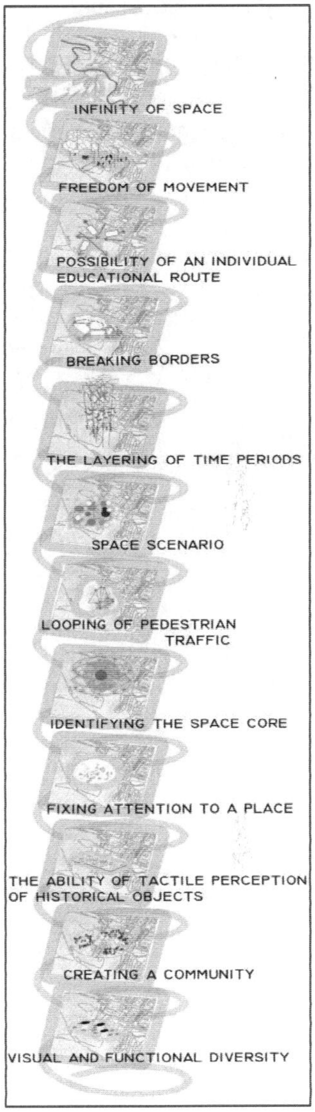

Figure 3. Impacting criteria for the project area.

Historical objects that do not have such valuable characteristics can enter as part of a single new architectural environment object, becoming the architecture of the narrative, giving an impulse and a clear vector to the future.

The designer, using the ideological platform of "urban memory", offers an environmental dialogue, immersing the visitor in an articulated plot, and changes the passive interaction with space to an immersive one [2,7].

The model being formed by the concept of contextual hybrid is aimed at identifying the list of requirements and restrictions. They are focused on achieving the harmonious integrity of the future environmental object as a space of image and form [1,5], where all the material components of the existing environment become resultant "characters" of the spatial mise en scene.

5. Acknowledgments

The main provisions presented in the article were tested in the framework of research work on the following contracts: "Development of forecasts and scenarios of spatial planning of the city Blagoveshchensk of the Republic of Bashkortostan" No. RB-04-19 and "Development of the concept of long-term spatial planning of the city Blagoveshchensk of the Republic of Bashkortostan" No. 15/12-19-MOBRB

References

[1] Rasuleva Y V, Kudasheva D R, Titarenko A V 2019 The concept of a multifunctional urban exoskeleton. *The Eurasian Scientific Journal.* **4(11)** 18SAVN419
[2] Rossi A 2015 *Architecture of the city* (Moscow: Strelka Press) p 264
[3] Paperny V Z 2012 The Spirit of time and the spirit of place in architecture *Speech: the spirit of place* **9** 44–49
[4] Malakhov S A 2011 Methods of compositional and contextual design as the most important elements of the binary opposition of an architect's professional activity *Vestnik VGASU: urban planning and architecture* **3** 107–112
[5] Lynch K 1982 *The image of the city* (Moscow: Stroyizdat) p 328
[6] Koolhaas R 2015 *Bigness or the problem of Large, Generic City, Junkspace* (Moscow: Strelka Press) p 84
[7] Donovan V 2015 Book review: Rites of Place: Public Commemoration in Russian and Eastern Europe *Anthropological forum.* **25** 178–180

IOP Conf. Series: Materials Science and Engineering **944** (2020) 012012 doi:10.1088/1757-899X/944/1/012012

Urbanization processes in the daily life of rural residents

T A Bulygina[1], E V Tufanov[2], S V Yanush[2], I N Kravchenko[2] and V A Ivashova[2]

[1] North Caucasus Federal University, 1, Pushkin street, Stavropol, 355017, Russia
[2] Stavropol State Agrarian University, 12, Zootechnicheskiy lane, Stavropol, 355017, Russia

E-mail: vivashov@mail.ru

Abstract. The article presents a vision of urbanization processes in the daily life of rural residents at the regional level. The methods of studying the processes of urbanization in the daily life of rural residents were: analysis of approaches to solving the problems of urbanization on the example of the Strategy for Socio-economic Development of the Stavropol Territory until 2035, taking into account a review of publications of scientists in scientometric databases; analysis of statistical data characterizing the vital activity of the urban and rural population of the Stavropol Territory; analysis of the results of a questionnaire survey of the population. The study found that a scientific discussion of various aspects of urbanization of rural residents confirms the relevance of the topic and shows approaches to solving problems of improving the quality, comfort and query of residents on the characteristics of the living environment. There is a certain lag in the comfort of the environment for the daily life of citizens living in rural areas compared to urban areas. This position is articulated at the level of understanding and the desire of residents to the best forms and models of environmental comfort. In the Strategy for Socio-economic Development of the Stavropol Territory in the period up to 2035 and other program documents for the development of the region, it is given much attention to the topic of complex transformations of rural settlements, both at the level of target indicators and in the concretization of measures to achieve them. The positive dynamics of the processes of urbanization in the life of rural residents is monitored in terms of satisfaction with the comfort of living in settlements and target indicators of strategic documents of the territories and the region as a whole.

1. Introduction

The relevance of research issue of rural territories and associated definitions occupies an important place in world research trends: food and territorial security of countries and regions, urbanization, depopulation, preservation of folk culture and traditions, eco and rural tourism, organic food and many others.

An important aspect in the complex of these thematic areas is urbanization (as the migration of the population from village to city). In search of the causes of this phenomenon, many researchers come to the conclusion that the living conditions of the rural population play a significant role. They are characterized by low territorial accessibility (long distances, lack or interruption in transport links, absence or poor quality of pedestrian sidewalks and hard road surface, etc.) to the services of education, culture, domestic services. The multifaceted nature of the investigated problem is confirmed in numerous studies covering a wide range of subject areas. The studies were conducted in different regions of the world, which confirms the high relevance of the stated topic.

IOP Conf. Series: Materials Science and Engineering **944** (2020) 012012 doi:10.1088/1757-899X/944/1/012012

The issues of harmonizing ecosystem services, expanding urban areas in China and rural economies were considered by the authors of the study Zhang K., Sun X., Jin Y., Wang R., Zhang S. [1]. Using the example of 30 villages, they considered 4 models of the provision of ecosystem services and the cost-effectiveness of these activities for local households. It has been established that ecotourism services are the most beneficial. Similar conclusions were drawn in a study of the authors Li H., Nijkamp P., Xie X., Liu J. [2]. Moreover, the situation is not unique to China. Researchers Romanenko Y.O., Boiko V.O., Shevchuk S.M., Barabanova V.V., Karpinska N.V. talk about EU countries [3], where there is a stimulation of the agricultural sector through the development of agritourism. For our study, an important conclusion is that, along with maintaining the uniqueness of the ecosystem characteristics of the area, it is necessary to provide tourists with a high level of quality of life during a visit to rural settlements. Accordingly, the infrastructure of rural settlements should have a qualitatively different level, close to urban conditions.

The conclusions important for understanding the requirements for infrastructure and quality of life in rural settlements are proposed by the authors Wang C., Gao B., Weng Z., Tian Y. [4]. The study found that the lack of standard public sector services for urban areas, in particular education, led to the abandonment of settlement in the village. When the opportunity arises within the framework of the Chinese Government-funded Resettlement Project to reduce poverty, the population chooses urbanization. The striving for a new quality of life and a new distribution in the structure of free time and employment is obvious.

The influence of urbanization processes on the climatic features of the area is indicated by the authors of a study conducted in Nigeria [5]. In their opinion, the resettlement of people from rural to urban is due to two main reasons: the search for a better paid job and a better quality of life. Maintaining the intensity of this process of population migration leads to negative consequences in the field of ecology: due to the expansion of urban construction, the traditional landscape of territories changes, forest areas decrease, which leads, among other things, to an increase in temperature on the earth's surface. For our study, it is important that the search for mechanisms to ensure the quality of life in rural areas gives a multiplier result, including the field of ecology. In continuation of the environmental aspect, the experience of regulatory regulation of land use processes in the conditions of increasing urbanization is offered by Polish scientists Kurowska K., Kryszk H., Marks-Bielska R., Mika M., Leń P. [6]. Providing institutional mechanisms for the conservation of rural ecosystems and, at the same time, developing conditions for comfortable living and high quality of life in rural areas is an urgent research issue for different types of economies and regions. Researchers Liu T., Huang D., Tan X., Kong F. emphasize the integration of various spatial management systems, which requires a fundamental transformation of the development planning philosophy [7]. The main impulse of the new philosophy should be a higher attention to the conjugation territories in urban and rural integration and strict observance of the legal norms of urban development. This position is also confirmed in publications of other authors [8, 9]. They emphasize the need to build a developed rural infrastructure and implement a strategy for the revival of rural areas, which will reduce the proportion of abandoned agricultural land and settlements.

The social aspects of urbanization and the importance of providing equal quality of medical services in rural and urban areas are considered by the authors of the article Ameye H., De Weerdt J. [10]. Using a comparative analysis of the health of children living in a rural city, scientists came to the conclusion that the lack of access to health care and sanitation is an important factor in reducing the health of rural children. This is especially important at the initial stage of children's life. Subsequently, all other factors ensuring the level of children's health are leveled. Similar findings were obtained by Lu C.L., Hsu Y.H., Su W.L., Tsai K.S., Li C.Y.: inequality between urban and rural areas was recorded on the basis of regression analysis of preventive health care data for children in Taiwan [11]. Thus, as in the case of education services, the development of health services at a qualitatively high level will contribute to reducing the outflow of the population from rural areas.

Modern agricultural production is undergoing significant changes, and for the traditionally farming regions of China, the issue of attention of the authorities to updating the life support systems of peasants is being updated. Authors Ge D., Long H., Qiao W., Sun D., Yang R. emphasize rural transformation. In their opinion, the creation of conditions for entrepreneurship will return rural migrants back [12].

Models of transport accessibility in the context of the infrastructural conditions of rural areas of China are considered by the example of a description of everyday practices of rural residents [13]. The authors note that despite significant advances in the construction of rural roads, the operation of public transport remains unsatisfactory, and for rural residents, safe conditions for traveling on personal light-weight economical transport are of particular importance. This factor is important for planning the characteristics of transport infrastructure projects in rural areas.

The multidimensionality and relevance of the study of urbanization issues in the daily life of rural residents were confirmed in a study of the authors Sampson L., Ettman C.K., Galea S. [14]. Based on a review of publications by the global scientific community, the authors' hypothesis found that living in rural areas reduces the level of psychological problems of the population. Depressive conditions are more widespread in more densely populated areas than in rural areas. The psychological aspects of the relevance of preserving rural territories require decisions at the state level and address issues of national security [15, 16].

2. Materials and methods

The methods of studying the processes of urbanization in the daily life of rural residents were: analysis of approaches to solving the problems of urbanization on the example of the Strategy for Socio-economic Development of the Stavropol Territory until 2035, taking into account a review of publications of scientists in scientometric databases; analysis of statistical data characterizing the vital activity of the urban and rural population of the Stavropol Territory; analysis of the results of a questionnaire survey of the population.

A survey of residents of the Stavropol Territory was carried out in 2019 on a regional stratified sample, in which different types of settlements were presented. In total, 1049 people took part in the study, among them urban residents made up 59% (619 people) and rural people – 41% (430 people). According to the results of a questionnaire survey of the population, we can confidently talk about a significant difference in the level of comfort of the living environment of the urban and rural population.

Table 1. Distribution of answers to the question
"Rate the level of comfort in the village of residence?", %

Answer options	Total according to the sample, %	Including	
		Urban dwellers, %	Villager, %
1. High level of comfort	12,6	15,9	9,3
2. Comfort level above average	26,4	35,8	15,1
3. Average comfort	41,0	36,9	46,0
4. Comfort level below average	8,8	7,5	11,0
5. Low comfort	11,2	3,9	18,6

Most respondents satisfactorily assess the degree of living comfort: 41.0% of respondents indicated an average level of comfort, with 36.9% of respondents in the city and 46% of respondents in rural areas. 51.7% of respondents rated higher than average and high in the city and 24.4% in rural areas. The difference in assessments of the comfort of living is obvious: the urban dwellers more highly assess the living conditions than the rural residents.

According to the Strategy of Socio-economic Development of the Stavropol Territory in the period up to 2035, the leadership positions of the region will be provided, first of all, with a unique standard of quality of living environment. It includes the combination of dynamism and diversity of large cities and life "on earth" in a big house, a big family, in proximity to nature and without harming nature. From the state of the region-donor population for the centre of Russia, the region should move to the state of the region - a magnet of highly qualified personnel.

Improving the quality of life in small towns and rural areas is a strategic task for the region as a southern region with a strong development of the agricultural sector. The complex of measures included in this task is aimed at reducing the migration outflow and includes measures aimed at the economic development of the village, as well as ensuring the availability of basic social services. First of all, these are measures in the field of agricultural development aimed at creating a viable competitive economy that provides jobs for the population, as well as in the field of development of road and municipal infrastructure (reconstruction and development of the water supply and sanitation system, maintaining the condition of the roadbed in a normative state). At the same time, a package of measures in the field of social development will have a significant impact on reducing migration outflows, namely: 1) improving medical support and access to medical services through the development of the network and material and technical base of primary health care institutions (hereinafter referred to as primary health care), a wider introduction of telemedicine technologies and the use of mobile teams and medical laboratories; 2) improving the quality and accessibility of education in rural areas through the introduction of digital technologies in education (digital school), the development of a regional network of youth creativity centers; 3) an increase in the number and variety of cultural and leisure activities, as well as an increase in the accessibility of services – the modernization of the material and technical base of cultural centers, the introduction of mobile forms – bibliobuses, mobile movie systems, car museums; 4) improving the quality and increasing the number of sports infrastructure in rural areas – reconstruction and construction of new sports and fitness complexes and sports grounds, including at the expense of private investors and as part of a program to support local initiatives. At the same time, it is important to create conditions in the countryside for playing sports not only for young people (playgrounds for playing sports), but also for people of middle and older ages (outdoor exercise machines, fitness rooms, gyms for physical therapy).

In the statistical survey "Stavropol Territory in Figures and Facts" there is an obvious lag in the provision of housing conditions for the rural population. The data are presented in Table 2.

Table 2. Comparative data on housing stock in urban and rural areas of the Stavropol Territory

Index	2015	2016	2017	2018
Average housing provision, m^2 per inhabitant	23.7	23.9	24.3	24.7
including:				
in urban areas, m^2 per inhabitant	25.6	25.9	26.3	26.8
in rural areas, m^2 per inhabitant	21.0	21.2	21.4	21.7

With the observed steady trend towards an increase in the average housing provision in urban areas, this indicator increased from 25.6 m^2 in 2015 to 26.8 m^2 in 2018. While in rural areas, an increase occurred from 20.0 m^2 in 2015 to 21.7 m^2 in 2018 without even reaching the 2015 level in urban areas. Improving housing conditions is becoming an important factor in ensuring the comfortable life of rural residents.

In a review of scholarly publications, an important approach was highlighted in providing a comfortable living environment for the rural population, which positively affects the quality of life and contributes to the consolidation of the population in rural areas. This is the development of the economy and agricultural production with the solution of infrastructure problems in rural areas. An analysis of the goals and objectives of the agricultural development program in the Stavropol Territory until 2021 shows that it is planned a comprehensive solution of infrastructure projects and activities related to agricultural production.

Within the framework of the program event "Comprehensive Equipping of Social and Engineering Infrastructure with Settlements Located in Rural Areas, Construction and Reconstruction of Roads", by 2021 it is planned to commission 72.00 kilometers of local water pipelines; 45.00 kilometers of gas distribution networks; 275 student places in general education organizations; 19.39 kilometers of paved public roads leading from a network of public roads to the nearest community facilities 8.0 thousand square meters of planar sports facilities in the countryside of the Stavropol Territory.

The level of area improvement of municipal district of the region territories has a significant impact on the living conditions of the Stavropol Territory population. Currently, in the territory of 176 municipal formations of the region there are 41829 apartment buildings and 10457 courtyard territories. 880 courtyards are improved. The share of improved courtyards in the municipalities of the region in the total number of courtyards in the municipalities of the region is 8.42 percent. 21 of the 776 public territories in municipalities of the region have been improved. The share of developed public territories in municipalities of the region in the total number of public territories in municipalities of the region is 2.71 percent. Out of 176 municipal formations of the territory, 18 municipal formations of the territory have implemented measures for the improvement of domestic territories and public territories with the labor participation of citizens and organizations. The share of municipalities of the region that have implemented measures to improve yard territories and public territories with the labor participation of citizens and organizations in the total number of municipalities of the region is 10.23 percent.

The strategic goal of the improvement policy implemented in the Stavropol Territory is to create conditions for systematically improving the quality and comfort of the modern urban environment in the Stavropol Territory by implementing a set of priority measures to form a modern urban environment in the municipalities of the region. The priority of the improvement policy in the Stavropol Territory aimed at achieving this strategic goal is to increase the comfort of living conditions for the population of the Stavropol Territory.

The share of municipalities in the Stavropol Territory implementing measures to improve yard and public areas with the labor participation of citizens and organizations by 2022 will amount to 52.84% of the total number of municipalities, including rural

3. Conclusion

A review of modern publications on the development of a comfortable living for rural residents, urbanization in their daily activities, an analysis of strategic policy documents implemented at the regional level and the results of a survey of residents of the Stavropol Territory allow a number of conclusions:

- a scientific discussion of various aspects of urbanization of rural residents confirms the relevance of the topic and shows approaches to solving problems of improving the quality, comfort and request of residents on the characteristics of the living environment;
- there is a certain lag in the comfort of the environment for the daily life of citizens living in rural areas compared to urban areas; it is articulated at the level of understanding and the desire of residents to better forms and models of environmental comfort;
- in the Strategy of Socio-economic Development of the Stavropol Territory in the period up to 2035 and other program documents for the development of the region, a significant attention is given to the topic of complex transformations of the territories of rural settlements, both at the level of target indicators and in the concretization of measures to achieve them;
- the positive dynamics of the processes of urbanization in the life of rural residents is monitored in terms of satisfaction with the comfort of living in settlements and target indicators of strategic documents of the territories and the region as a whole.

References

[1] Zhang K, Sun X, Jin Y, (...), Wang R and Zhang S 2020 *Ecological Indicators* **115** 106363
[2] Li H, Nijkamp P, Xie X, Liu J 2020 *Sustainability (Switzerland)* **12(8)** 3148
[3] Romanenko Y O, Boiko V O, Shevchuk S M, Barabanova V V and Karpinska N V 2020 *International Journal of Management* **11(4)** 605-613
[4] Wang C, Gao B, Weng Z, Tian Y 2020 *Land Use Policy* **95** 104627

[5] Ahmed H A, Singh S K, Kumar M, (...), Dzwairo R and Lal D 2020 *Urban Climate* **32** 100600
[6] Kurowska K, Kryszk H, Marks-Bielska R, Mika M and Leń P 2020 *Land Use Policy* **95** 104614
[7] Liu T, Huang D, Tan X, Kong F 2020 *Land Use Policy* **94** 104498
[8] Qu L, Li Y, Feng W 2020 *Ecological Indicators* **112** 105783
[9] Wang Y, Peng P, Li Q, Chen Z and Tang W 2020 *Sustainability (Switzerland)* **12(8)** 3356
[10] Ameye H, De Weerdt J 2020 *World Development* **130** 104950
[11] Lu C L, Hsu Y H, Su W L, (...), Tsai K S and Li C Y 2020 *Public Health* **182** 102-109
[12] Ge D, Long H, Qiao W, (...), Sun D and Yang R 2020 *Journal of Rural Studies* **76** 85-95
[13] Ao Y, Zhang Y, Wang Y, Chen Y and Yang L 2020 *Journal of Transport Geography* **85** 102708
[14] Sampson L, Ettman C K, Galea S 2020 *Current opinion in psychiatry* **33(3)** 233-244
[15] Baklanov I S , Baklanova O A , Erokhin A M , Ponarina N N, Akopyan G.A 2018 Myth as a means of ordering and organizing social reality, *TarihKültürveSanatAraştırmaları* T. 7.- **2**. pp 41-47
[16] Ivashova V, Goncharov V, Erokhin A., Kolosova O, Berkovsky V 2018 Life strategies of graduates of regional universities: the reality of Russian society, *International Journal of Civil Engineering and Technology* T. 9 **13**. pp 398-404

IOP Conf. Series: Materials Science and Engineering **944** (2020) 012013 doi:10.1088/1757-899X/944/1/012013

Comparative analysis of urban and rural environmental quality: the opinions of residents of the region

I S Baklanov[1], O A Baklanova[1], E E Nesmeyanov[2], V A Ivashova[3] and L A Kabardokova[4]

[1] North Caucasus Federal University, 1, Pushkin Str., Stavropol, 355017, Russia
[2] Don State Technical University, 1, Gagarin Square, Rostov-on-Don, 344000, Russia
[3] Stavropol State Agrarian University, 12, Zootechnicheskiy Lane, Stavropol, 355017, Russia
[4] Branch of MIREA - Russian Technological University in Stavropol, 8, Kulakova Av., Stavropol, 355035, Russia

E-mail: vivashov@mail.ru

Abstract. The article presents the results of a comparative analysis of the opinions of the rural and urban settlements' population of the Stavropol Territory on the qualitative characteristics of the living environment. The quality of life in the village, which to a small extent depends on the state of the environment (transport accessibility of infrastructure and social facilities, road and pavement, lighting, etc.), plays a significant role in the choice of permanent residence. Currently, the rural population has high demands on the living environment. Rural settlements have territorial-landscape, climatic features, the historically established traditional spatial organization of rural settlements and the social lifestyle of the villager. Therefore, the urban model of improvement in rural areas cannot be used. It is important to study the opinions of local residents to understand the directions of transformation and improve the quality of the environment of rural settlements. Directions for the development of the environment of a settlement significantly depend on the characteristics of traditional socio-cultural practices of the population. This study allows us to more accurately formulate the desired contours of the model of the environment of settlements. On the example of a comparative analysis of the problems of transport accessibility of social infrastructure by residents in different types of settlements, we have shown the actual discrepancy in their significance. To plan directions for the development of rural transport infrastructure, it is necessary to solve the issues of hard cover for light-duty vehicles that are designed to carry small cargoes necessary for personal subsidiary plots.

1. Introduction

The improvement of rural settlements is relevant for a number of reasons. Firstly, the acceleration of scientific and technological progress and related innovations that increase the quality of life of the population of the city, all this is an important factor in choosing the life of the city rather than the village. Secondly, the difference in the improvement of the territories of settlements that are under the jurisdiction of municipalities deliberately puts citizens in unequal conditions, updating the issue of raising living standards in rural settlements. Thirdly, the territorial-landscape, climatic features, the historically established traditional spatial organization of rural settlements and the social lifestyle of a rural resident don't allow the direct use of the urban model of rural areas improvement. Therefore, it is important to study the opinion of residents of rural settlements in order to understand the directions of transformation and improve the quality of the environment of rural settlements

The problems of this study are relevant not only for rural Russia and, in particular, the Stavropol Territory. A review of modern publications confirms the widespread interest and relevance of information regarding possible directions for the modernization of the rural environment.

Scientists Peña-García A., Sędziwy A. in their study cited the results of a comparative analysis of different lighting models for rural areas [1]. It is worth noting that the capabilities of LED equipment have great advantages in energy consumption over traditional lighting. For the spatial position of rural settlements and small municipal budgets, such a proposal for coverage is very relevant. At the same time, the issue of light pollution is becoming relevant. Scientists emphasized the multidimensional nature of the problem of light pollution and proposed not to reduce it to the formal prohibition of white LED.

Scientists Khadiyanto P., Soetomo S., Hadi S.P. in their article rightly raised the issue of the environmental quality of rural settlements [2]. They referred to already confirmed data on the direct relationship between environmental quality and the quality of life of people. Using the example of an Indonesian village, it is shown that the expansion of industrial zones negatively affected the environmental situation of this village. Infrastructure facilities, in particular water supply, are not designed for industrial level. The authors of the study note that the quality of the environment is determined by the condition of buildings, structures and infrastructure, location and accessibility, as well as the convenience of the physical environment. The quality of life is determined through health, economic, educational and social relations. The results showed that not in all regions with poor physical environmental conditions, people assess low quality of life as low. The decrease in environmental quality requirements is explained by the way of life of the rural population formed over many generations.

A definite confirmation of the need to take into account the lifestyle features of the rural population is the study of the author Kobi M. [3]. The author shows the significance of the needs of residents, analyzing the contours of urban architectural anthropology, which are characterized by social and material practices that are distinct from rural settlements in the winter season, the wide use of the artificial environment and climate control. He considers urban architecture as a social sphere, which has its own characteristics, including ethnographic ones. The study of the features of socio-cultural practices allows us to more accurately formulate the desired contours of the model of the environment of settlements.

The authors of Salamone F., Belussi L., Danza L., Ghellere M., Meroni I. developed a model for assessing the current energy and environmental behavior of the built-up environment and possible scenarios for updating the coverage of settlements [4]. An integrated approach, implemented in the energy-ecological analysis of the urban environment, is useful for the modernization of the environment of rural settlements.

The authors of Shao T., Jin H. in their publication consider the qualitative characteristics of traditional and new housing in rural areas [5]. The relevance of this study is dictated by the integration processes between the city and the village in modern China, which led to the active construction of housing. The results of the generalization of traditional approaches to the construction of rural housing and the additional features of modern energy-saving technologies, as well as the requirements for comfortable living, have identified improving factors. These include zoning of the thermal environment, optimization of heating modes, achieving energy conservation through additional thermal insulation in the construction of buildings. Important findings for our study are: the increasing demands of rural residents for housing comfort and the articulated need to take into account traditional approaches to rural development.

The authors of the study of the relationship between the temperature environment and thermal insulation of clothes for rural residents of China Zhao W., Chow D., Sharples S. came to a conclusion about the imbalance in the level of economic development between urban and rural areas [6].

IOP Conf. Series: Materials Science and Engineering **944** (2020) 012013 doi:10.1088/1757-899X/944/1/012013

Differences in incomes, living conditions and lifestyle affect the perception of thermal comfort by residents of urban and rural settlements. Thus, they emphasized significant differences in the perception of a comfortable environment and lifestyle of the rural and urban population. This statement aims us to study the differences in the perception of the environment in different types of settlements and emphasizes its relevance.

The theme of the climatic differences between urban and rural environments is relevant for many regions of the world [7, 8]. The authors of the study Taher H., Elsharkawy H., Newport D. analyzed the influence of urban green spaces on improving the microclimate in the urban environment using the example of large cities in England [9]. The mayor of London approved the London strategy until 2050, in which the improvement of urban green systems is planned through an increase in the number of trees, green walls and green roofs. The comparative characteristics of urban and rural settlements highlighted by the authors remain characteristic. Rural development is less dense, open spaces prevail, few impenetrable surfaces (roads, sidewalks, buildings, etc.) [10].

Thus, it is confirmed the assumption of the difference in the perception of the comfort of the living environment in different types of settlements; the characteristic differences between the urban and rural environments, such as buildings, lighting, landscaping pavements, etc. are determined. A study of the opinions of residents of the Stavropol Territory shows the relevance of this study in the Russian regions [11, 12, 13].

2. Materials and methods

In the Stavropol Territory, in 2019, a population survey was conducted on a random stratified sample. In total, 1187 people took part in it. The sample size allowed the segmentation of opinions of residents by type of settlements: the regional capital, cities, rural district centers, villages with a population of over 3 thousand people, villages with a population of less than 3 thousand people.

The formation of the population's request for the qualitative characteristics of the living environment can be based on an assessment of the level of satisfaction with existing conditions. Areas of dissatisfaction determine the possible direction of the transformation of the settlement environment.

Table 1 presents comparative data characterizing the assessment of transport accessibility of social infrastructure by residents in various types of settlements.

A survey of residents of the Stavropol Territory was carried out in 2019 on a regional stratified sample, in which different types of settlements were presented. In total, 1049 people took part in the study, among them urban residents made up 59% (619 people) and rural people – 41% (430 people). According to the results of a questionnaire survey of the population, we can confidently talk about a significant difference in the level of comfort of the living environment of the urban and rural population.

Table 1. Comparative analysis of the assessment of transport accessibility of social infrastructure by residents in various types of settlements, %

Type of location	Convenient to reach	Not convenient to reach
Regional Center	64,8	35,2
City	66,7	33,3
Rural district center	54,1	45,9
Village over 3 thousand people	21,1	78,9
Village less than 3 thousand people	71,4	28,6
Total for the surveyed population	60,2	39,8

IOP Conf. Series: Materials Science and Engineering **944** (2020) 012013 doi:10.1088/1757-899X/944/1/012013

In the segmentation by type of settlements, the most satisfied with the transport accessibility of the social infrastructure facilities were residents of small villages (with a population of less than 3 thousand people) – 71.4% said that it was convenient for them to get there. The level of satisfaction with this factor in the quality of the living environment is also above average for residents of cities and the regional capital: 66.7% and 64.8%, respectively. This situation is due to several reasons. Segmentation is presented in the following table.

Table 2. Comparative analysis of the reasons why residents gave such answers, %

Type of settlement	Walking (by transport) takes a lot of time	Rare appearance of (does not run) public transport	Public transport stop is far away	Parking problems	Poor pedestrian area	Poor condition or lack of ramps
Regional Center	33,3	11,1	0	33,3	22,2	0
City	34,6	23,1	7,7	26,9	0	3,8
Rural district center	50,0	25,0	0	6,3	6,3	0
Village over 3 thousand people	73,3	6,7	0	0	20,0	0
Village less than 3 thousand people	25,0	25,0	50,0	0	0	0
Total for the surveyed population	44,3	17,7	5,1	17,7	10,1	1,3

For villages with a population of over 3 thousand people and less dense development characteristic of rural areas, the main problem is the constant need to overcome great distances. Residents do this on foot or by transport, but in any case it takes a lot of time. The same problem, only slightly less acute is observed in rural district centers. It can be solved by more developed public transport options. For small villages, the most significant problem is the far-reaching public transport stops, as indicated by 50.0% of the survey participants.

There are 2 obvious interconnected vectors of the needs of rural residents: light motor vehicles for moving and transporting small loads and hard road surface, designed specifically for this type of transport.

For the regional capital and cities, parking problems are characteristic, which requires systemic solutions in planning the functional load on the urban space.

3. Conclusion

Based on a comparative analysis of the quality of the urban and rural environment, taking into account the opinions of residents of the southern region of Russia, a number of conclusions can be made:

1. Directions for the development of the environment of a settlement significantly depend on the characteristics of traditional socio-cultural practices of the population, the study of which allows us to more accurately formulate the desired contours of the model of the environment of settlements.

2. On the example of a comparative analysis of the problems of transport accessibility of social infrastructure by residents in different types of settlements, we have shown the actual discrepancy in their significance.
3. For planning directions for the development of rural transport infrastructure, it is necessary to solve the issues of hard cover for light-duty vehicles that are designed to carry small cargoes necessary for personal subsidiary plots.

References

[1] Peña-García A, Sędziwy A 2020 *Journal of Illuminating Engineering Society of North America* **16(2)** 147–156
[2] Khadiyanto P, Soetomo S, Hadi S P 2020 *IOP Conference Series: Earth and Environmental Science* **409(1)** 012042
[3] Kobi M 2019 *Social Anthropology* **27(4)** 689–704
[4] Salamone F, Belussi L, Danza L, (...), Ghellere M, Meroni I 2019 *IOP Conference Series: Materials Science and Engineering* **609(7)** 072018
[5] Shao T and Jin H 2019 *IOP Conference Series: Materials Science and Engineering* **609(7)** 072063
[6] Zhao W, Chow D, Sharples S 2019 *IOP Conference Series: Earth and Environmental Science* **329(1)** 012023
[7] Pitts A, Gao Y and Le V T 2019 *Sustainability (Switzerland)* **11(19)** 5515
[8] Lau K K-L, Chung S C and Ren C 2019 *Building and Environment* **154** 227–238
[9] Taher H, Elsharkawy H and Newport D 2019 *IOP Conference Series: Earth and Environmental Science* **329(1)** 012046
[10] Pelevin S I, Taubaev B D and Baklanov I S 2018 I*nternational Journal of Civil Engineering and Technology* **9(11)** 2437–2443
[11] Goncharov V N, Nesmeyanov E E, Kolosova O U, Arutyunyan V V and Ivashova V A 2019 *Journal of Physics: Conference Series* **1353(1)** 012135
[12] Ivashova V A, Tokareva G V, Agalarova E G, Nadtochiy Yu B and Yushchenko I V 2020 *IOP Conference Series: Materials Science and Engineering* **775(1)** 012020
[13] Baklanov I S , Baklanova O A, Erokhin A M , Ponarina N N, Akopyan G A 2018 *TarihKültürveSanatAraştırmaları* **7(2)** 41–47

IOP Conf. Series: Materials Science and Engineering **944** (2020) 012014 doi:10.1088/1757-899X/944/1/012014

Strategy for the urban environment development of the regional capital: prospects and challenges

V A Ivashova[1], **G V Tokareva**[1], **G A Narozhnaya**[2], **L A Zvereva** [3] **and N G Bondarenko**[3]

[1] Stavropol State Agrarian University, 12, Zootechnicheskiy Lane, Stavropol, 355017, Russia
[2] Branch of MIREA - Russian Technological University in Stavropol, 8, Kulakova Av., Stavropol, 355035, Russia
[3] North Caucasus Federal University, 1, Pushkin Str., Stavropol, 355017, Russia

E-mail: vivashov@mail.ru

Abstract. The article discusses approaches to strategic planning of the urban environment of the regional capital and the challenges that need to be overcome in the course of preparing a strategic development project. Strategic planning of the urban environment of the regional capital has its own important distinctive features. This is due to the status and mission of the administrative centre of the territory. A review of the literature shows the relevance of the research topic and social practice of solving current and future challenges. The strategic planning of the urban environment of the regional capital should be based on such criteria as the involvement of citizens in strategic planning and assessment of the effectiveness of strategies, harmonious development of the urban ecosystem, sustainable social and economic development of the city, infrastructural and administrative support for innovation, increasing social capital, safety of the urban environment in all its manifestations and ensuring the democratic rights and freedoms of citizens. We show the key challenges and vectors of development of the urban environment on the example of the analysis of the development strategy of the capital of the southern region of Russia – Stavropol. The results of a survey of citizens and the business community give an idea of the comfortable urban environment in terms of living and doing business, they are an important addition to understanding the problem areas of the regional capital development.

1. Introduction

Strategic management is a way to combine all actions and decisions in the field of strategic planning into a single management system. The main goal of the implementation of strategic management is to provide the local authorities of the city of Stavropol with a unified system of goals and objectives for the development of the city of Stavropol, relevant and updated through feedback mechanisms, which will help reduce the risk of decision-making. Federal Law No. 172-FL of June 28, 2014 "On Strategic Planning in the Russian Federation" establishes the legal framework for strategic planning in the Russian Federation, coordination of state and municipal strategic management and budget policy, as well as the powers of local governments and the procedure for public, scientific and other organizations in the field of strategic planning.

Strategic planning of the urban environment of the regional capital has its own important distinctive features. This is due to the status and mission of the administrative centre of the territory. A review of the literature shows the relevance of the research topic and social practices for solving current and future challenges.

In the study of Mandeli K., the author has shown the importance of reflecting local values in the formation of public zones of urban space, using the example of the second largest city in Saudi Arabia, Jeddah, which is its "economic capital", as well as the largest city in the administrative district [1]. In the course of radical re-planning of Jeddah, modern approaches to planning and design in developing cities were used, which made it possible to form a quality urban space. The development of proactive urban design was tailored to the needs of residents and businesses and, in general, improved the quality of the environment. The design of public spaces was carried out taking into account the traditions and values of the regional community and was aimed at improving the image of the economic capital of the state. For our research it is important to focus the environment on comfortable living and its functionality not only from the point of view of the local community, but also from business communications: a developed high-class hotel and restaurant network, discussion and exhibition grounds and complexes, wide representation of financial institutions, etc. In addition, among the mandatory parameters of a comfortable urban environment aimed at the economic prosperity of the capital of the region, there should be territorial and transport accessibility of all infrastructure facilities and the modern level of digital services and their high specifications.

The authors of the article Lim Y., Edelenbos J., Gianoli A., based on a wide review of scientific publications, determine the target settings for the strategic innovative development of the city through sustainability and high quality of life [2]. Current approaches to strategic planning of urban development, according to the authors, are ensuring citizens' involvement in strategic planning and assessing the effectiveness of strategies, harmonious development of the urban ecosystem, sustainable social and economic development of the city, infrastructural and administrative support for innovation, and increasing social capital. Also, the safety of the urban environment in all its manifestations and ensuring the democratic rights and freedoms of citizens should not be left without attention. All approaches are interrelated and at the level of formulating the target settings of a particular city are interpreted depending on the context of the territory: the achieved level of socio-economic development and contemporary regional and world challenges.

An important aspect of the strategic development of the urban environment, associated with modern energy consumption practices, is considered by the authors of the study Natanian J., Aleksandrowicz O., Auer T. [3]. The study highlights the importance of urban density and configuration for optimizing energy consumption, especially in hot climates. For the strategic planning of the urban environment of the capital of the region, it is important to regulate the energy load not only depending on the type of development (residential or office), but also the effects of the building model (area and materials of walls, windows, roofs) and parameters of their location (distance between buildings, orientation, etc.).

The authors Skalicky V., Čerpes I. propose a comprehensive methodology for assessing a livable urban environment as a basis for strategic planning. In their opinion, the system of criteria for assessing the quality of the living environment should be focused, first of all, on the suitability for human life [4].

Researchers Shaw A., Saharan T. in their work consider the problems of resettlement of the population from areas of dilapidated and hazardous housing in connection with the development of cities and the growth of the quality of life of citizens [5]. The authors' comparative analysis of the satisfaction of residents who refused the resettlement package and continued to live in their huts, and those who accepted the package and moved to new apartments provided by the government, showed the benefits of resettlement: improved sanitary and living conditions, social well-being and quality of life. For our study it is important that the strategic orientation of the regional capital development program is aimed to improve the quality of life of the population and the gradual resettlement of residents, including the historical centre of the city, from dilapidated to modern residential buildings.

IOP Conf. Series: Materials Science and Engineering **944** (2020) 012014 doi:10.1088/1757-899X/944/1/012014

The harmonious development of the urban ecosystem is an important criterion for the planning and design strategy of the urban environment. The authors Veinberga M., Zigmunde D. in their scientific work emphasize the need for new approaches that will provide a qualitative improvement in the aesthetic and environmental characteristics of public urban areas [6]. On the basis of an expert survey they analyzed the criteria for balancing aesthetics and ecology in the urban environment. These include compositional unity, uniqueness, consistency with architecture, condition of built elements, decorative quality of plants, care and maintenance, naturalness, typical landscape, wildlife, presence of wildlife, vegetation structure and use of local plants. For our study it is important to note that Stavropol, being the regional capital, has repeatedly, since Soviet times, been among the best cities for greening public urban areas and has not only high-quality traditions, but also effective practices for organizing the process of landscaping and landscape design. All this creates a unique image of the urban environment and contributes to the development of communicative, cultural, and leisure zones inherent in the regional centre.

The authors Vitkova L., Silaci I. introduce an important concept – a creative city, which helps in determining the strategic vectors of the urban environment development [7]. The basis for the forward movement for a creative city is the socio-cultural potential of organizations and the population: a favourable social climate, a high level of tolerance, the atmosphere of a living city, a rich social life, informal social activity. All these characteristics are important for the strategic development of the regional centre and aim the administrative resource at the correct, effective use of the socio-cultural potential of the territory.

For a city with a rich historical heritage in the process of strategic development, it is important to rely on the existing traditions of the urban environment, giving them new, modern and demanded functions. And here it is appropriate to speak of revitalization as a phased restoration of the urban environment, in which it becomes more habitable. This topic is devoted to the publication of the research results of the authors Fross K. and Guminska A. [8]. They rightly point out that changing consumer expectations regarding public spaces in connection with technological progress and changing lifestyles presupposes a timely update of the quality of the urban environment. The purpose of such changes is to eliminate spatial and ecosystem conflicts of areas and objects and to strengthen the multipurpose potential and sustainable development of territories.

Among the criteria for strategic planning of the urban environment of the regional capital, an important place is occupied by the involvement of the local community in making key decisions. Social practices for the implementation of this approach are presented in the study of the authors Kasala V., Smatanova K. [9]. They note that public participation is necessary for two reasons: to identify the real needs of local residents as an urban environment and to incorporate proposals into urban development programs.

An interesting aspect of designing the quality of the urban environment is associated with the colour coloration of the surrounding architecture and landscape design. In the article, Kraus M. shows the results of the influence of colour on the perception and assessment of the environment; it is reflected in the health and well-being of residents [10].

There are also gender differences in the perception of environmental quality. The authors Kraus M. and Novakova P. come to the conclusion that this knowledge is necessary and useful for developing appropriate strategies for creating and maintaining a stable and comfortable environment [11].

Thus, the review of publications that determine approaches to developing a strategy for the development of the urban environment of the regional capital shows the relevance of the topic under consideration and establishes important markers of its success: citizen involvement in strategic planning and assessment of the effectiveness of strategies, harmonious development of the urban ecosystem, sustainable social and economic development of the city, infrastructural and administrative support for innovation, increasing social capital, safety of the urban environment in all its manifestations and ensuring the democratic rights and freedoms of citizens [12, 13, 14, 15].

IOP Conf. Series: Materials Science and Engineering **944** (2020) 012014 doi:10.1088/1757-899X/944/1/012014

2. Materials and methods

To determine the approaches to strategic planning of the urban environment of the regional capital we analyzed the documents of the Strategy for the development of the city of Stavropol, which set out the prospects and modern challenges. We conducted the comparison of the key positions of the developed Strategy with significant criteria for the success of the development of the centre of the region, which were identified in the literature review. We analyzed the results of a survey of residents of Stavropol (427 people), conducted by the City Administration in 2019. They provide information on the satisfaction of residents with the quality of the urban environment and characterize the areas of its improvement. The results of a survey of entrepreneurs (43 people in 2019), as an expert group, were also considered to understand the image of Stavropol as a centre for regional socio-economic development and the compliance of infrastructure components with the expectations of this category of consumers. All stages of the study of the strategy for the development of the urban environment of the capital of the region allow us to draw well-founded conclusions and proposals for its improvement.

Strategic planning in the city of Stavropol is an activity on goal-setting, forecasting, planning and programming of socio-economic development, aimed at solving the problems of sustainable socio-economic development of the city of Stavropol. When developing and implementing strategic planning documents, all documents are agreed on goals, objectives, priorities and indicators of achieving goals. These include: strategy of social and economic development of the city of Stavropol; action plan for the implementation of the Strategy of social and economic development of the city of Stavropol; forecasts of the socio-economic development of the city of Stavropol for the medium and long term; budget forecast of the municipality of the city of Stavropol, Stavropol Territory for the long term; municipal programs of the city of Stavropol.

The mechanism for implementing the Strategy provides for the use of different means and methods of influence: legal regulation, administrative measures, financial instruments (tax, property, tariff, debt, etc.), mechanisms of organizational and information support.

The goals and objectives of the Strategy, the mechanisms for their implementation are taken into account in the development of draft municipal legal acts of the city of Stavropol, including the budget of the city of Stavropol for the next financial year.

Analysis of the developed tasks aimed at the integrated development of the urban environment shows that there are positive coincidences (results) and areas for improvement in terms of the optimal criteria for strategic planning. Well-articulated criteria include the harmonious development of the urban ecosystem, the safety of the urban environment in all its manifestations.

Insufficient attention in this section is paid to promoting the image of Stavropol as the capital of the region and business activity. The urban environment is not functionally shown from the point of view of business and public communications. In general, the tasks have not been formulated for developing the urban environment of the regional capital aimed at sustainable social and economic prosperity of the city, increasing social capital, and ensuring the democratic rights and freedoms of citizens.

We considered the assessment of the quality of the urban environment through the satisfaction of residents with its infrastructural functions. Population survey data in absolute terms are presented in Figure 1.

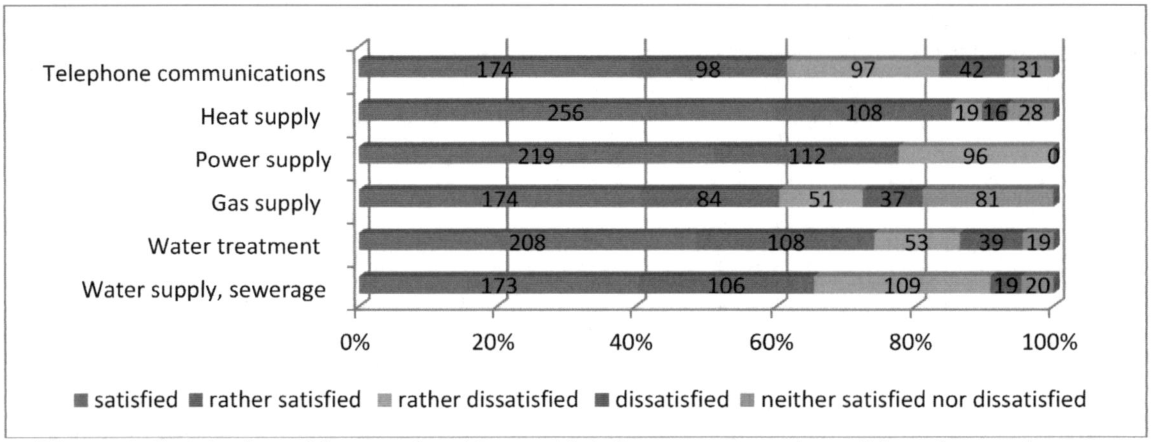

Figure 1. Satisfaction of Stavropol residents with infrastructure services of the urban environment.

The highest level of satisfaction is caused by the organization of heat, power supply, and water treatment services. The area for improvement is water supply, sanitation, and communication services.

The quality of the infrastructure characteristics of the urban environment was assessed by representatives of the business community of Stavropol in the framework of an expert survey of the city administration.

Table 1. Expert assessment of the quality of infrastructure characteristics of the urban environment (in points on a 5-point scale)

Infrastructure subsystem of the urban environment	Estimation of the terms of obtaining access to services, average score	Evaluation of the efficiency of documentary registration, average score	Assessment of the objectivity of the cost of connecting business objects, average score	Average value of the quality of work of the infrastructure subsystem of the urban environment, average score
Water supply, sewerage	3,2	3,3	3,4	3,3
Gas supply	3,2	2,7	2,7	2,9
Power supply	3,1	3,5	3,1	3,2
Heat supply	4,0	3,5	3,3	3,6
Telephone communications	3,9	4,0	4,1	4,0
Average of the scores	3,5	3,4	3,3	3,4

According to the business community, the most favorable situation in the infrastructural subsystems of the urban environment is in the field of communications – on a 5-point scale, the average assessment of access to services, the promptness of documenting, and the cost of connection was 4.0 points. The worst situation is in the gas supply subsystem, where the average score was 2.9 points. Thus, areas for improving the work of infrastructural subsystems of the urban environment in the development strategy of the regional capital are associated with ensuring sustainable processes in the field of gas supply, electricity supply, water supply, and sanitation. Assessment of the quality of work of infrastructure subsystems can be improved by increasing the objectivity of the cost of services and increasing the efficiency of documenting.

3. Conclusion

The study allows us to draw a number of conclusions aimed at improving the strategy for the development of the urban environment of the regional capital.

Firstly, important criteria for the success of a strategy are: citizen involvement in strategic planning and assessment of the effectiveness of strategies, harmonious development of the urban ecosystem, sustainable social and economic development of the city, infrastructural and administrative support for innovation, increasing social capital, safety of the urban environment in all its manifestations, and ensuring the democratic rights and freedoms of citizens.

Secondly, the analysis of the developed tasks aimed at the integrated development of the urban environment shows that there are positive coincidences (results) and areas for improvement in terms of the optimal criteria for strategic planning. Well-articulated criteria include the harmonious development of the urban ecosystem, the safety of the urban environment in all its manifestations. Insufficient attention in this section is paid to promoting the image of Stavropol as the capital of the region and business activity. The urban environment is not functionally shown from the point of view of business and public communications. In general, the tasks have not been formulated for developing the urban environment of the capital of the region aimed at sustainable social and economic prosperity of the city, increasing social capital, and ensuring the democratic rights and freedoms of citizens.

Thirdly, according to the results of a survey of residents of the regional capital and representatives of the business community, areas have been identified for improving the quality of the infrastructure characteristics of the urban environment. For the population, the area for improvement is water supply, sanitation, and communication services. For the business community – gas supply, electricity supply, water supply and sanitation. Assessment of the quality of work of infrastructural subsystems can be improved by increasing the objectivity of the cost of services and increasing the efficiency of documenting.

In general, we can talk about a satisfactory current state of the infrastructure functions of the urban environment, which is a good basis for capacity building and development of all significant components of the environment. In addition, to ensure the objectivity of public opinion data, independent studies of the satisfaction of residents with the comfort of the urban environment are needed, it will increase the involvement of citizens in the processes of strategic planning and decision-making.

References

[1] Mandeli K 2019 *Cities* **95** 102409
[2] Lim Y, Edelenbos J and Gianoli A 2019 *Cities* **95** 102397
[3] Natanian J, Aleksandrowicz O and Auer T 2019 *Applied Energy* **254** 113637
[4] Skalicky V and Čerpes I 2019 *Cities* **94** pp 44–54
[5] Shaw A and Saharan T 2019 *Environment and Urbanization* **31(2)** 597–614
[6] Veinberga M and Zigmunde D 2019 *IOP Conference Series: Materials Science and Engineering* **603(4)** 042016
[7] Vitkova L and Silaci I 2019 *IOP Conference Series: Materials Science and Engineering* **603(3)** 032072
[8] Fross K and Guminska A 2019 *IOP Conference Series: Materials Science and Engineering* **603(2)** 022088
[9] Kasala V and Smatanova K 2019 *IOP Conference Series: Materials Science and Engineering* **603(3)** 032063
[10] Kraus M 2019 *IOP Conference Series: Materials Science and Engineering* **603(5)** 052097
[11] Kraus M and Novakova P 2019 *IOP Conference Series: Materials Science and Engineering* **603(5)** 052084
[12] Pelevin S I, Taubaev B D and Baklanov I S 2018 I*nternational Journal of Civil Engineering and Technology* **9(11)** pp 2437–2443
[13] Goncharov V N, Nesmeyanov E E, Kolosova O U, Arutyunyan V V and Ivashova V A 2019

Journal of Physics: Conference Series **1353(1)** 012135
[14] Ivashova V A, Tokareva G V, Agalarova E G, Nadtochiy Yu B and Yushchenko I V 2020 *IOP Conference Series: Materials Science and Engineering* **775(1)** 012020
[15] Baklanov I S, Baklanova O A, Erokhin A M, Ponarina N N and Akopyan G A 2018 *TarihKültürveSanatAraştırmaları* **7(2)** pp 41–47

Communication organization issues through the design of visual space in a digital environment

A S Khabibova[1], E A Khizhnyak[1]

[1] Moscow City University, Institute of Culture and Arts, 129226, 2nd Agricultural Passage, 4, building 1, Moscow, Russia

E-mail: alevtina.khabibova@gmail.com; alias.07@mail.ru

Abstract. The article reveals the main types of visualization to support complex activities, specifies the forms of using methods of information visualization by means of computer technologies. The main principles of teaching communicative design are described, with account for modern visualization technologies and technologies for exhibiting design-engineering products in the digital space. The article reveals various aspects of the design teaching methodology, focused on the comprehensive use of visualization tools in the learning process. Practical recommendations are given that allow you to build a training program taking into account a wide range of new digital media technologies.

1. Introduction

Modern trends in the development of society with the advent of new information technologies entailed the accelerated development of new communication links and contributed to a change in the means of communication, which led to the rapid development of a number of areas of computer technology-based design. Visual communication has become the most effective way of problem-solving. It has been proven by science that verbal and visual information is perceived and processed differently by the brain. Visual experience is thinking in images that gradually accumulate and thus increasing the conceptual base of a person.

Complex structural relationships require a lot of words and sentences to explain the details. However, they are more effectively learned by the means of visual communication. The language of visualization is universal, thus it's preferable to display different kinds of information graphically.

With the spread of digital technologies, the means of transmitting communication messages began to develop and improve more actively, and their organization changes.

The organization of visual material through the screen involves the use of various communication technologies, including the following plans:

1) an informational plan, which involves the accuracy of the visual message transfer;

2) presentation plan, which necessitates the creation and usage of images accounting for the characteristics of the target audience;

3) a communication plan, involving support for joint actions and connections of participants in virtual communication.

At the present, specialists in the creation of visual communications are mainly graphic designers whose professional training is aimed at developing creative thinking, obtaining versatile knowledge

IOP Conf. Series: Materials Science and Engineering **944** (2020) 012015 doi:10.1088/1757-899X/944/1/012015

and practical skills in the field of composition, color theory, typography, photography, etc., as well as mastering specialized computer programs.

2. Materials and methods

In order to study the content of the use of communication technologies, let us focus on the visualization of the communication support of social practice.

When discussing the presentation plan, it's necessary to take into account the expositional environment that influences the creation and formation of visual material. According to Marshall McLuhan, the parameters of this environment drastically affect the development of design, while video and interactivity provide for openness and dynamics, where the work area is larger than the window through which we look [2]. “Computer technologies are an intermediary, a medium between virtual models of presentation and exhibition environments, which can be both real and virtual” [3, p.116].

Due to the fact that exhibition environments are needed for complex activities, they include various types of resources. Let's consider the capabilities of the primary means of information visualization in various situations.

Information visualization in scientific and educational spheres, as well as business these days is highly dependent on infographics - otherwise known as information graphics, which is "data journalism" (set, catalog) of graphic signs, images, diagrams, various types of graphs, used for the purpose of organizing various information, achieving the greatest accessibility for understanding the topic in question, its better visual presentation.

Scientific research using the eye tracking method has shown that the subjects pay attention primarily not to the text, but to the image. This is where they search for the main meaning of information [1, p. 229]. Infographics usually combine text and graphic information. They complement each other organically.

Modern infographics are very diverse. Each type has its characteristics, pros and cons. Infographics are divided into two main types by to the method of presentation - static and dynamic, according to their functional purpose - illustrative, cognitive and communicative. Common types of classification are also 2D and 3D infographics, video infographics. Let's dwell on the essence of each of them.

2D infographics are flat images that have dimensions in two planes (height and width), which can be either still or animated.

3D infographics are a tool of data transferring by the means of creating 3D models in special computer programs and presenting images that are more realistic, palpably three-dimensional and at the same time animated, i.e., moving and transforming in space and time. Thus, 3D infographics are animations made in special 3D computer programs. Layered graphics make it more impactful, engaging and artistic.

Video infographics are an effective way to present static data. Information in the form of animated graphics is distinguished by a clear sequence and structure, due to which it is easier and more interesting for the audience to perceive.

Additionally, one can add sound effects, music and/or a voiceover track.

One of the most actively used means of modern communication is presentation, the purpose of which is to accompany spoken information (such as reports, lectures, messages, etc.).

Presentation infographics as data visualization are extremely popular these days. It is impossible to imagine any presentation without them. Graphic elements or objects, especially appealing from the point of view of their design solution, quickly and effectively convey the necessary information to the listener.

The visual design of the presentation, the use of all available graphic solutions in the design of its slides play an essential role for the sensory and aesthetic perception of information. The means by which the designer helps to convey information are color, fonts, textures, modular grid, etc.

The most common presentation programs today are PowerPoint, StarOffice Impress, Google Slides and Keynote. A large number of themes, applications, templates, designs of graphs, diagrams,

diagrams, etc. have been developed for them. Infographics can be easily exported from PowerPoint files into other applications to create more creative presentations, such as Google Slides, just like content in other applications is imported into PowerPoint. Presentations can be animated, for which there are several modes or animation options for slide elements, as well as for transitions from one slide to another. It's possible to expand the selection of templates. For example, in PowerPoint you may do so by contacting one of the sites dedicated for this purpose. "The current trend in infographics is not the traditional use of templates or standard icons, but the transformation and generalization of infographics. Innovative technologies make it possible to create visual storytelling with a clear understanding of the conveyed visual information" [4, p. 143].

For the visualization of complex information in presentations, you can define general requirements, the fulfillment of which can make the presentation interesting, creative and informative, not to mention easily understood. Let us highlight the main ones.

1. The presentation should not be overloaded with text, graphics or images, especially small ones.
2. Presentation slides should have a consistent design that helps convey its style.
3. The font should be compatible with the composition and the graphic form, which, as a rule, has a more significant volume and information density.
4. Slide text should be short, clear and structured. In recent years, designers have often made font the basis of a slide composition, highlighting it with color gradation, scale, color or texture.
5. The color scheme should be uniform, not overloaded with different colors. It's best to select no more than 3 basic colors. The color combinations must follow the well-known rules of color theory. Contrasting colors enhance the artistic expression of the presentation. When it comes to the trends of the recent years, bold color combinations are especially prevalent.
6. The composition of the slides should be balanced, integral, built with account for the psychology and physiology of human visual perception.

The aesthetic appeal of presentations is one of the important parts of their design, especially for graphic design students.

It should be noted that in the modern world, animation and video are the most important means of communication in the digital environment. The use of video format is well illustrated by the organization of TEDx conferences. The model of an 18-minute presentation on a hot topic, as a rule, is accompanied by a presentation displayed on a touch screen. Graphic outline, made to visualize reports, creates the effect of visualizing the style and manner of each speaker.

The next most commonly used format at the moment is a combination of interactive elements of video and animation. Animated graphics are gaining more and more popularity in the context of remote communication. To design this kind of visualization, the drawing process is filmed, which is edited with added sound and special effects. Icons and illustrated text accompanying the story gradually appear on the screen. This approach does not require expensive video processing. At the same time, experience in graphic facilitation is required from the person creating such a presentation. The term "facilitation" originates from the English verb "facilitate" - to promote, help, simplify, direct. The purpose of such a presentation is to convey the idea, to facilitate understanding of the relationships between elements. As practice shows, such an animation stimulates the imagination and creates a higher level of involvement of the communication participants.

Ways to effectively use visualization when working online include the following components:

1. Well-crafted images, videos, and animations that highlight key points.
2. The use of graphic templates in a digital environment that allow participants to engage in a common interaction simultaneously.
3. Creation of virtual meeting rooms and enabling of virtual meeting rooms visual components during web conferences that allow to involve large groups of people in communication.

According to D. Sibbet, visualization is an extremely complex phenomenon, which includes several key areas of the world of visualization: cognitive visualization, graphic design, data visualization, visual facilitation [7].

IOP Conf. Series: Materials Science and Engineering **944** (2020) 012015 doi:10.1088/1757-899X/944/1/012015

For the optimal organization of communication in a digital environment, a special approach is required. One must take into account the ways to complete the following tasks:

- orientation of communication participants in a virtual environment;
- the use of accurate visual images, specifically chosen for the target audience;
- designing support for participants and their joint activities.

Thus, a specialist who provides interactive visualization in organizing communication is faced with a wide range of questions - what kind of exhibition environment to choose, what conditions to create for the participants, which templates and metaphors to choose, how to prioritize and structure a presentation scenario.

According to a number of researchers, the successful implementation of training for specialists in the field of graphic design is only possible under the condition of versatile professional training. Thus, S.A. Semin, considering various aspects of communication, writes that the activity of a modern designer requires professional thinking, that combines the figurative and systemic principles, brings new socio-cultural meanings into reality [6, p.16].

3. Results

Clarifying the relationship between various aspects of training in the education of students in the direction of Communicative design, we tested the system of educational creative tasks. The inclusion of students in design practice was carried out in several stages. At the very beginning of their training, the students were offered tasks geared for developing figurative thinking. By building their own "image libraries", students begin to confidently operate with images, which may include both simple and abstract concepts. In parallel with this, it became more expedient to master the simplest techniques of animating graphic elements. Adding animated text or pointers to a presentation or poster develops an understanding of the interactive composition.

As a result of such tasks, it was unconditional that a schematic representation of an object allows them to accurately formulate the main properties, make them understandable to others, reveal hidden connections, show the problem under discussion as a single integral system [8].

Animated service elements and graphics are used to create a unified composition, with the key points highlighted by a whole range of means: color, location, shape, and dynamics. Thanks to modern computer graphics programs and graphic tablets, such tasks are of great interest to students.

At the next stage, students begin a more complex task that includes analyzing information in terms of building a hierarchy. While studying articles, books or speeches, they create graphic notes. Based on the previous knowledge in the field of animation, the elements of this outline can be dynamic and influence the viewer's perception of significant moments. Over the course of this work, future designers reflect on and learn to build correlations.

Key messages and correlations are brought to light and contribute to a clear understanding of the hierarchy within information blocks.

This kind of work prepares students for a stage that allows them to work with more complex information and create infographics. At this stage, interactive visualization of various graphic elements takes on a more professional look.

Further work in the training of a specialist in the field of graphic design is aimed at addressing the more complex aspects of organizing information. These include the creation of scenarios for complex presentations, including text blocks, graphics and diagrams. At this stage, in parallel with figurative thinking, systematization skills are developed. Over the course of training, there are tasks introduced to create diagrams that demonstrate difficult situations, complex concepts and ideas.

Systematic work in this direction contributes to the development of such necessary qualities for a designer as creativity, flexibility of thinking, perception of information and the ability to combine disparate facts into a coherent picture.

The practice of using visualization tools in the learning process has shown that students develop the following skills and abilities:

1) structuring information and identifying relationships and hierarchy between elements;

2) understanding the capacity of digital technologies in the field of compositional shaping;
3) development of scripts for presentation materials;
4) the ability to select the most successful kinds of font, color and stylistic solutions to convey the meaning of information of varying complexity.

This approach to training is, in our opinion, aimed at comprehensive professional training of a modern specialist in the field of graphic and communicative design.

4. Conclusion

Therefore, the modern development trends, which has become dynamic and personalized, have led to a significant change in the transmission of information, which is clearly observable with the example of digital products of the Internet industry, digital educational environments, the organization of video broadcasts and many other forms of digital visual communication. The changes place new demands on the visual organization of content in communicative design. At the moment the most relevant is the organization of visual material through a screen. In the changing situation, it is necessary to look for new ways of creating visualization, which include nonlinearity, editing, blurred transitions, multimedia and heterogeneity.

Analysis of this type of material serves as the basis for the development of a number of courses and assignments to train designers in the design of visual space in a digital environment.

Possession of computer technology and the ability to design the visual component of information in a digital environment is the key to success and demand for a specialist in various areas of modern communication. Such an orientation in the methodology of teaching computer graphics is of certain scientific and methodological interest for the professional training of graphic designers and has great prospects in the educational process in higher education.

References

[1] Bulatova E V, Alekseeva A S and Lomtatidze O V 2015 Experimental methods of analysis of perception of creolized media texts *Ait-tracking in psychological science and practice: collection of articles. articles Resp. ed.* VA. Drummers (Moscow: Kogito-Center) pp 228–233

[2] McLuhan M 2018 *Understanding media. External extensions of a person* (Moscow: Kuchkovo field)

[3] Boyur R, Ishchenko E, Semin S and Shchitova V 2003 What and how designers are taught in Togliatti *Design Problems*, collection of articles, ed V L Glazychev *Design Problems* ed V L Glazychev (Moscow: Union of Designers of Russia) pp 100–149

[4] Radionov S I, Radionova O R, Pilyugin R V and Khizhnyak E A 2017 Graphic design: modern development trend*s World of Education - Education in the World* (Moscow: Izvestia) Number 4 (68) pp 140–147

[5] Roehm D 2012 *Visual thinking* (Moscow: Mann, Ivanov and Ferber)

[6] Semin S A *Communication time (Electronic resource)* URL: https: //www.fondgp.ru/old/lib/chteniya/27/46/prezentatciya_Semin_itogovaya.pdf (date accessed: 22.07.2020)

[7] Sibbet D 2016 *See the solution: Visual methods of business managemen*t (Moscow: Alpina Publisher)

[8] Khabibova A S 2019 The use of graphic facilitation as a means of increasing the efficiency of student learning within the elective module "Digital technologies in art" *Problems of modern pedagogical education: collection of scientific papers* (Yalta: RIO GPA) Issue 64 Part 4 pp 250–252

IOP Conf. Series: Materials Science and Engineering **944** (2020) 012016 doi:10.1088/1757-899X/944/1/012016

Choice of style solution for a design project

I V Khristoforova [1,2], T I Krasikova [1], L V Silcheva[1], E N Sorokotyagina[3]

[1] Leonov Moscow Region University of Technology, Gagarina st. 42, Korolev, 141017, Russia
[2] Financial University under the Government of the Russian Federation, Leningradsky ave. 49, Moscow, 125993, Russia
[3] Kosygin Russian state University, Malaya Kaluzhskaya st. 1, Moscow, 119071, Russia

E-mail: hristo@list.ru

Abstract. The publication reveals the role of marketing research in the choice of style solutions for design concepts in educational institutions. The results of the analysis of styles used in the design concepts of educational institutions, conducted with the involvement of 159 analog objects, are presented. The problem of conflict between the creative and technological components in the design profession is discussed. The role of the digital component in the effectiveness of work on a design project is emphasized. The list of popular professional computer programs used in the course of design engineering. Based on these programs, the student project presented as an example was executed. Discussed issues of digitization of design in conjunction with the assessment of the author's contribution and the impact of the creativity of a particular designer

1. Introduction

1.1 The Role of pre-project marketing research in design

Design, as a sphere of artistic design, solves a wide range of tasks aimed at creating new consumer values in a wide variety of areas of activity: urban planning, architecture, and various industries. Publications of scientists and practicing designers in applied areas of design reveal the role of shape and color [1, 2], ergonomic and stylistic characteristics in the design of products, objects, complexes, and space [3-5]. All these elements should be defined based on the needs of future users. Therefore, marketing research plays a special role in creating new products (goods and services) [6]. At the same time, experts talk about the need to have certain qualities: a sense of empathy and a special professional design thinking that allows you to identify these needs and meet it.

1.2. Features of interior design of educational institutions

The main goal of the environment designer in the field of education is to create the most comfortable, convenient and multifunctional space. The influence of the environment on the quality of the educational process is not in doubt among modern specialists in the field of environment design [7-8]. Professionally designed design can help increase motivation and performance of students [9].

A designer working with educational institutions should take into account, first of all, the influence of the environment on the learning process. When designing a modern space for educational purposes, the following principles are taken into account: openness, privacy, multi-functionality,

IOP Conf. Series: Materials Science and Engineering **944** (2020) 012016 doi:10.1088/1757-899X/944/1/012016

transformability, interactivity, inclusiveness and efficiency [10].

1.3. Creative and digital components of modern design

Designed products-novelties and newly introduced objects in the timely world are created with the direct participation of designers, their creative skills and artistic taste. Equally important is their knowledge and skills in the field of digital technologies that allow you to visualize and translate ideas into specific computer images, drawings, two-and three-dimensional digital projects. These two most important components often act not only as partner competencies, but sometimes come into conflict, which will raise the question of the advantage of the technocratic aspect over the creative one, which gives rise to a number of problems [10].

To create a project designer resorts to the help of professional software. In the process of creating a project, the following software products are used as auxiliary resources: Autodesk AutoCAD, 3Ds Max, Corona Renderer, Photoshop, Corel, and others. The authors of the article are convinced that these tools are auxiliary elements that can not be effectively used without the creativity of a particular person and artificial intelligence can not fully replace it

2. Materials and methods

The preparation of this publication were: secondary information obtained in the course of Desk research, analysis was carried out by methods of grouping and systematization, goal setting and design. We used materials from the laboratory of Construction and applied design engineering of the Moscow technological University of the Moscow region (MGOTU) and 9 student final qualifying works (WRC) of the 2019-2020 academic year performed within its activities. Pre-project studies were conducted in each of these WRC, which determine the basis of the design project concepts chosen by the authors. The results of grouping and classifying content information are presented in the next section.

3. Results

To obtain data, the authors systematized information based on the results of the analysis of 159 analog objects presented in the WRC of students graduating in 2020. In them, the pre-project analysis examined the applied styles presented in the table. 150 existing objects were studied, analyzed in 9 WRC, performed at the department (schools, colleges, technical schools, universities, leisure and educational centers, including art, music, technical sports development and foreign languages) and developed projects in the WRC. The results of the research can also be clearly demonstrated by the results of work on one of the projects described in this publication. The author of the project A.A. Yakovleva, student, department of design photo, supervisor Ph. D., associate professor A.B. Demenkova, consultant, candidate of architecture U.V. Sazonov. Our example shows the algorithm of a design project: from the idea and marketing analysis to implementation in sketches, drawings and computer visualization of the winter garden in the courtyard of the University. As we can see from the table data, the most popular styles in education during the analysis period were classical, minimalistic, and ecological (Table 1).

Table 1. Analysis of characteristics and frequency of styles in design concepts of educational institutions

№	Name of the style	Description of the stylemain characteristics	Projects analogs, %	Projects WRC, %*
1	avan gard	The Use of modern materials, unusual furniture, the use of a bright and contrasting color palette.	9	11
2	art deco	geometric lines. Bright inserts and materials (metal, mirrors, glass). Restraint and monochrome	4	11
3	clas sic	Strict and restrained style and geometry of rooms and furniture. The colors used are pastel, beige, and cream	17	35
4	conte mporary	Modern, democratic, finishing materials-inexpensive, artificial. The color is varied from light shades for floors and walls to bright or dark and rich tones for furniture.	8	11
5	construct ivism	A style that has nothing superfluous, clear geometry and high functionality of rooms and furniture, movable partitions, natural lighting, soft colors.	6	-
6	loft	A style that adapts industrial enterprises to modern premises, combining the rough (concrete, beams, pipes) and elegant (upholstered furniture, mirrors, latest technology) into a modern ensemble.	2	-
7	mini malism	Simplicity and brevity, lack of decor. Neutral color palette. Cabinet furniture, simple forms, strictly sealed.	29	55
8	art nouve au	Smooth lines and intricate outlines in the decor of rooms and furniture, ornaments in the form of climbing plants, asymmetry of forms, arches, openings, stained glass Windows, decorative wrought-iron inserts	1	-
9	high-tech	Comfortable light and without decor made of modern synthetic, glass and metal furniture and the latest technology in the equipment, which are located in spacious rooms. Abundance of light from various sources. The Windows have blinds or roll-up curtains.	2	-
10	punk style	punk style is a fantastic style that combines vintage furniture and all sorts of mechanisms. Interior colors are dark: slate, wood-brown, cherry. Furniture in dark or rich colors, chairs and sofas with copper rivets, low tables and cabinets with drawers	-	-
11	techno	Strict geometry and rough materials, pipes, communications. Combination of living room and kitchen. Sliding furniture. Cool, urban colors. Lighting - spotlights, lamps on brackets and tripods.	1	-
12	futurism	Modern and creative style with a strong emphasis on ergonomics, functionality, the latest materials and technologies. Availability of free spaces. Walls, ceilings, and furniture may have unusual shapes. The monochrome of the light interior is diluted with bright accents.	2	-
13	eclecti cism	The composition of items from the past, present and future is balanced by the selection of color, shape or texture. Lamps and decor items are just as diverse. An abundance of decor and Souvenirs.	-	-
14	eco	The use of natural and natural materials (wood, cork, raw stone, ceramics), an abundance of living plants, a lot of light. Use of paints of natural colors and shades.	14	33
15	ethical	A style that reflects the culture of a particular ethnic group. Finishing materials and a range of shades characteristic of the home of a particular people. Authenticity of the environment and the presence of furniture, items, and decor typical of certain cultures.	5	22

Source: Systematized by the authors based on the analysis of 159 objects.

* Two styles were allowed to be combined in WRC projects.

IOP Conf. Series: Materials Science and Engineering **944** (2020) 012016 doi:10.1088/1757-899X/944/1/012016

A practicing designer uses drawing as a "process of thinking on paper". The ability to present a creative idea on paper is very important for the development of the project, so the skill of sketching is a very powerful tool in the Arsenal of the designer.

Quick sketches and technical drawings contribute to the emergence of new ideas and reflect the overall picture of the future project. During the implementation of the described project, a number of sketches were made (Fig. 1, 2) the Plans that need to be created, as a rule, are technical drawings, layouts (Fig.3). A 3D visualization of the project is created based on sketches and drawings (Fig. 4).

Figure 1. Sketch of the winter garden.

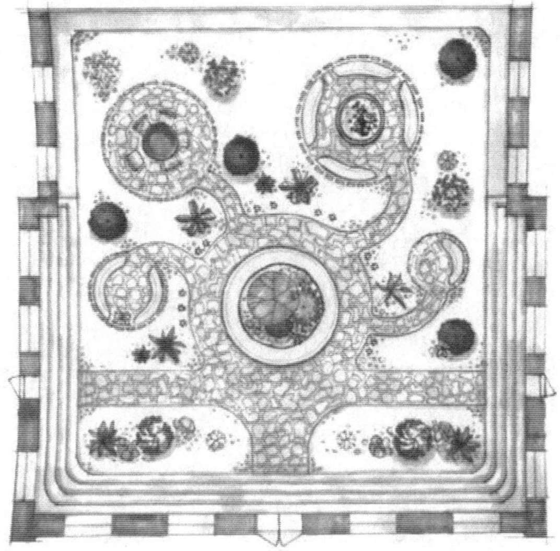

Figure 2. Plan of the winter garden fitting plan.

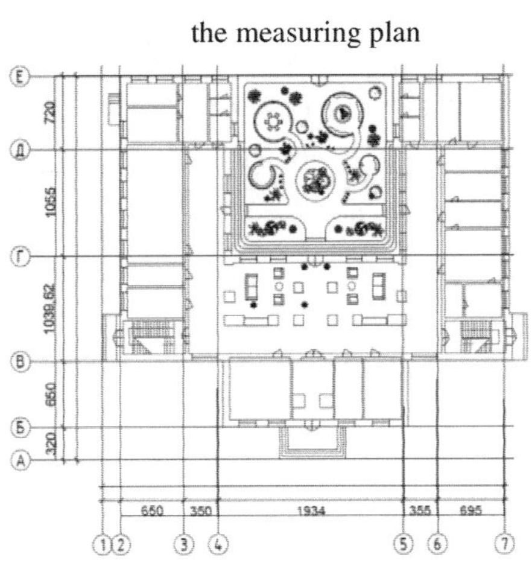

Figure 3. Plan of the 1st floor and winter garden.

Figure 4. 3D Visualization of the winter garden.

4. Conclusion

Thus, it can be stated that the project concept is created based on the choice of style, on the basis of pre-planned marketing research, built by conducting content analysis in the course of Desk research, analyzing analogues, developing sketches, preparing drawings and 3D visualization, agreed with the customer.

The problems of applying academic drawing and painting by modern designers, as well as the use

IOP Conf. Series: Materials Science and Engineering **944** (2020) 012016 doi:10.1088/1757-899X/944/1/012016

of digital technologies that significantly increase the effectiveness of the designer's work, but also create problems of the author's contribution and influence of a particular individual's creativity, remain debatable issues.

It is necessary to emphasize once again that marketing approaches are the basis not only for making management decisions, but also for creating design projects.

References

[1] Arkhipova A A, Syreyshchikova O A, Khristoforova I V and Arkhipova T N 2014 Variability of educational environment design *Materials of the XIII International scientific and practical conference. Theory and practice of modern science* 155–159

[2] Moon G 2018 Conflict colors. Design and Culture *The Journal of the Design Studies Forum* **10(3)** 369–377

[3] Loy J, Canning S and Little C 2015 Industrial design digital technology *Procedia Technology.* **20** 32–38

[4] Savelyeva Zh, Bunina K and Pavlov A 2018 Design features for educational institutions *[Electronic resource]. URL:https://edu design.ru/design_components (date accessed*: 09.07.2020)

[5] Petrosova I A, Guseva M A and Lopasova L V 2013 Preparation of initial information and design of haddresses in three-dimensional SAPR *Design and Technology* **36(78)** 42–47

[6] Khristoforova I V, Demenkova A B and Arkhipova T N 2019 Digital technologies in the design field: from marketing research to product development *II International Scientific Conference GCPMED. Global Challenges and Prospects of the Modern Economic Development*, 871-876 *[Electronicresource]*.URL:https://www.europeanproceedings.com/proceedings/EpSBS/volumes/gcpmed2019 *(date of the application* 09.07.2020)

[7] Osipova N V 2002 Aestheticization of the educational space of the school. Monograph. Publishing house of Moscow state University 188

[8] Osipova N V and Zubkova M F 2010 Typological variative features of educational institutions as the basis for purposeful estheticization of educational space *Bulletin of the Moscow state regional University. Pedagogy* **3** 48–53

[9] Bartashevich D A 2019 Market Research of foreign language schools in the Metropolitan metropolis: Development trend *Scientific notes of young researchers* 24–31.

[10] Khristoforova I V, Kovalev V G, Arhipova T N, Sireischikova O A and Makeeva D R 2015 The processes of technocratization in the profession of the designer *Asian Social Science* **11(7)** 138–144.

IOP Conf. Series: Materials Science and Engineering **944** (2020) 012017 doi:10.1088/1757-899X/944/1/012017

Historical and architectural monuments of wooden architecture in the territory of historical settlements in Belgorod region

L I Kolesnikova[1], V V Aksenov[2], M V Brakovenko[2]

[1] Belgorod State Technological University named after V G Shukhov, 46 Kostukov str., Belgorod, 308012, Russia
[2] Belgorod State Institute of Culture and Arts, 73 Korolev str., Belgorod, 308033, Russia

E-mail: arh.lik@yandex. ru

Abstract. The study analyzes the origin and development of wooden architecture in Belgorod region in connection with the development of the Wild Field on the southern borders of Russia. Volumetric-planning and architectural and compositional features of the preserved wooden heritage buildings were considered: church buildings and residential buildings in large settlements both for poor population and for more prosperous one. The study describes the architectural and compositional construction of temples and residential buildings of different centuries. The authors analyzed the construction periods and the impact of architectural styles that were embodied in the space-planning solution and the design of facades of both wooden churches and residential buildings. Particular attention was paid to cultural heritage sites of regional significance – architectural monuments located in cities of Belgorod and Stary Oskol and villages of Belgorod region. The article presents the research work related to the conversion of a cultural heritage site into a hotel and a museum.

1. Introduction

Historical and natural features of Russia determined the specificity of wooden architecture. Russian craftsmen created a relatively small number of original typical elements and structures to solve different construction problems find to make each building look individual. The main tool of the carpenter was an ax. They made all types of buildings: religious, residential and outbuildings. The scope of carpentry work throughout the entire Russian Middle Ages was really enormous: the chronicles report a lot about construction of wooden fortification cities, starting with the note written under 988: '... And men and the best of the Slovenes from the Chrivitches and from the Chuditches, and from the Vyatiches started to ax, and thus inhabited cities emerged' [1]. This continued until the very end of the 17th century. Many outstanding monuments of wooden architecture, which have always been the most fragile part of the architectural heritage, have not survived to our time. Some of these disappeared due to people's negligence or misunderstanding of the value of these buildings, some of these were beyond the age limit, and others burned down in numerous fires [2].

IOP Conf. Series: Materials Science and Engineering **944** (2020) 012017 doi:10.1088/1757-899X/944/1/012017

2. Materials and Methods

The materials and methods of the study are based on a comprehensive study of the adaptation processes of wooden heritage buildings in the context of changing socio-cultural priorities. The materials studied were archival sources and cartographic materials, materials from museum funds, the current regulatory and recommendatory documents related to assessment of architectural monuments and regulation of their use, including the fundamental one: Federal Law of 25.06.2002 No. 73-FZ (as amended on 13.07. 2015) On Objects of Cultural Heritage (monuments of history and culture) of the peoples of the Russian Federation. The study employed a field survey of historical architectural objects. Retrospective analysis and complex assessment of the territory were performed to achieve the objectives of the study.

3. Main Part

The 1580s is the time when the construction of fortresses on the lands of the so-called Wild Field, on the southern borders of Russia started in Russia. The construction of the Belgorod line began in 1635 and lasted twenty years. The lands of the Wild Field were abundant with forests, which provided the main building material. Servants and settlers who carried out all construction works added the features and traditions of their native places into the architecture. Folk architectural traditions determined the specifics of residential and religious architecture in the new fortified cities and adjacent settlements. The fortifications of the fortresses were made of wood, as were all the buildings inside the fortresses, which often functioned as both a city and a prison [3]. Wooden buildings went rapidly to rack and ruin, and were replaced by new buildings. In this continuous renewal, the architectural demands of society also changed, and new architectural and compositional, and volumetric-spatial solutions appeared.

3.1. Temple architecture

In the 18th and early 19th centuries, the architectural styles of stone architecture had a definite impact on wooden architecture. Wooden temples appeared in villages, and their architectural and planning, and volumetric-spatial solution was based on compositional techniques and decorative design of facades borrowed from classicism. The era of eclecticism also was reflected in both cult and residential architecture. At the end of the 19th century and early 20th century, under the influence of Art Nouveau, elements of stone urban architecture became predominating in wooden architecture, in both volumetric planning of buildings and design of facades, which caused the shift from traditional compositional techniques in both religious buildings and residential and public buildings.

A striking example in the list of preserved church buildings in Belgorod region is the most ancient wooden temple of the Prophet Elijah in Verkhossna village built in 1762 in the former border fort that was built in 1637 and 9 years later became the fortress city of Verkhosensky. In 1991–1992, Temple of the Prophet Elijah was restored. Full disclosure of the main structures of the monument during field surveys revealed two main construction periods of the temple building. The first period (from 1762 to the 80s of the 19th century) is characterized by volumetric planning composition that includes a quadrangle with a light lantern illuminated from the north and south. From the east, there was a pentahedral apse, and from the west it was a square frater with beveled corners to the cardinal points. The original volume of the church had a classic compositional three-volume structure. A similar composition of three timber blockings is a distinctive feature of Ukrainian wooden architecture of the 17th–18th centuries. The roof lantern makes it possible to open the space of the temple to the very top, and this is an imitation of stone architecture. The second construction period dates back to the end of the 19th century. During this time, the building was extensively reconstructed, which expanded the temple volume, changed the plan, facades and interiors. The Temple of the Prophet Elijah reconstructed at the end of the 19th century significantly lacked its both architectural and artistic merits of the volumetric composition and the characteristic details in the decoration of the facades (Fig. 1). As a result of all the changes in the temple architecture, it fully resembled a stylized architecture of the 19th century [4].

IOP Conf. Series: Materials Science and Engineering **944** (2020) 012017 doi:10.1088/1757-899X/944/1/012017

a b c

Fig. 1. a – plan of the temple after restoration of the late 19th century; b – the south-west view of the temple; c – eastern facade. Photo by L.I. Kolesnikova

Imitation of the architecture of stone buildings made wooden buildings look stone by means of board siding, profiled cornices and fillets, columns, chiseled balusters, and other elements of urban architecture, as well as plastering and painting of facades.

Typical examples of imitation of classicism in temple architecture are such temples as the temple in honor of the Kazan Icon of the Mother of God in the village of Cossack Fox, Church of the Nativity of Christ in the village of Gorki, and Church of the Nativity of Christ in the village of Kryukovo (Fig. 2). The temple in honor of the Kazan Icon of the Mother of God in Cossack Fox village is the most interesting example in both the volumetric planning construction and in the facade design. Built in 1780, it was rebuilt twice in 1834 and 1907. The facades, cruciform in the plan, are covered with wooden siding. Windows are decorated with lining provided with large key stones, in the altar, rectangular windows are completed with archivolts of the arched windows, the light drum is cut through by arched windows, and the facades of the north and south side-chapels are divided into spinners by pilasters. The main (western) entrance to the temple is decorated with a columnar portico. For the temple of the Nativity of Christ in the village of Gorki and that in the village of Kryukovo, characteristic features are both a cruciform volumetric planning composition and the design of facades, which used wooden siding, profiled cornices, imitation of key stones in the design of window openings and column porticos. Column porticos adorn the western facades in the temple in honor of the Kazan Icon of the Mother of God in Cossack Fox village and in the temple of Nativity of Christ in Gorki village. In the temple of the Nativity in Kryukovo village, porticoes are used not only for the western facade but also for the northern and southern facades.

a b c

Fig. 2. a – temple in honor of the Kazan Icon of the Mother of God in the village of Cossack Fox, b – temple of the Nativity of Christ in the village of Gorki, c – temple of the Nativity of Christ in the village of Kryukovo. Photo by L.I. Kolesnikova

IOP Conf. Series: Materials Science and Engineering **944** (2020) 012017 doi:10.1088/1757-899X/944/1/012017

3.2. Residential architecture

The oldest buildings of residential architecture that have survived to this day in Belgorod region date back to the 19th century. These houses are mostly four-sided or five-sided, in which the fifth wall is located not along the house, but across.

They are typically shingle plastered and whitewashed; in the villages, houses covered with reeds can be found nowadays (Fig. 3). The houses of wealthy people in the period of classicism were built with columnar porticoes and mezzanines. In the era of eclecticism, the houses were large in area, two-storey, often the first floor was stone, and the wooden floor was shingle plastered and painted or covered with a wooden siding. Under the impact of Art Nouveau, original residential houses of wealthy citizens appeared that adorned the city in the second half of the 19th century.

a b

Fig. 3. Residental buildings: a – four-sided house; b – five-sided house

The preserved and most interesting buildings of the classical heritage include the residental house of the magistrate Kurchaninov in Belgorod. The house located on the relief with a slight slope is built on a high stone basement, and there are domestic premises due to the relief difference in the southern part of the basement plan. The house features shingle plastered and whitewashed wooden walls. Large rectangular windows are framed with lining. In the center of the main facade there is a four-column Doric portico with a wooden entablature and a triangular pediment with a large window. High bases are erected under each column. Heating stoves, molded cornices and sockets have been preserved in the house. A fence and massive curbstones for gates with shallow exedra niches flanked by semi-columns are made of brick in stylistic unity with the house.

The house of the magistrate Kurchaninov is an example of 'wooden' classicism and is part of the building that formed the spatial urban environment of the 19th century (Fig. 4).

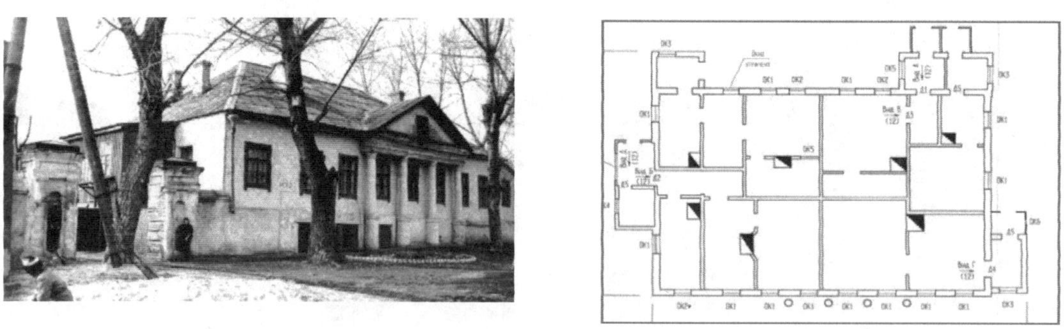

a b

Fig. 4. Residental house of the magistrate Kurchaninov: a – general view; b – house plan. Photo by L.I. Kolesnikova.

IOP Conf. Series: Materials Science and Engineering **944** (2020) 012017 doi:10.1088/1757-899X/944/1/012017

Wooden houses with mezzanines are rare in Belgorod region. One of these houses has survived on the Les na Vorskla site located in the vicinity of the village Borisovka, Borisovsky district, on the right bank of the river Vorskla. The history of the house is connected with the Tikhvin Monastery of the Mother of God Dormition, which was founded in 1714 by Count Field Marshal B.P. Sheremetev, the owner of the village. To the north of the monastic complex, across the road, there are several 19th century buildings that belonged to the monastery. Some of them were residental houses of the monastery priests, including the cultural heritage site *The house where academician and geobotanist V.N. Sukachev worked in 1934–1941.* The house was built in the first half of the 19th century near the road to the monastery. The southern front facade is facing the Tikhvin Monastery of the Mother of God Dormition. The house is a log house without a trace, it was originally plastered and painted. In the 20th century, accretion was made to the northern facade, which distorted the volumetric spatial composition of the house.

Fig. 5. *The house where academician and geobotanist V.N. Sukachev worked in 1934–1941.* Draft design for the reconstruction of a house with conversion of the heritage building into the hotel and museum of academician V.N. Sukachev

The initial volumetric-spatial composition of the house was a symmetrical rectangular building with a mezzanine located above the central part, oriented along the longitudinal axis from north to south. Gable roof mezzanine with triangular pediments to the north and south. The mezzanine is decorated with two balconies: one is at the level of the second floor across the width of the mezzanine, and a smaller one is at the level of the attic. Slightly extended balconies are fenced with carved wooden rapports. The pediment of the mezzanine has a semicircular window, the shape of which is replicated on the front surface of the canopy above the balcony supported by wooden columns with capitals. The canopy over the balcony of the second floor rests on the same columns. There is a large open terrace on the ground floor. Lateral one-story volumes on the western and eastern floors are completed with three-pitched roofs (Fig. 5).

At present, the house is partially used, since the major part of the building is under the threat of collapse. In 2018, the specialists of the Department of Architecture and Urban Planning of BSTU proposed a draft design for reconstruction of this house with conversion of the heritage building into a hotel and museum of academician V. Sukachev [5].

A two-storey wooden house with a mezzanine is a typical example of late classicism, which has survived until the present day, and is an integral part of the building of the monastery complex.

The heyday of the architectural style of Art Nouveau fell on the years 1890–1914. A new trend in architecture embraced not only the capital cities of the Russian Empire but also the cities of the peripheral provinces. Regional differences were manifested in the number of storeys and the choice of materials: in the capital, multi-storey buildings were built of stone, in other places, there were mansions with two or three floors made of wood with a stone basement. The rejection of straight lines and angles in favor of more 'natural'; lines results in fluid silhouettes and accentuated decorativeness. Art Nouveau house designs employed historical elements selectively in favor of more expressive fashion solutions. In Belgorod region, a few stone houses built in this style have survived, wooden architecture, as less durable, is nevertheless represented by the extraordinary residential house of Kobzev in the city of Stary Oskol (Fig. 6).

a

b

Fig. 6. Residential house of Kobzev: a – front facade; b – south-eastern side of the house. Archive photos

The house was built by the merchant M.S. Kobzev for his daughter in 1868 and named *Caprice of Margarita*. The author of the project was N.P. Voikova. Carpentry works were carried out by the Pilyugin brothers from Kazatzkaya, and fretwork was performed by the Gerasimovs, father and son. The hostess of the house, Margarita, organized a salon in the house, and art and cultural figures were invited to the house. The house, located on the south side of the former Lower Trade Square, on a relief sloping to the west, attracted attention not only by its unusual shapes and rich wooden carvings, but also by its bright colors as the walls were painted light green, the decor was lilac, and the roof was covered with orange copper sheets. The house, built on a steep relief, has two floors – a stone basement and an upper wooden one. It is rectangular in plan with a suite-hall layout. The roof is complicated by three volumes in the form of truncated pyramids located above the street corners of the house: the pyramid on the left focuses on the main entrance, the central one shows the location of the 'great hall.' All the pyramids are decorated with carved dormer windows. The street facades are divided into three parts by wreaths completed with basement scarps. The scarps are covered with carved pilasters. The windows are decorated with keeled kokoshniks with carved overlays. The facades are completed with a richly decorated frieze. A baluster fencing is arranged along the perimeter of the roof. To date, the upper floor has been dismantled due to the protracted restoration works [6].

The Kobzev house is a unique architectural monument, where the forms of the emerging Art Nouveau are combined with the techniques of Old Russian architecture. It is the only example of a wooden house built during the transition period from the Russian Revival to the Art Nouveau style that has survived in Belgorod region.

4. Conclusion

The study showed that the majority of historical and architectural monuments in the historical settlements of Belgorod region that have survived until the present day were built in the period from the 19th century to the beginning of the 20th century [7]. Thus, the study of wooden heritage buildings in Belgorod region yielded the following conclusions:

1. The oldest peasant and city houses survived until the present day were built not earlier than the first quarter of the 19th century.
2. A great number of four- and five-sided houses were shingle plastered to insulate the walls, since by that time Belgorod region was short of good timber for mass construction.
3. Many of the houses of this type are still covered with reeds in the villages.
4. Residential houses of prosperous citizens built from timber were insulated not only with shingle plastering but also with felt. Roofs were covered with tiles, which have now disappeared.

5. At the end of the 19th century, copper or black sheet steel was used as roofing material in the houses of prosperous citizens, which was usually painted green to give buildings an elegant and festive look.
6. The majority of the wooden churches date back to the 19th century, much less number – to the 18th century.
7. In course of time, the buildings of the temples were restored, which often changed their original appearance, the facades were decorated with elements characteristic of fashionable style trends borrowed from stone architecture, which brought features not typical of wooden architecture.

For preserving heritage sites, the study of wooden heritage buildings and their restoration are of practical importance to consider the local architectural features, which must be taken into account during monument restoration. The analysis of volumetric-planning, volumetric-spatial solutions and methods of decorative design of facades of wooden heritage buildings made it possible to systematize them, identify characteristic changes in their historical development, designate their characteristic features and, in this regard, to prepare the ground to solve problems of modern restoration.

Acknowledgements
The study was carried out as part of the Program of flagship university development at Belgorod State Technological University named after V G Shukhov, using equipment of High Technology Center at BSTU named after V G Shukhov.

References
[1] Shopina E V and Drobot A S 2019 Russian wood architecture as an example of household buildings *Technical Aesthetics and Design Research* **13** pp 45–53
[3] Kolesnikova L I 2012 *Orthodox churches in Belgorod and Stary Oskol* BSTU Publishing House, Belgorod 145
[4] Kolesnikova L I 2015 *Regional features of wooden temples in Belgorod region on the example of the temple of the prophet Elijah in the village of Verkhossna, Biryuchensky district, Belgorod region* BSTU Publishing House, Belgorod **5** pp 21–25
[5] Perkova M V, Roshupkina O E, Kolesnikova L I and Vaytens A G 2019 The urban conflicts identification and resolution on the Belogorie reserve territory *IOP Conference Series: Materials Science and Engineering*
[6] Kolesnikova L I 1999 Kobzev House *Belgorod Encyclopedia* Regional Publishing House, Belgorod 119
[7] Perkova M V, Kolesnikova L I, Tribuntseva K M and Sementsov S V 2019 Architectural and planning features of educational institutions of the late 19 – early 20 century as a characteristic of identity of the historic town center *B coll.: Urban Form and Social Context: from traditions to newest demands proceedings of the XXV* ISUF International Conference. Siberian Federal University, Institute of Architecture and Design pp 61–67
[8] Perkova M V and Kolesnikova L I 2013 Holy Trinity Monastery: features volumetric-spatial composition *Bulletin of BSTU named after V G Shukhov* **6** pp 70–74
[9] Kolesnikova L I and Sementsov S V 2018 Belgorod city estates of the 18th century *Bulletin of BSTU named after V G Shukhov* **3** pp 65–68
[10] Korenkova G V, Chernysh N D and Mityakina N A 2016 Commitment to the traditions of Russian Orthodox church building in the formation of special knowledge of future designers *Bulletin of BSTU named after V G Shukhov* **6** pp 105–08

IOP Conf. Series: Materials Science and Engineering **944** (2020) 012018 doi:10.1088/1757-899X/944/1/012018

Identification of a tool for evaluation of lighting solutions by users on the example of architectural lighting

I V Koval[1], N V Bystryantseva[1], S V Kolgushkina[1], E U Lekus[1]

[1] ITMO University, Kronverksky Pr. 49, bldg. A, St. Petersburg, 197101, Russia

E-mail: koval.yan@bk.ru

Abstract. This article devoted to description of investigation citizens perception the lighting solutions of architectural lighting at the historical centers of Saint-Petersburg and Copenhagen. The main purpose is recognition the factors of lighting solution attractiveness for further implementation in lighting design. The experiment where every respondent should take evaluation of one or the other lighting solutions has been done for research this area. The questions creation and data analysis occurred by using the Mehrabian and Russell psychological model (PAD) also method S-O-R (stimuli, organism, response). As the result such factors of attractiveness as *luminance, color, artistry (imagery), contextual correlation* have been determined. This research has been conducted within master thesis at CLD ITMO University

1. Introduction

Architectural lighting (AL) is one of the main factors of the city light environment formation also its touristic and financial attractiveness. However, a high level of global urbanization and increasing city quantity has led that a city lighting process has become more typical and massive, in this case citizens' opinion and local city infrastructure feature are not be taken into the account.

The main purpose of the research - is recognition the factors of lighting solution attractiveness, toolkit allowing to make a connection between result expectations and architectural lighting design itself. Lighting is the most essential tool which with combination of architectural features create the unique artistry look [1]. Lighting designers transfer semiotic image, officials oriented by already existing solutions in other countries, trying to achieve same result, but it is total dangerous for city light environment in general [2, 3]. Whatever ideas and concepts of architectural, lighting designers or officials have, there is an ordinary user, who lives there, understands this territory and responses to changes [4].

The task this investigation was analysis and observation user's interest, construct feedback relatively AL. The idea of construct such dialog occurred based on the global trend of space humanization (Human Centered Design, Democratic Lighting), where personal and community values, social aspects, dynamic of the social interactions are the basic of design [5, 6].

1.1. Comparing and description architectural lighting feature in Saint-Petersburg and Copenhagen

For more suitable object analyzing the embankments' facades in Saint-Petersburg and Copenhagen (Denmark) had been chosen. A part of the investigation was in Copenhagen, where the one of tasks was research capital light environment and its features. Copenhagen is differing a lot from Saint-Petersburg: the historical part in fact does not have AL. The city can be described as "the dark capital", however

IOP Conf. Series: Materials Science and Engineering **944** (2020) 012018 doi:10.1088/1757-899X/944/1/012018

there is a logical explanation. A point of that it is a consequence of Copenhagen's municipality plan, where the purpose is decreasing a level of carbon dioxide emission [7]. The major tasks such vision was: decrease level of light pollution; lighting design in accord with nature; decrease of power consumption; increase a role of daylighting. At the result, step-by-step Danish capital municipality realized a strategy of humanization of the city and light environment. Another feature, which likely has influence on city light environment by indirectly, is the Denmark historical past. There is an unofficial law "Law of Jante" by Askel Sandemose that affects people's worldviews. So, the culture factor in light preferences formation reflects in user's demand and the attitude towards solutions, form their own understanding of lighting quality.

2. Methodology

The experimental part of the research is to use the survey method as a tool to obtain information from people (users) with combination of two psychological models, S-O-R [8] and Mehrabian-Russell (PAD) [9]. The main idea is comparing the primary reaction of a person and any criteria (e.g. luminance, contrast, color). In this investigation these are tools for correlation of lighting parameters and subjectively assess users. The questioning was conducted as an electronic one in social networks.

The practical part of the research itself consists of three stages: selection of lighting solutions based on comparison of light parameters, conducting surveys, data analysis [10]. For exploration possible perception features of lighting solutions in the inhabitants of different countries, the experiment was conducted as part of a research internship in Denmark, the survey was divided into two options: for Russian-speaking and English-speaking users. For each selected lighting solution 4 questions were formulated using the Likert evaluation system [11] and one question with an open answer. The questions have been drafted in such a way that each answer can be interpreted in the methods described above. The first two questions refer to the first psychological model of PAD to get an emotional response from users. The next two are for correlation of the received data with such parameters as luminance and color. The last question complements the information from the answers to the first four questions, but at the same time is an important source of data such as the nature of luminance distribution, the contrast of facade elements with each other, and the interaction of the entire building with its surrounding context.

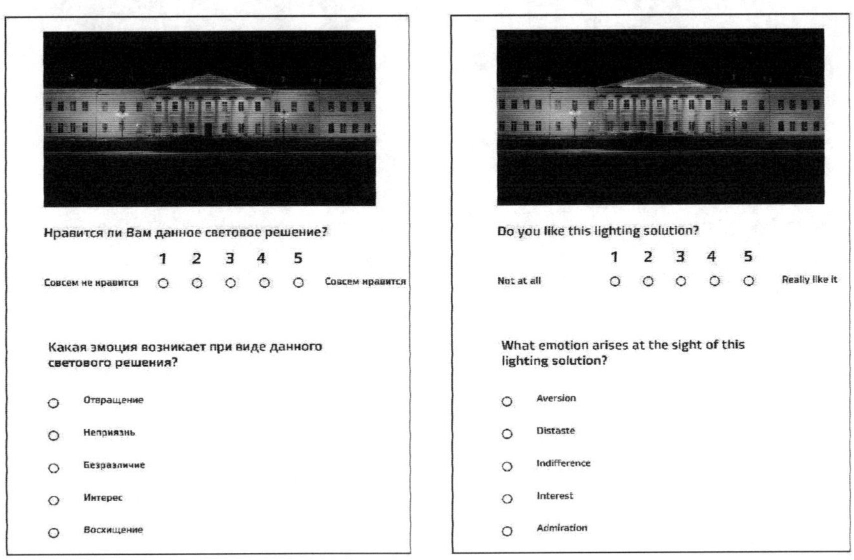

a b

Figure 1. Examples of questions from the bilingual survey: a – for Russian-speaking, b – for English-speaking.

IOP Conf. Series: Materials Science and Engineering **944** (2020) 012018 doi:10.1088/1757-899X/944/1/012018

Respondents aged between 18 and 55 years participated in the online survey. The total number of participants is 206, including 100 men and 106 women. All respondents were initially familiarized with the procedure without describing the experiment itself. It's necessary for achieve more clear results. The buildings of the classical system of form formation located on the embankment of St. Petersburg and Copenhagen were chosen as the necessary incentives. At the same time the lighting solution was analyzed in terms of such criteria as: luminance (intensity), luminance distribution, contrast (luminance or colorful) between the object and the background, as well as additional information was taken into account the category of the building, its function and status. An example of output information is illustrated in Figures 2 and 3. Figure 4 shows the selected seven lighting solutions.

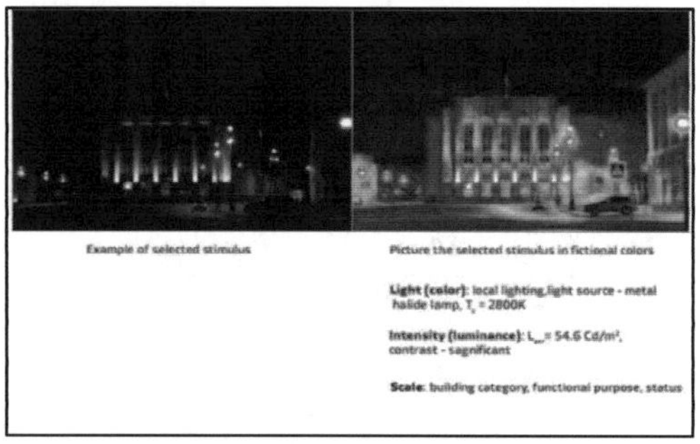

Figure 2. Example of lighting solution attributes.

Figure 3. Example of photo fixation of a lighting solution.

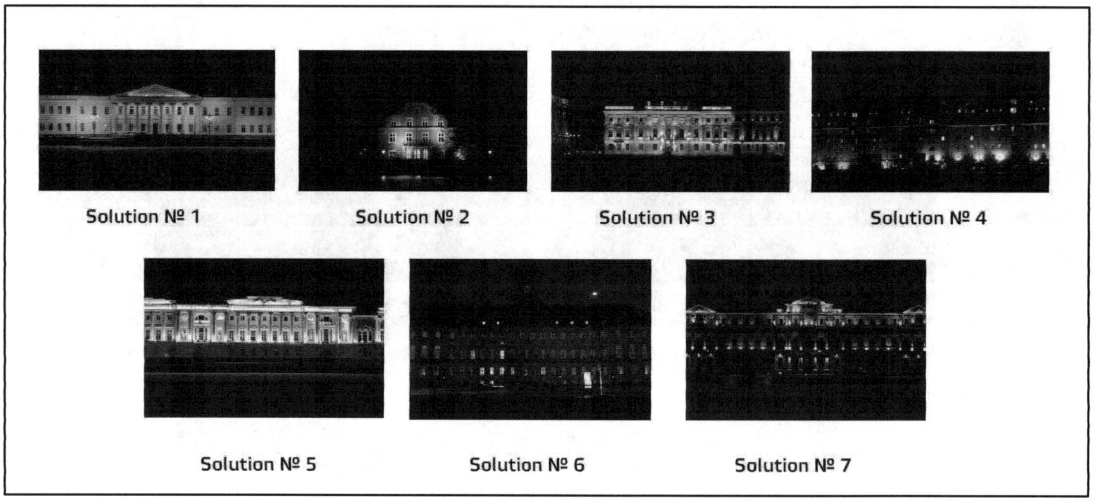

Figure 4. Selected lighting solutions.

2.1. Mehrabian-Russell Model Results Processing

At the result of questioning all responses have been analyzed with using PAD model and represented through cycling diagrams:

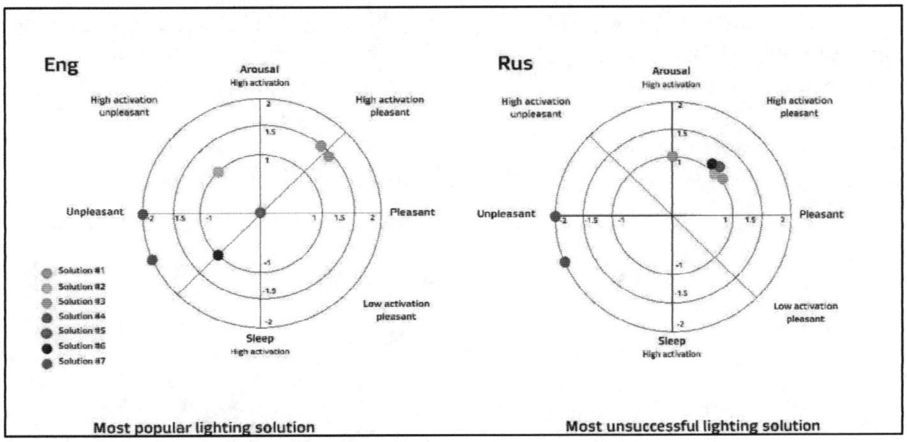

Figure 5. Cyclic diagrams of emotional perception of lighting solutions in two variations.

In additional of identifying most attractive lighting solutions from users' demands point of view, the international perception difference has been discovered. Diagrams in Figure 5 show more expressive emotional response from Copenhagen's citizens. At the result two most and least successful solutions have been chosen based on respondents' replies. Lighting projects are further depicted at Figures 6 and 7.

a

Figure 6. Most successful and popular solution.

b

Figure 7. Least successful solution.

For illustrating more clear perception difference two disputable lighting solution have been selected whose analysis is presented below. All diagrams and indexes on a right match with English-speaking survey, on a left - Russian-speaking. Also, worth to note that parameters on the horizontal axis with rating scales diagrams illustrated below meet from the worst responses (not at all) to the best ones (really like it). In the end of data analysis about the Solution №2 from English-speaking respondents it has been defined that this lighting project arised controversial feeling. On the one side, a warm flood light makes a cozy and comfortable atmosphere. On the other hand, the high lighting contrast between the center and the boundaries of the building, according to people' replies, caused fear, discomfort and suspicion. At the same time Russian-speaking users had this very contrast expressed by the appearance of such words as "secret" and "mystery", expressing emotions more positively.

Comparison of the results on Solution № 6 was evaluated by two groups in diametrically opposed way. Such bright illumination turned out to be not usual for English-speaking respondents. There were a lot of negative responses where frequent words of description were: "too bright", "discomfort for eyes", "strong contrast". In opposite, the Russian-speaking citizens, in addition to comments on the high luminance, noted and commented on the interaction of lighting used with the architecture features, any advantages and disadvantages. Such diagrams clearly illustrate the difference in lighting preferences between cultures, and the answers can also be used to draw conclusions about a greater awareness of lighting among foreign users.

2.2. S-O-R model results processing

The second stage of user assessment analysis was already conducted using the "stimulus-organism-response" model (Figure 8).

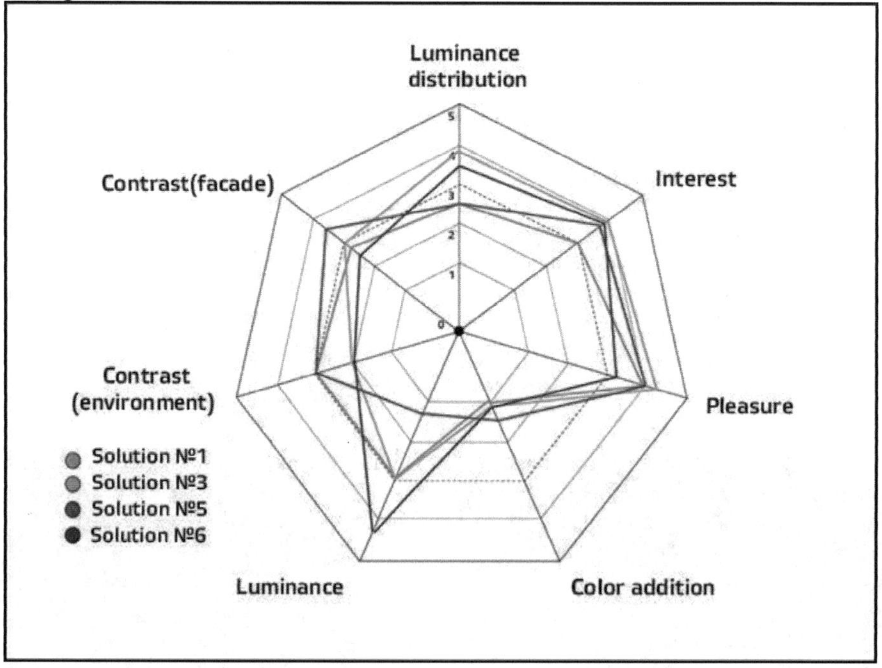

Figure 8. Cyclic S-O-R diagram.

Judging by the diagram, there is a general tendency of changes in the emotional state and increasing interest in the lighting solution at significant luminance indicators (in two of the four values the luminance indexes exceeds the normal values of 30 Cd/m2), but in cases with Solutions № 5 and № 6 strong overestimation and strong understatement of this parameter leads to ambiguous and even negative reaction. Other parameters in each light project brought both positive and negative contribution. Also, negative estimation concerning addition of colour light is visually shown, however in several solutions' addition of luminaries with certain colour, as an option of leveling of lacks the general AL was offered. A brief conclusion on each lighting solution based on user comments. From the point of view of foreign residents, high luminance of 45.29 cd/m2 (Solution № 6) is a fundamental factor of negative assessment. luminance 25.3 cd/m2 (Solution № 3) caused the greatest positive response, luminance 13 cd/m2 (Solution № 5) caused the most emotional response in the form of discussion, where opinions were divided between "mysterious" and "dull" evaluations.

3. Conclusion. Factors of lighting solutions attractiveness in terms of user evaluation

Based on the results of the experimental study of preferences of urban residents regarding architectural lighting, we can highlight the factors of attractiveness of the lighting solution: *luminance, color, artistry (imagery), contextual correlation*. Luminance is the most important human factor because it forms the basis for the image of the object. In St. Petersburg city dwellers prefer lighting solutions with increased luminance, as it forms a certain sense of "understanding" the object. The second factor is color. The overwhelming number of respondents considered unacceptable the use of coloured light in the AL of

IOP Conf. Series: Materials Science and Engineering **944** (2020) 012018 doi:10.1088/1757-899X/944/1/012018

the presented facades. Artistry (imagery), is a set of several indicators: Contrast (luminance or colorful) between object and background, luminance distribution, and concept idea. By this we mean the artistic task that lies at the heart of the project when it comes to lighting. Such a task should be thought out, justified and informative for the user of the environment. Contextual connection means the interaction of the surrounding light background and the light architecture of the structure.

In conclusion, within of the research work carried out to increase the attractiveness factor is recommended: to use luminance in the range of values 25-30 cd/m2; refrain from using coloured light on classic architectural buildings. Preferred color temperature is 3000-3500 K. Color rendering index of the light device Ri not less than 80. With working with the solution to achieve a uniform distribution of brightness by aligning the average value of contrast to the ratio 1/3. When solving the conceptual connection with the environment not to use the ratio of 1/7 luminance of the object and the background (in accordance with the norms is characterized as significant). For "alignment" of contrast it is necessary to work with the surrounding objects. It should be noted that the objects used in the study have a lighting solution based on their daily image. In order to create a complete lighting solution for the perception of the embarkments panoramas, it is recommended to stay close with such a concept. To confirm the accuracy of the results of the investigation should further study the formulated factors of attractiveness of light solutions. The study of social aspects of lighting based on a qualitative method makes it possible to identify both positive and negative features of AL. Such research can make an educational contribution to the formation of the residents' perception that the city lighting, and architectural lighting, particular. Also, it plays an important role in connecting the social life of people and urban space. Carrying out such "educational", research has the ability to improve the level of light culture of users, which may later have an impact on the formation of a more demanding request for urban lighting, increasing the criteria for the professional level of the light environment of the city and its developers.

4. Acknowledgments

We would like to thank Michael Mullins and Ellen Kathrine Hansen for their research assistance and equipment providing.

References

[1] Gusev N M and Makarevich V G 1973 *Light architecture* Moscow. Stroyizdat
[2] Szczepetkov N I 2006 *Lighting design of the city.* Architecture-C
[3] Bystryantseva N V 2015 *Complex approach in creation of the evening city light environment* (Moscow: Moscow Institute of Architecture | MARKHI) 21
[4] Gooch D, Kortuem G and Wolff A 2015 *Reimagining the Role of Citizens in Smart City Projects*
[5] Brandi U, Geissmar-Brandi C 2007 *Light for cities: lighting design for urban spaces.* Switzerland. A handbook. Springer 168
[6] Berntzen L and Johannessen M R 2016 The role of citizen participation in municipal smart city projects: Lessons learned from Norway *Smarter as the new urban agenda* Springer, Cham, pp 299–314.
[7] Ravn K B 2017 Copenhagen municipality *Big profits from LED lights on the streets* [Electronic Resource] Access mode: ing.dk - https://ing.dk/artikel/koebenhavns-kommune-stor-fortjeneste-ved-led-lys-paa-gaderne-199225
[8] Nunthiphatprueksa A 2017 The application of stimulus-organism-response paradigm: The role of social media in Thailand's destination image and behavioral intentions Thailand. National Institute of Development
[9] Mehrabian A and Russell J A 1974 *An approach to environmental psychology* (The MIT Press)
[10] Casciani D and Rossi M 2012 ELSE Experience of lighting sustainability in the environment *Cumulus Conf. Northern World Mandate –Towards Open and Participative Cities* 24–26
[11] Likert R, Roslow S and Murphy G A 1934 Simple and Reliable Method of Scoring the Thurstone Attitude Scales *Journal of Social Psychology* **5(2)** 228–238
[12] SP 52.13330.2016 Daylighting and artificial lighting

IOP Conf. Series: Materials Science and Engineering **944** (2020) 012018 doi:10.1088/1757-899X/944/1/012018

[13] EN 2004. EN 13201–2:2004. Road lighting. Part2: Performance Requirements

IOP Conf. Series: Materials Science and Engineering **944** (2020) 012019 doi:10.1088/1757-899X/944/1/012019

Arctic design methods: Developing technology for onboarding life in the extreme environment

S G Kravchuk[1], **N P Garin**[1], **N V Klyusov**[2]

[1] Ural State University of Architecture and Art, 23 K. Liebknecht St., Yekaterinburg, 620075, Russia
[2] Institution of Additional Education "Technopolis", 4/1 Melik-Karamova St., Surgut, 628400, Russia

E-mail: svetlana_usenyuk@mail.ru

Abstract. The paper reports the results of a 3-yr long research project (2017-2020) into Arctic design and its potential to contribute to the development of the Russian Far North, with emphasis on mobility in conditions of underdeveloped infrastructure and remoteness from economic centres. By way of employing the case study approach, we conducted several expeditions to explore DIY transport vehicles created by inhabitants of the Russian hinterland. We elaborate on the results of data collection and analysis to propose a triad of design methods aimed at working with local makers. The overall approach is illustrated with conceptual sketching. In the final discussion of the results, we outline potential directions for further research.

1. Introduction

The Russian Far North with its thinly scattered population, long distances and lack of roads and infrastructure presents a major challenge in terms of sustainable development and, in particular, strategic transition from 'conquering' the northern frontiers to 'onboarding' life in the North. A team of design researchers carried out a three-year research project (2017-2020) by way of contributing to the solution of this national priority based on the research and design expertise of the Arctic Design School (ADS) set up within the Department of Industrial Design at the Ural State University of Architecture and Art (USUAA) in the early 1980s, now the university's major lab for design research. Within the framework of this project, the existing principles and practices in the area of design for extreme northern environments were reviewed. One of the outcomes is a triad of design methods aimed at developing artifacts and technologies that would better serve people in remote/isolated areas with severe climatic conditions.

1.1. Thematic Focus

The main design and research challenge stems from the basic characteristics of human-technology interaction in the extreme environment of the Arctic/North: in such conditions, man relies on technology, which is more vulnerable than in other climates and is more difficult to maintain and repair, resulting in growing psychological tension in the user. Thus, the research question is: what kind of man-made environment should there be for the Arctic, and what technological and aesthetic characteristics would it have? With reference to the professional specifics of design: how should the special "Arctic materiality" be designed, including industrially produced objects which should support and protect

IOP Conf. Series: Materials Science and Engineering **944** (2020) 012019 doi:10.1088/1757-899X/944/1/012019

human life while causing no damage to the nature?

For design professionals, this means that "arctic adventure" cannot be regarded a "specific case" for design practice; it is an independent area in its own right. Hence, there is little rationale in using existing design principles, methods and techniques developed and tested in more "civilized" environments and moderate climatic conditions. According to Nigel Cross, improving the design process implies the development of new design methods [1, p. 46]: in our case, this means that design for an extreme environment requires developing its own set of tools and systematic procedures that would ensure the expected tangible outcome.

Having studied a historical array of scientific and technical experiences in the creation of "object-spatial environment" for the conditions of the Russian North [see, for example 2–4], we discovered a rather scarce toolkit based primarily on the modernization of existing prototypes of either dwellings and clothing or machinery and other equipment developed for other conditions (more the comfortable climate of the midlands) and for performing other functions, with minor adjustments for the harsh climatic conditions of the North. The inability of the industry to meet the need for reliable, convenient and affordable transport engendered a spontaneous amateur design movement. These specimens and practices of "amateur design" – particularly in the field of transportation – inspired us to develop special design methods, homemade craftsmen being priority target users of the practical results of our work.

2. Data and methods

The development and testing of a set of design methods for the Russian North/Arctic extreme environment envisaged in the project required the use of the case study approach outlined by Yin as a "linear, but iterative process" [5] that appears to be especially useful when research is in its early probing stage. The main source of information for the study of appropriate arctic design technologies and practices was ethnographic explorations: during the period of 2017-2020, we made five expeditions (Arkhangelsk Oblast – 1, Ulyanovsk Oblast – 1, Perm Krai – 2, and Yakutia – 1) involving participant observation and participation/co-creation with the user [6].

Based on the array of empirical data obtained, we focused on the transport sector for two reasons: (1) the paramount importance of the theme of mobility in an extreme environment with poorly developed infrastructure and long distances from economic centers; (2) opportunities for testing design methods developed on live material from the regions under study, with subsequent possible applications in housing and equipment design. In order to strictly and unambiguously interpret our observations, hypotheses and proposals, we focused on a specific scale – movement travelling to meet personal domestic needs and micro-mobility implemented in small-size "transport units" of 1-2-seat ATVs.

The case study we present in the section below is grounded in the village of Pozhva in the Perm Krai, a center of user community innovations in the field of transport that was discovered and surveyed twice: in March 2018 and July 2019. The fieldwork included participant observation, semi- and unstructured interviews with 14 participants, age ranging from 45 to 63, and a co-design experiment with the proposed set of methods applied to the development of all-terrain vehicles by teaming up amateurs (3 persons), design students (4 persons) and a design professional (1 person).

3. Results

3.1. Methods

The triad of design methods developed in the course of the study encompasses various scales and "spheres of influence" in design for extreme environments: progression from external superficial transformation of a one-off thing (*method of composition keys*) to the development of a geographically confined and locally rooted system of objects (*factor method*) and, ultimately, facilitation of locally appropriate and accepted technologies and equipment improving human life (*co-authorship method*). These methods may be used separately or constructed into a "road map" to a new, more attractive lifestyle and a new culture of living in remote, sparsely populated areas with a severe climate. Sticking to the report style of this paper, we provide only a brief portrayal of the methods, while a detailed description with illustrations is presented in the collective monograph [7] and on the project website arcticdesignresearch.ru.

3.1.1 Method of composition keys

Objective of use:

- For targeted improvement of usability in various cases, for example, where a transport vehicle needs to be raised to the level of a commercial product, particularly where the object to be improved presents a product in which engineering solutions have been implemented – from reducing friction in individual components to increasing energy efficiency – without due regard for its aesthetics/design.
- For upgrading objects which have their functional qualities preserved but aesthetically are outdated and require visual/stylistic reincarnation.

The essence of the method implies translating the image of the initial object into the language of a formal (non-pictorial) composition and developing a "composition key", i.e. a two-dimensional abstract composition containing in laconic form concentrated information about the visual characteristics of the object – the compositional center, rhythm, symmetry, etc. The next stage involves analyzing and reformatting the initial composition key allowing for up-to-date requirements and new trends in various areas of science, technology, and culture (for example, enhanced safety in operation, protection against adverse natural and climatic elements, eco-friendliness, latest technological achievements and modern materials, imagery and stylistic characteristics of the brand, etc.), and creating a new image of the product based on the modified composition key.

Expected outcome: the method is aimed at facilitating psychological and emotional adaptation to the environment achieved by means of ensuring the unity of functional, structural and aesthetic qualities of man-made things. A conscious application of the law of artistic composition results in a product with "ideal" consumer characteristics – utmost correspondence to the intended use and satisfaction of all aspects of sensory perception.

3.1.2 Factors method

Objective: to develop an informative and comprehensive brief for designing a mass-produced/batch product: perform an analysis of the form and technical structure of a DIY off-road vehicle in order to elicit the local factors that are important for developing a light-weight off-road transport. In design, these factors are called form-generating – they make up the basis of the design process to resolve the issue of travelling where there are no roads.

The essence of the method: it is a sequence of analytical actions of browsing (mentally) through all links between the form of the product and the conditions of its functioning in the environment. The list of actions is characterized by step-by-step transition from identification of explicit (obvious, visible) factors to concealed ones, from passive contemplation of the form to its active apprehension.

Expected outcome: the method reveals an array of environmental factors that are most characteristic of a particular area. In so doing, there is no need to describe or note or take into account individual factors (cold, damp, wind, etc.); of greater importance is the unique combination of factors, their inter-relations and respective form-generating techniques that have become traditional for the indigenous culture. Knowledge of these combinations provides the designer with exhaustive information for handling design problems, with a ready-made program/algorithm of form-building. The tangible embodiment of this algorithm is a product with a clear/ideal destination. It should be noted that for each remote region of the country there is its own combination of factors and respective system of form-generation techniques.

3.1.3 Co-authorship method

Objective of use: to create conditions for a creative contribution from the user to the development of an absolutely new object/product for oneself. The co-authorship method presents construction of a context for designing while part of the design work is realized by the user from the set of components, or "semi-finished products", using the instructions developed by designers.

IOP Conf. Series: Materials Science and Engineering **944** (2020) 012019 doi:10.1088/1757-899X/944/1/012019

The essence of the method: it is based on the natural ability of man to improve and tailor things and technologies to "fit" his needs by identifying "weaknesses" as a potentiality for improvement, find new uses, etc. The method itself is realized as an asynchronous sequence of actions/iterations: designers suggest a certain provocative solution, which is then modified and improved by users; the designers then get feedback on how it was implemented in the object, resolve problems or propose new provocations; the users pick them up and continue, etc. The iterative process of working, as a rule, happens in real settings and is not limited in time.

Expected outcome: the method results in a thing which is still unfinished while the process of endless modifications and adaptations of the object to the user's needs and to the environment has and can have no final solution. In this case, the product "grows out" beyond being a commodity and turns into a "live" object, literally animated by its owner and co-author.

3.2. Case study: Pozhva "Jeeps"

Pozhva is a workers settlement established in the mid-18th century at Stroganovs' iron-making mill on the river Kama in Perm region. Later, the mining business declined but in the 20th century its human resources and technologies found use at Pozhva Machine Engineering Factory manufacturing fire-fighting equipment. In the 1990s, the settlement became known for its production of light-weight off-road vehicles on low-pressure tires (Fig. 1) – the locals called them "jeeps" despite the fact the vehicles did not have a cab. The general "jeep" concept was borrowed from technical periodicals and adapted to local potentialities and conditions. The design that became classical was a four-wheel rear-drive arrangement with low-pressure tires; it was equipped with the Izhevsk motorcycle engine. The frame was welded at the factory using what was available at that time, and then it was fitted out in private garages. Amateur mechanics also assembled four-wheelers, articulated vehicles and vans, but these were one-off specimens. Pozhva thus grew into an off-road vehicle making center in the Kama region, reaching a peak output in 2000-2007 when the Internet helped promote it across Russia; where the choice of mass-produced cross-country vehicles was still limited. The owners tried repeatedly but failed to get their products legalized by the Gostekhnadzor technical oversight agency to enable their use "on-road", which only served to have the "off-road" label getting stuck to them.

Figure 1. Examples of Pozhva "jeeps". Images by I. Abramov, 2018.

3.2. Conceptual sketching

The basic set of essential characteristics which an ideal transport vehicle should have for the northern (roadless, climatically severe) conditions includes: autonomy (minimal requirements as to infrastructure needed), maneuverability, eco-friendliness (low ground pressure), economic efficiency, maintainability, and customizability.

This section presents a series of conceptual sketching proposals, a designerly reflection on the structure and visual appearance of the Pozhva "jeep" and a "rough" idea for a light-weight small-size ATV. The rectangular-section metal frame of this ATV is arranged longitudinally. The rider sits on it as on a motorcycle. The wheels are tubes. The frame allows a cage or overhead structure to be mounted. There is no suspension. Thanks to large shock-absorbing wheels, the roughness of the terrain is dampened effectively. Drive is through the rear wheels, the front wheels being steerable. The ATV is capable of fording shallow water and can carry small weights. With minor modifications of the wheels, this updated "jeep" could find applications in various areas – from flat and swampy to hilly and snowy terrains of the Russian North.

Figure 2. Modifications of the Pozhva "jeep" for different conditions and user needs. Images by N. Klyusov, 2020.

4. Discussion

In the project reported, we focused on considering small-size transport within the systems of geography and culture by way of expanding professional designer sensitivity to the context of technology creation and use. Our results thus contribute to the discussion on the geographical construction of technology (G-COT) and the potential of remote areas to evoke people's creativity as technology users [see more in 8–10].

While aimed at solving mobility problems in remote areas with extreme climatic conditions, the proposed triad of design methods is not limited to the geographical borders of the Arctic / Far North. In this vein, a comparably vast, remote, and extreme African continent, i.e. a space for makeshift solutions to everyday basic needs [11], can provide yet another "testing ground" for further research and co-design practices with lay designers. With these methods – based on a competent design interpretation of local knowledge embodied in objects and technologies – we aim to contribute to the ongoing debates on decolonizing design [12].

Based on the analysis of the collected empirical material, we claim that inventiveness, self-organization and cooperation are the hallmarks of the northern communities defined by remoteness and harsh climatic conditions, which is also reflected in design as an activity to organize autonomous life support in an isolated/extreme environment [13]. Even at this stage of initial formulation, which definitely needs to be tested, this statement represents a discovery in the field of design and, in a broader sense, in the field of human-technology relations.

5. Conclusion

The paper presents an alternative view of the phenomenon of the extreme environment as a space of new opportunities for designing locally appropriate solutions, particularly all-terrain vehicles. By exploring DIY means of transportation created by inhabitants of the Russian hinterland, the work provides insights into local practices of physical development and expansion of geographical space. These means and practices are presented as innovatory resources for designing customized small-sized vehicles. To use this resource efficiently, we have developed a unique set of design methods that combines sensitivity to socio-cultural aspects with artistic and aesthetic tools. These methods are unique to the Arctic Design field and, at the same time, available for use in other extreme contexts. We have arrived at this result by addressing the theme of territorial distance, isolation, and transportation as a means of overcoming them. Roadlessness and transport irregularity are, of course, key characteristics of the Arctic in the context of its industrial development and resettlement, but quite similar extreme manifestations of isolation and roadlessness and the need to overcome these problems can be found literally a hundred kilometres from any major city across Russia.

6. Acknowledgments

This work was supported by the Russian Science Foundation (grant № 17-78-20047).

References

[1] Cross N 1989 *Engineering design methods* (Chichester [England]; New York: Wiley)
[2] Dibbern J S 1977 Soviet and Japanese oversnow vehicles *J Terramechanics* Dec; **14(4)** 227–36
[3] Derunov G, Kirindas A and Ksenofontov I. 2011 *Mashinnaya tyaga saney: istoriya russkogo snegokhoda* [Machine pull sledges: The history of Russian snowmobile] (Rybinsk: Mediarost)
[4] Kalemeneva E 2018 Arctic Modernism: New Urbanisation Models for the Soviet Far North in the 1960s. In: Wormbs N, editor *Competing Arctic Futures Palgrave Studies in the History of Science and Technology* (Cham, Switzerland: Palgrave Macmillan) 213–41.
[5] Yin R K 2009 Case study research: design and methods. 4th ed. (Los Angeles, Calif: Sage Publications) (Applied social research methods).
[6] INUSE Codesign Journey Planner [Internet]. http://codesign.inuse.fi. Available from: http://codesign.inuse.fi
[7] Garin N, Kravchuk S. 2020 Design for roadlessness: sketches on methods of designing small-size all-terrain vehicles (album-monograph) [Electronic resource] (Ekaterinburg: USUAA)
[8] Norcliffe G. G-COT The Geographical Construction of Technology. Sci Technol Hum Values. 2009 Jul 1;34(4):449–75.
[9] Ardener E "Remote areas": Some theoretical considerations [reprint 1987]. HAU J Ethnogr Theory. 2012;(2 (1)):519–33.
[10] Usenyuk S, Hyysalo S and Whalen J. Proximal Design: Users as Designers of Mobility in the Russian North. Technol Cult. 2016 Nov 16;57(4):866–908.
[11] Campbell A D. Lay Designers: Grassroots Innovation for Appropriate Change. Des Issues. 2017 Jan;33(1):30–47.
[12] Schultz T, Abdulla D, Ansari A, Canlı E, Keshavarz M, Kiem M, et al. What Is at Stake with Decolonizing Design? A Roundtable. Des Cult. 2018 Jan 2;10(1):81–101.
[13] Korgin N, Kravchuk S. Arctic design theory: Methodological issues and interests' concordance problems. In: Proceedings of the International Research and Practice Conference "THEORY OF ACTIVE SYSTEMS - 50 years" (TAS-50, Moscow) [Internet]. Moscow: Institute of Control Sciences RAS; 2019 [cited 2020 May 25]. 464–72. Available from: http://www.mtas.ru/search/search_results.php?publication_id=22775.

IOP Conf. Series: Materials Science and Engineering **944** (2020) 012020 doi:10.1088/1757-899X/944/1/012020

Visual comfort in the organic architecture of an individual residential building

S V Ilvitskaya[1], T V Lobkova [1], V A Lobkov[1]

[1] State University of Land Use Planning, 15, Kazakova Str., Moscow 105064, Russia

E-mail: Tanya.ten.z@mail.ru

Abstract. Organic architecture is based on unity with the natural context. Moreover, the natural environment in all its diversity has a positive effect on humans. Visual comfort is one of the necessary components of the sustainable architecture of an individual home. And visual auspiciousness as one of the aspects of environmental friendliness is gradually beginning to be included in the number of components of "green design". The article discusses and analyzes the relationship of the principles of organic architecture with techniques for ensuring visual comfort of the architectural space of an individual residential building.

1. Introduction

The modern design of an individual residential building is distinguished by the breadth and diversity of architectural creativity. However, when developing design solutions with a high level of comfortable functioning of the space of a residential building, often not given due attention to visual comfort. Nevertheless, the visual organization of space is constantly in interaction with a person, exerting a certain influence on him.

Visual comfort is one of the necessary components of the sustainable architecture of an individual home. And visual auspiciousness as one of the aspects of environmental friendliness is gradually beginning to be included in the number of components of "green design" [1].

Based on the analytical comprehension of modern "green" trends, focusing on the favorable visual interaction of architecture and the surrounding space, a significant role should be given to organic architecture. This direction arose at the end of the 19th century, and finally established itself in the 30s - 50s of the 20th century. Currently, the principles of organic architecture continue to be embodied in modern methods and techniques for the formation of environmentally friendly buildings, combining high technological effectiveness and close interconnection with the visual natural aspect.

2. Materials and methods

The main research method in this work is a comparative analysis of the principles of organic architecture and methods for creating a favorable visual context in the architecture of a residential building.

A comfortable visual environment is an environment with a wide variety of elements in the surrounding space. The main requirement for visual comfort is the lack of homogeneous and aggressive fields. The visual environment of an architectural object should not have large smooth planes devoid of any visible elements, or an excessive number of similar details. The environment with nature-like elements is considered the most comfortable for perception. Consequently, the characteristic features of a visually positive environment should be diversity in silhouettes and colors, diversity in remoteness and sparseness of visible objects. Also, in the architectural solution, it is necessary to provide a direct relationship with the surrounding natural space [2-4].

The demand for organic architecture by the beginning of the 21st century is associated with the development of technology and, at the same time, a craving for individual housing, in which architectural forms can be a continuation of nature. According to the plan of the architect F.L.Wright, the founder of the organic trend, the dwelling was supposed to combine modern achievements and tradition, be an integral part of the environment, organically fit into the natural environment and be intended for the landscape in which it was created [5-7].

Modern organic architecture also has at its core the interaction of a residential building with a natural context. The conceptual model of such a house provides for alternative energy supply systems, the use of ecological materials and the provision of a comfortable architectural environment in conjunction with nature. An example of an individual residential building in the style of new organic architecture can be the Villa Shell in Karuizawa by ARTechnic architects. Interesting examples of organic architecture in Russia can be an individual house in the village of Vartemyagi, a house in the village of Aleksandrovskaya, Wright Style House in Moscow (IK-architects), a house in the village of Zhukovka (Fig. 1).

Fig. 1. Individual residential buildings in the style of organic architecture.

IOP Conf. Series: Materials Science and Engineering **944** (2020) 012020 doi:10.1088/1757-899X/944/1/012020

3. Research results

From the point of view of visual comfort and the relationship with the natural environment in organic architecture, the following principles should be distinguished:

- **Interaction with the landscape** is one of the most important techniques, which provides for a strict fit of the building into the natural environment, which allows you to create a single composition of the architectural object and the surrounding space.
- **The use of natural materials** - materials should be used with a minimum of finishes, preserving the natural look as much as possible. This technique provides interconnection and integration into the natural environment.
- **The unity of architecture with nature** - architectural compositions with the placement of terraces, open interior spaces, a large glazing area allows you to visually connect the interior and exterior.
- **A variety of decorative solutions** – is aimed at enriching the architecture of the house with various decorative elements.
- **The individuality of the appearance of the house.** This technique is manifested in the uniqueness of each building, reflecting the visual image of the area, traditions, inner world and preferences of the inhabitants of the house.

The table illustrates the role of organic architecture principles in providing visual comfort to the individual home environment (Fig. 2).

Principles organic architecture	The value of principle in providing visual comfort
Interaction with the landscape	interaction with the natural environment
The use of natural materials	Variety, interaction with the natural environment
The unity of architecture with nature	Visual expansion of spatial boundaries
Variety of decorative solutions	Positive visual perception of the architectural space
Individuality	Interconnection of worldview and aesthetic aspects in architecture

Fig.2. The principles of organic architecture in providing visual comfort

4. Conclusion

At the present time, when environmental problems are aggravated and a person is often cut off from nature, it is especially important to form a favorable natural environment in architecture. The relationship with nature provides visual comfort and a positive visual perception of the architectural space, which is absolutely necessary for a modern person and should be an integral part of modern design (Fig. 3).

The organic direction in architecture successfully solves the problem of ensuring a favorable perception of the space of an architectural object. This architectural style is quite modern and interconnected with environmental design. Thus, this study of the relationship between the principles of organic architecture and methods of ensuring visual comfort in design can help to increase the level of environmental comfort in an individual residential building.

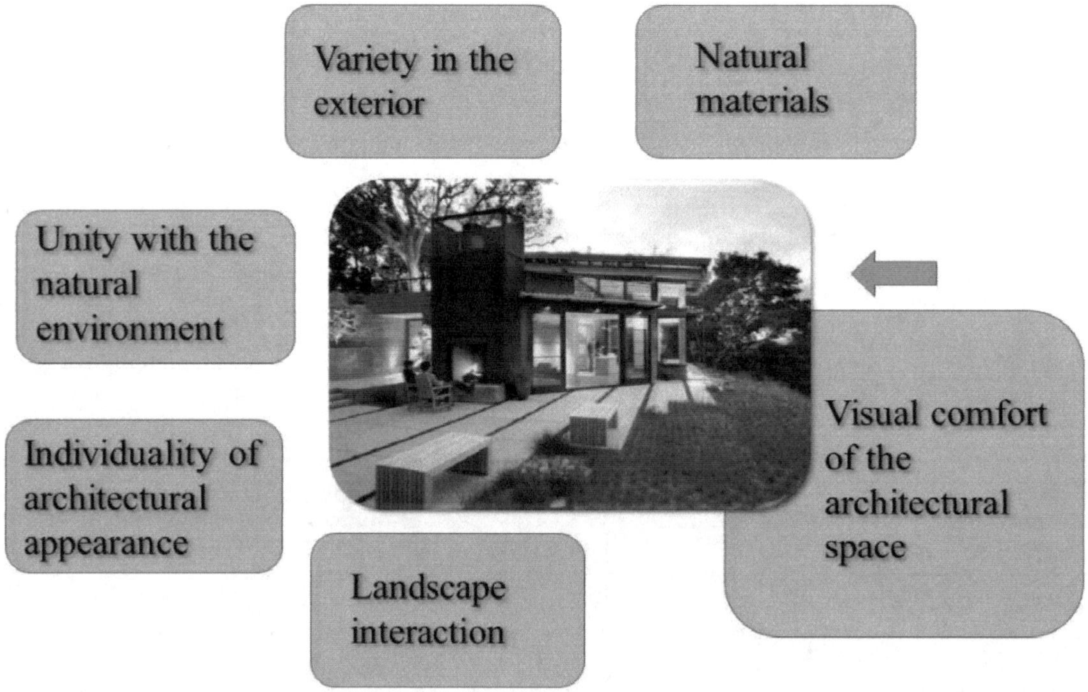

Fig.3. The scheme of methods for providing visual comfort in the architecture of an individual residential building

References

[1] Ilvitskaya S V and Lobkova T V 2018 Philosophy of unity with nature as basis of energy-efficient house architecture *IOP Conf. Ser.: Mater. Sci. Eng.* g451 (2018) 012161
[2] Krasilova L A 2015 *News of univ. Invest. Build. The prop.* The interaction of the architecture of an individual residential building with the natural environment (late XIX - early XXI centuries) **4** 15
[3] Ilvitskaya S V, Lobkov V A and Lobkova T V 2019 Natural materials in sustainable architecture building system *IOP Conf. Ser.: Mater. Sci. Eng.* 687 055030
[4] Shuvalov V M, Komarova JA, Golovataya OV and Kenich O 2016 *Vestnik RUDN* Organic architecture: in harmony with man and nature **4** 50
[5] Smirnova S N 2014 Environmental responsibility of the architect and its impact on ensuring environmental safety of architectural solutions *Priv. Sci. J* **4** 199
[6] Lobkov V A, Ilvitskaya S V and Lobkova T V 2019 *International Journal of Applied Sciences and Technologies Integral* Sustainable architecture in the aspect of environmental management **2** 17
[7] Ilvitskaya S.V., Lobkov V.A., Lobkova T.V.2019 *Academia. Architecture and construction* Natural materials in "green" architecture **2** 130–133

IOP Conf. Series: Materials Science and Engineering **944** (2020) 012021 doi:10.1088/1757-899X/944/1/012021

Relevant forms of youth suit and transforming accessories and ergonomic approach for their development

V A Maltseva[1], L N Maljavko[1]

[1] Bunin Yelets State University, 28, Kommunarov ave., Yelets, 399770, Russia

E-mail: diztory@mail.ru

Abstract. The article presents data on the design, composition and design work on the implementation of the design project of a suit and accessories-transformers for young people under the motto "Cold Desert". The article presents the result of the analysis of fashion trends for the season "autumn-winter 2020-2021", the choice of a source of creativity, the development of a series of sketches of the collection, an artistic and technical description of the appearance of products, determining the most accurate and modern method of designing a suit, studying the properties of the materials used, making drawings basic and model designs of the project and the choice of processing technology for products. The studies have shown that the ergonomic approach improves consumer qualities of products and contributes to the adaptation of their imaginative solutions and designs for everyday use.

1. Introduction

In our opinion, the issue of using an ergonomic approach in the works of fashion designers, textile designers and costume designers is of paramount importance for modern design (Haider Ackermann, Stefano Pilati, Tom Ford, Simon Port Jacques, Greg Lauren, Raf Simons, Kim Jones, Hiroki Nakamura, Denis Shevchenko, Alisa Ushakova, Nino Shamatava and other costume designers). The development of collections of clothing and accessories for mass production should be carried out taking into account the capabilities and characteristics of potential consumers, contribute to the creation of comfort, a positive outlook and evoke many bright, positive emotions.

In this article, the design of a suit and accessories is considered as an innovative activity aimed at achieving a specific goal. This goal is formulated in the form of a project task, namely, the creation of relevant accessories for a promising youth collection that departs from the usual forms and meets four requirements: ergonomics, environmental friendliness, aesthetics and economy.

The significance of the ergonomic approach using in the designing process of transforming accessories in our study is determined by:

- the need for agreement between the requirements for the design project and the psychological and physiological characteristics of potential consumers;
- comprehension of the influence of human behaviour, activity and well-being on the formulation and solution of problems of design and production of a suit and accessories;
- integration of the requirements and characteristics of a person into a design project [1];
- the ability to analyze and structure the ongoing changes in human life, to highlight the main characteristics of the "man-suit-environment" system to create fundamentally new constructive volume-plastic solutions.

IOP Conf. Series: Materials Science and Engineering **944** (2020) 012021 doi:10.1088/1757-899X/944/1/012021

Organization of the productive work on the design project of the suit and accessories needs to consider this process in different aspects: conceptual approaches (O.I. Genisaretsky, G.V. Tolmacheva, M. Heidegger); psychological and physical approaches (V.M. Bekhterev, L.Ya. Dorfman); creative approaches (F. Barron, D. Feldman) scientific and technical approaches (St. Garner, T.O. Berdnik, O.I. Genisaretsky), artistic approaches (A.S. Kvasov, S.A. Mednik), systemic approaches (M. Chikszentmihalyi, D. Feldman) and ergonomic approaches (K. Marell, V. Yastrzhembovsky) [2 – 8]. The scientists mentioned above consider various aspects of the issue we are investigating, but do not offer ways to solve it.

A hypothesis of the study: the process of designing a suit and accessories will be more productive if an ergonomic approach is used as a methodological basis.

The following sections present methods of organizing the process of designing a suit and accessories and the results of their approbation at the Yelets State University named after I.A. Bunin.

2. Materials and methods

Following the hypothesis of the study and the degree of development of the problem, the following tasks were set:

1. to substantiate the productivity of the design of a suit and accessories-transformers from the standpoint of an ergonomic approach;
2. to develop a collection of suit models and transformer accessories from the standpoint of an ergonomic approach and following the fashion trends of the fall-winter 2020 / 2021 season;
3. to approve a methodology that contributes to the production design of a suit and transforming accessories from the standpoint of an ergonomic approach.

The tasks can be solved by organizing the process of designing a suit and accessories-transformers on a theoretical and methodological basis, and the effectiveness is checked using the selected criteria and indicators.

The study aims to determine the effectiveness of the ergonomic approach in the development of relevant forms of a suit and transforming accessories for young people.

Following the purpose of the study, the following tasks were solved:

- fashion trends in themes, colour palette, fabrics, trends and accessories in the fall-winter 2020 / 2021 season were analyzed,
- a source of creativity was selected (silhouette form, colour scheme, texture),
- a series of collection sketches were developed, artistic and technical description of the appearance of the products,
- the most accurate and modern method of designing a suit was determined,
- the properties of the materials used were studied,
- drawings of the basic and model designs of the project were made,
- the technology for processing the products was selected.

The questions of the influence of modern fashion on the formation of personality were also considered [9].

The choice of material influences the protective, aesthetic and utilitarian function of the suit, allowing for comfort and confidence.

For design project, the following materials were used: multi-textured artificial leather, raincoat fabric, denim, the knitted fabric of different texture and density, artificial suede, chiffon, lining fabric and dublerin (Figure 1). Modern technologies make it possible to improve materials, making them wear-resistant, durable, able to cope with low and high temperatures, hypoallergenic and safe for the human body. The materials used in the design project are pleasant to touch, have high wear resistance, strength, excellent breathability, and are easy to clean. The most ergonomic knitted fabrics "kashkorse", "footer", "noodles", "mesh" with an original weave of natural, artificial and (or) synthetic fibres. By their properties, they are antistatic, hygroscopic, hygienic and hypoallergenic.

a b c d

Figure 1. Collection materials: a. artificial leather, b. denim, c. jersey, d. knitted mesh

In the development of the design project, the method of analysis, the method of associations, the combinatorial method of design, the method of transformation were used. The calculation and graphic methods of designing garments were considered (the Unified method of designing clothes of the countries of the members of the Council for Mutual Economic Assistance, the methodology of the Central Experimental and Technical Sewing Laboratory, the methodology of the Central Research Institute of the Garment Industry, the method "Mueller and Son" and the dummy method. Since not all systems and methods for designing garments take into account the deforming properties of fabrics and the class of deployment accuracy, a unified method of designing of Central Experimental and Technical Sewing Laboratory was chosen for the designed collection. This technique made it possible to unify calculations, detail and classify all allowances, simplify the graphic construction of individual nodes. Within the framework of the ergonomic approach, the technique allows working with an individual human figure, taking into account all the features of the physique, to accurately select increases for free fit, which guarantee an excellent fit.

This technique is especially relevant when using modern computer-aided design of garments in the development of original design techniques, taking into account the new properties of modern materials.

3. Determining the effectiveness of an ergonomic approach in the development of actual forms of costume and accessories-transformers for young people

The continent of Antarctica, celebrating the 200th anniversary of its discovery by Russian seafarers F.F.Bellingshausen and M.P. Lazarev, served as the source of creativity for the development of a collection of costume and transformer accessories under the motto "Cold Desert".

According to the author's intention, the collection of models should reflect the beauty and purity of the nature of Antarctica, its global ecological significance. The idea is achieved through a colour scheme, the use of materials of different textures, and sophisticated structural modelling. Asymmetry and geometric articulations repeat the individuality and massiveness of the pointed peaks of rocks and glaciers. Moreover, the different texture of materials conveys the state of nature of the continent: the smoothness and fluidity of waters, fogs and ice sheets, natural relief.

Clothing that protects people in the harsh climate of Antarctica must meet all ergonomic criteria. This requirement also influenced the design project of the costume for young people. The ergonomic approach is the most effective in modern urban fashion design, which requires practicality and comfort in all areas of human activity.

An analysis of fashion trends for the fall-winter 2020-2021 season showed that a modern person chooses products that have a free silhouette that allows free movement, soft shapes to give calm and comfort, as well as a creative cut that brings some novelty and individuality. This style of clothing not only matches the urban environment but also complements it with its unusual shapes (Figure 2).

a b

Figure 2. Key themes of the fall-winter 2020-2021 season: a. Global Heritage Sports, b. "Urban survival"

The creation of the collection provides for a systematic development of the source of creativity, namely, the nature of Antarctica, which, while transforming, preserves the general idea and systemic unity (Figure 3).

a b c

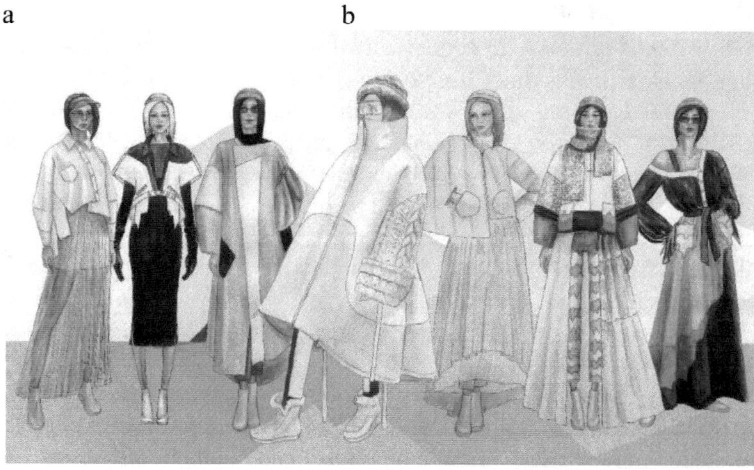

d

Figure 3. Transformation of the source of creativity and drawing up a sketch series: a. source of creativity, b, p. transformation of the source of creativity, d. sketch of the collection "Cold Desert" (author – O. Golikova, the student of Bunin Yelets State University)

The collection's peculiarity lies in its sophisticated constructive modelling (Figure 4). It includes a wide assortment range: trousers, skirts, sundresses, sweaters, blouses, parkas, coats, dresses, realizing the idea of an urban wardrobe for every day. The collection is created, taking into account the principle of interchangeability, which corresponds to the ergonomic approach to the suit design.

IOP Conf. Series: Materials Science and Engineering **944** (2020) 012021 doi:10.1088/1757-899X/944/1/012021

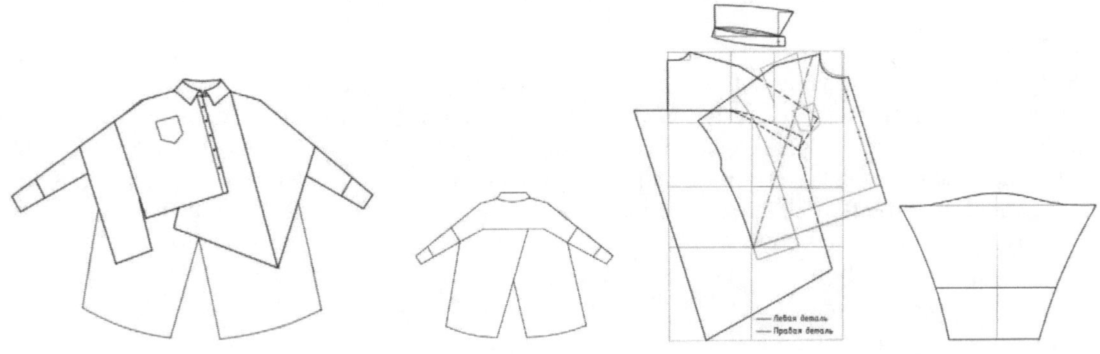

Figure 4. Modelling and designing a shirt. Model 1.

Numerous transformer accessories complement the collection: bags, glasses, hats, scarves.

According to fashion trends in the fall-winter 2020 / 2021 season, bags are distinguished by functionalism, practicality, a variety of shapes and sizes. Whether it is tiny structured cases and micro bags, or padded backpacks and bags made from premium quilted nylon with a satin finish, functional, detachable pockets adorned with trendy detailing.

Stylish transformer bag (Figure 5) is the embodiment of design ideas from the standpoint of an ergonomic approach. This bag is a phenomenon that adapts to any life situation in a few seconds. The apparent advantage of the developed transforming bag is its versatility. A multifunctional accessory for a modern person is becoming a necessity. Depending on the needs of a person, a transforming bag can easily modify four main parameters: shape, dimensions, number of compartments and appearance. The most popular option is the transformation of shape and size. The oversized rectangular bag can be easily transformed into a bolero if folded in half and secured with fittings. A neat oval shape, the bag acquires after fastening the tape "zipper" on the sides.

Figure 5. Convertible bag

The bag capacity can be increased by changing the number of pockets. Some models have detachable wallet pockets that can be used separately from the bag. Complete modification of the texture and colour of the bag occurs when it is turned inside out. Opening the zipper allows easily changing the design of the bag and use it as a sundress, skirt and even a sleeping bag. Detachable straps and hiding shoulder straps provide maximum comfort while moving. Another valuable advantage is saving money because instead of buying several things, we buy one that performs the functions of five or even more models.

Modifying the transforming bag makes it possible to experiment with Pearson image. Such a bag does not fit into the standard rules of wearing, so it can be ideally combined with many elements of the suit.

4. Conclusion

Design and compositional and design work was carried out to complete the design project of a suit and accessories-transformers for young people under the motto "Cold Desert". It was experimentally found that the ergonomic approach significantly increases the consumer qualities of products, allows them to adapt their imaginative solution and design for everyday use. A survey of potential consumers (girls 14-22 years old) after the demonstration of the collection showed that convenience, comfort and practicality in combination with aesthetic appeal were noted by 96% of respondents, modernity and quality of materials – 84% of respondents, originality of the created image – 80% of respondents. The design project can be implemented in small-scale industrial production.

5. Acknowledgments

We express our gratitude to the design students of Bunin Yelets State University for their participation in the experimental work References

References

[1] GOST R ISO 26800-2013 Nacional'nyj Standart Rossijskoj Federacii 2014 *Ergonomics. General principles and concepts* OKS 13.180; 11.180.30 (Moscow Standartinform)

[2] Genisaretskij O I 1995 *Deyatel'nost' proektirovaniya i proektnaya kul'tura* [Design Activities and Design Culture] (Moskva: Progress)

[3] Tolmacheva G V 2015 Konceptual'nyj dizajn: ot idei k produktu [Conceptual design: from idea to product] *Mezhdunarodnyj zhurnal prikladnyh i fundamental'nyh issledovanij* **(10-4)** pp 728–730

[4] Dorfman Y A 1997 *Ehmocii v iskusstve: Teoreticheskie podhody i ehmpiricheskie issledovaniya* [Emotions in Art: Theoretical Approaches and Empirical Studies] (Moscow: Smysl)

[5] Bruno C 2018 De los no lugares al espacio basura: diseño de los espacios de globalización *Arte, Individuo y Sociedad* **30(2)** 261–273

[6] Barron F and Harrington D 1981 *Creativity, intelligence and personality. Ann. Rev. of Psychol* **32** 439–476

[7] Garner St. 2005 Revealing design complexity: Lessons from the Open University *CoDesign: International Journal of CoCreation in Design and the Arts* **1(4)** pp 267–276

[8] Berdnik T O 2007. Rozhdennye letat' [Born to fly]. DIZAJN-REVYU, (1). [Online] Available: *http://design-review.net/index.php?show=article&id*

[9] Victoria Toribio-Lagarde, Dolores Álvarez-Rodríguez. 2019 La moda y su influencia en la identidad cultural adolescente. Arte, Individuo y Sociedad **31(2)** pp 409–423

IOP Conf. Series: Materials Science and Engineering **944** (2020) 012022 doi:10.1088/1757-899X/944/1/012022

Humanistic approach to environmental design: modern principles of organizing an accessible environment

I V Naidenova

Institute of Architecture of Belgorod State Technological University named after V.G. Shukhov, 46 Kostyukova str., Belgorod, Belgorod region, 308012, Russia

E-mail: Irinanaidenova1992@gmail.com

Abstract. The article deals with certain design trends that characterize the humanistic approach to environmental design, as well as ways of interaction between urban planners and designers within this approach. The theoretical basis of the article is the results of research by specialists in Russia and abroad of social urban planning conflicts that arise in the process of creating public spaces and their use by various groups of residents, including those with limited mobility.

1. Introduction

The processes of changing society in the twentieth century contributed to the emergence of conflicts in cities. Specialists in Russia and abroad are currently engaged in establishing the nature of these conflicts and searching for ways to solve them. In our country, until recently, little attention was paid to the inconsistency of urban development in urban planning [1], since the Russian urban planning theory of the twentieth century is generally characterized by the predominance of the functional paradigm. Conflicts of interests and needs between participants in urban development activities are becoming particularly complex and large-scale in modern conditions of a market economy and a tense environmental situation [2]. At the stage of society's entry into the information age, all residents should have equal access to a comfortable and aesthetically attractive urban environment.

According to the world Health Organization, about 15% of the world's population has one form of disability. This indicator has grown by 5% over the past 40 years. This is due to an increase in the life expectancy of the population and an increase in the quality of health care. In Russia over the past ten years, there has been a positive trend towards reducing the number of disabled people (from 13.08 million in 2013 to 11.88 million in early 2020, according to Rosstat). The largest percentage of disabled people are older women, which can be explained by the still low life expectancy of men. However, the fact that the share of children with disabilities has increased from 568 thousand to 688 thousand people over the same billing period is alarming. Young people are more in need of realizing their rights to education, work, leisure, and personal potential, which is why significant resources are being allocated to address the problem of organizing an accessible environment in the country. In this situation, it is important to support the collaboration of specialists in the field of urban planning theory and design practice to implement an interdisciplinary approach to environmental design, taking into account local territorial features.

2. Materials and methods of scientific research

The object of the research is functional conflicts in the use of urban territories by low-mobility groups of the population. The search for ways to resolve these conflicts is a promising area of cooperation between specialists in the field of urban planning theory and designers. The method is a comparison of approaches to the organization of urban space that characterize different periods of the twentieth century. as a result of the study, it is concluded that the predominance of "humanistic" ideas in the design theory of the last decades and the need for wider application in domestic design of new design trends that have developed in the conditions of increasing attention to the problems of people with physical disabilities.

3. The main part

According to the authoritative European urbanist Jan Gale, "until 1960, cities around the world developed primarily on the basis of centuries-old traditions. Living in urban space was an important part of this experience, and no one doubted that cities were built for people." [5] In his research, he focuses on the current problem of reducing public urban spaces in favor of a growing number of cars and the development of transport networks, which was first pointed out in 1961 by the American representative of new urbanism, D. Jacobs.

For four decades, Gale has studied the positive experience of cities in resolving conflicts of interest between motorists and pedestrians, considering the city as an environment that should be, first of all, comfortable and safe for people traveling on foot or by Bicycle. In the aspect of modern design, a fairly simplified division of residents into "motorists", "pedestrians" and "cyclists" does not give a detailed picture of a safe and "friendly" city, since it does not cover the interests of people who have difficult access to public spaces due to physical restrictions. In this regard, it increases the value of deepening theoretical knowledge about the nature of urban planning conflicts in order to further resolve them, including those that arise at the social and everyday level between different categories of citizens.

Even 20 years ago, urban planners, in an effort to "return cities to people", mostly focused on the average parameters of the human figure. With the advent of the new Millennium, new trends are emerging in the design of urban spaces, based on a deep systematic study of the needs that determine the comfort of all residents. This is made possible, on the one hand, thanks to technology, on the other hand, the accumulation of experience of interaction with experts from related fields, fostering interdisciplinary approaches to design. Design, which existed in the format of practical activity at the stage of formation of the first human civilizations, since the first half of the twentieth century. Acts as a driving force for the socio-cultural development of a new type of society [6]. In General, modern design approaches can be described as human-oriented or humanistic.

Problems of organizing the accessible environment in Russia. The problem of poor accessibility of cities for low-mobility groups is currently receiving much attention at the legislative level. In 2008 Russia has ratified the UN Convention on the rights of persons with disabilities; the state program "Accessible environment" has been implemented since 2011. However, the effectiveness of existing approaches is still insufficiently studied by urbanists in the long term. This category of citizens is extremely diverse in composition, which makes it difficult to calculate statistics. In addition to the disabled, the low-mobility groups include pregnant women, parents with strollers, temporarily disabled citizens – people who are also difficult to move for a certain period of time and need quality improvement of the environment. Among the reasons that complicate the implementation of the state program, we can highlight:

- insufficient number of interdisciplinary studies, especially at the regional level, including specific medical knowledge about the features of restrictions in people with physical disabilities;
- incomplete regulatory framework, imperfect tools for strategic planning and mechanisms for monitoring the implementation of urban development projects [5];
- lack of practical recommendations, lack of practical experience of specialists outside of the millionaire cities.

Current design trends for creating an accessible environment. Everyone should have the right to easily accessible open spaces, just like the right to clean water. Everyone should be able to see the tree from their window, sit on a bench next to the Playground at home, or walk to the Park in ten minutes

IOP Conf. Series: Materials Science and Engineering **944** (2020) 012022 doi:10.1088/1757-899X/944/1/012022

[6]. Analyzing this statement, we can conclude that the low-mobility category of citizens in practice often experiences communication difficulties, for them the urban environment is not only uncomfortable, but also unsafe. The process of organizing an accessible environment is still insufficiently studied from the point of view of urban planning theory, and, for the most part, is the subject of designers ' activities. It is no accident that design is directly related to the scientific and practical discipline of ergonomics, which is based on the principle of correlation of objects that fill the environment with the anthropometric characteristics of a person. Experts note that in domestic design, “there is still a large gap between ergonomic knowledge and its application” [8], which generally confirms the opinion that urban planners do not pay enough attention to the scale of the human figure in urban space, and this is especially acute in the case of deviations of human anthropometric characteristics from the average parameters. Trends in "humanistic" design of recent decades are aimed a t creating an environment accessible to low-mobility groups of the population, and this process in foreign practice is often associated with the concepts of “universal design” and “inclusive design”.

Concepts of universal and inclusive design in project theory and practice. In contrast to, for example, ErgoDesign, the concepts of universal and inclusive design are not sufficiently covered by domestic specialists. They are often seen as synonymous, with the exception that " inclusive design "is an English term, and" universal design " is an American one. In this regard, it is useful to consider opinions of foreign experts, representing both of these concepts as independent. Yu Treviranus, a Canadian Professor and founder of the research center for inclusive design, has quite accurately defined that universal design creates a universal size for everyone, while inclusive design is an individual size for one person.

Universal design is usually presented as the design of an architectural environment that is accessible to most people without additional adaptation. The universal environment can be described as simple and intuitive to use. It is considered that universal design appeared in the first half of the twentieth century, largely due to representatives of architectural modernism. During this period, there is a systematization of knowledge about the features of the organization of external and internal space of buildings to improve working conditions and everyday life of residents of fast-growing cities. Le Corbusier, a follower of the theory of anthropocentrism in architecture, declared "...the house is a snail shell. Therefore, we need it to be made according to our standards" [9]. With the help of a system of proportions-Modulor, he created universal approaches to the design of the average proportions of the human figure.

Inclusive design originates in the 70s of the last century, thanks to the achievements of technological progress, with the introduction of technologies such as subtitles for the hard of hearing and audio books for the visually impaired categories of citizens. There are three main trends in the development of inclusive design over the past decades, which has contributed to the public recognition of people with disabilities:

1) legislative – expansion of the legal framework;

2) urban planning and environmental design-creating an accessible environment in the context of new construction or reconstruction;

3) technological-improvement of rehabilitation technologies due to progress in the fields of computer design and engineering.

The spread of new design solutions to remove barriers between people and the environment abroad was preceded by changes in legislation. For example, universal access to urban spaces for wheelchairs in the United States was organized after the passage of the Americans with disabilities act by Congress in 1990. In 1998, an amendment to the rehabilitation act of 1973 came into force, which provided for access to all electronic and information technologies for the disabled. In turn, the legislative changes were caused by the growth of the social movement in defense of veterans of numerous military conflicts since the mid-twentieth century. And the need for social and psychological adaptation of people who were injured during these conflicts. The final consolidation of the international legal framework for the social integration of persons with disabilities took place in 2006 with the adoption

IOP Conf. Series: Materials Science and Engineering **944** (2020) 012022 doi:10.1088/1757-899X/944/1/012022

of the UN Convention on the rights of persons with disabilities.

Computer tools for interactive design in virtual space allow [10] to design objects with mathematical accuracy that take into account individual anatomical features of a person. Thus, with the help of modern technologies, the designer can equip residential and public spaces for people with various forms of physical restrictions. Despite the fact that many developments have a fairly narrow range of applications, inclusive design as a direction of project activity has a great commercial potential. This opinion can be justified taking into account the above statistics, according to which the largest share of disabled people in Russia and in the world as a whole is made up of elderly people. In the context of an "aging population", the experience of specialists who are familiar with the physical characteristics inherent in different age categories and the features of the organization of the environment for them will be more and more in demand in mass design in the future. However, even now there are examples of developments that were originally intended for a narrow circle of users, which have also gained popularity among people without physical characteristics. The product line "Good Grips", developed by the American brand OXO specifically for people with arthritis since 1990, is such an example-OXO tableware with thick rubberized handles, especially the rotary can opener and vegetable peeler, even after two decades, finds a place in kitchens around the world.

4. Conclusions

Despite the fact that research by foreign urbanists in practice demonstrates the effectiveness of the humanistic approach to design, in Russia it will take 20-30 years to study the economic profitability of implementing inclusive developments. This does not reflect well on the investment attractiveness of social projects of the organization of an accessible environment, despite the increased attention to the problems of low-mobility groups at the state level over the past 10 years. In General, the positive dynamics that have developed in the domestic legislation allows us to expect significant changes in the issues of public recognition of people with disabilities, as well as in the practice of resolving conflict tensions in the urban environment through the use of an interdisciplinary approach to design and regional exchange of experience.

References

[1] Perkova M V Conflictological approach in urban planning 2018 *Architecture and construction of Russia (Moscow)* **2 (226)** pp 92–99
[2] Perkova M V 2019 Urban development of the regional settlement system and its elements (on the example of the Belgorod region)
[3] World report on disability [Electronic resource] World Health Organization URL: https://www.who.int/disabilities/world_report/2011/report/ru/ 05.06.2020
[4] The Number of disabled people for reasons of disability in the Russian Federation by region [Electronic resource] Federal register of disabled persons URL: https://sfri.ru/analitika/chislennost/chislennost/chislennost-po-prichinam?territory=1 05.06.2020
[5] Naidenova I V The European experience of reorganizing coal mines into cultural and business centers on the example of Moscow Essen and Genk 2018 *Vestnik of the Belgorod state technological University named after V G Shukhov* **6** pp 65–70
[6] Ian Gale Cities for people 2012 Alpina publisher *(Moscow)*
[7] Naidenova I V Interior design and architecture through the prism of time: from conflict to cooperation 2019 *Technical aesthetics and design research* **1** pp 32–38
[8] Runge Yu F, Manusevich Yu P Ergonomics in environment design 2005 Architecture-S (Moscow)
[9] Le Corbusier Architecture of the XX century 1977 TRANS with FR.; ed. by K T Topuridze Progress (Moscow)
[10] Schumann H Dynamische Raumgeometrie 2005 *Beiträge zum Mathematikunterricht* pp 533–536

IOP Conf. Series: Materials Science and Engineering **944** (2020) 012023 doi:10.1088/1757-899X/944/1/012023

Cycling development in Moscow as part of the policy of sustainable development of the metropolis

P P Nemov[1], **A I Novik**[1]

[1] Genplan Institute of Moscow, 2/14, the 2nd Brestskaya Street, Moscow, 125047, Russia

E-mail: anastasijanovik.7@yandex.ru

Abstract. Bicycle transport is a popular and promising means of transport in the largest cities, as well as the least developed in terms of business travel purposes. The paper substantiates the importance of developing bicycle transport infrastructure in the largest cities through the example of Moscow. The development of bicycle transport is inextricably associated with sustainable urban development and meets the goals of the VizionZero program, leads to the decrease in the use of individual motor transport, increases the physical activity of the population and also has a positive effect on the development of urban environment. The paper highlights the main purposes of the cycling movement of the population and emphasizes the importance of the development of business movement. The characteristics of the development of bicycle-transport network are given. The importance of the integrated approach in the development of a city is substantiated, including the development of a bicycle-transport network with provision of connectivity and safety during its construction. The problems of the development of cycling in Moscow are determined on the example of foreign and Russian cities, as well as strategies for the development of transport infrastructure. The value characterizing the quality of the network development in the city is determined presented by the length of cycle road network per 1 inhabitant. The main problems of the development of cycling transport in the city of Moscow are found and the prospects for the development of the bicycle transport network in Moscow are outlined through the example of the work of Moscow Transport Department and Moscow General Planning Research and Project Institute. The time spent on cycling along the selected routes is analyzed in comparison with car driving during rush hours in road network. As a result it is found that the time spent on cycling is lower compared to traveling by car. In addition, the possibility of developing a bicycle transport network in the red lines of Moscow road network is found. The authors put emphasis on the relevant character of the problem of social perception of cycling transport among the population and determine the directions for increasing the social approval of this type of transport.

1. Introduction

Nowadays more and more cities start to apply GreenMobility principles in their development. The main idea of this concept is to use environmentally friendly and safe means of transport, one of which is a bicycle. Therefore, the development of cycling infrastructure is the most important factor in the sustainable development of the city. At the same time, the Swedish program VizionZero is spreading in many countries, due to which the death rate on the roads is reduced. According to VizionZero program one of the main principles for reducing the the rate of road deaths is to reduce the infrastructure for

IOP Conf. Series: Materials Science and Engineering **944** (2020) 012023 doi:10.1088/1757-899X/944/1/012023

movement by individual transport and develop a comfortable environment for pedestrian and non-motorized movement, the main means of transport is a bicycle.

In Russia there is also a "Road Safety Strategy for 2018-2024", which provides a "zero tolerance policy for road deaths" [1], as well as a national project "Safe and High-Quality Roads" [2]. The purposes of the adopted strategies are similar to the VizionZero program, but the methods of their achievement differ significantly. The main objectives of this strategy are: improvement of control over road safety, development of first aid measures and security from road accidents, development of a system for fixing violations and improvement of road network. Thus, the basic principles of sustainable transport development and GreenMobility in the design of a new and operation of the existing road network are not included in the Road Safety Strategy of the Russian Federation, although road safety is inextricably associated with transport planning and urban environment design.

2. Materials and methods

The concept of sustainable city development is inextricably associated with the development of transport infrastructure. The integrated approach to its development includes the development of urban space not only for the movement of pedestrians, individual, public and truck transport, but also for cyclists. The integration of the bicycle infrastructure into the city transport framework will create a comprehensively developed transport system using alternative means of transport, as well as increase the physical activity of city residents and the comfort and attractiveness of urban environment.

There are the following **purposes** of cycling [4]:

- Leisure – the route length of such trips is 10-50 km;
- Business: can be both for the movement of the population from places of employment to residential areas at distances of 5-15 km and for small freight transportation.

In Moscow there are bicycle paths in almost every park, the development of a bicycle ring is icnluded in the sectoral scheme of the bicycle road network of Transport Department [Figure 5], and there is already a pedestrian "green ring" in the parks of Moscow with a length of 160 km. This meets the need of Moscow residents in cycling trips for recreational purposes. Business routes are developed mainly in the central part of the city, it was there that the first bike paths and bike sharing stations appeared after the implementation of My Street program. However, in a city with a centripetal motion of the bulk of the population, combined cycling routes are practically not developed. This is due to the actual absence of bicycle road network in the middle and outskirts of the city, the lack of safe bicycle parking on the territory of the transport interchange hub and in places where people change from a bicycle to public transport, and in the territory of residents (courtyards, residential areas, residential buildings). Combined cycling have an untapped growth potential in Moscow and can become attractive for people outside of walking distance from high-speed off-street transport stations (more than 0.8-1.5 km).

According to Strava heat map of the activity of cyclists, we can conclude that the need for the development of cycling infrastructure is observed throughout Moscow. This approach was included during the creation of the branch scheme for the development of the bicycle transport network of Moscow (Figure 1).

IOP Conf. Series: Materials Science and Engineering **944** (2020) 012023 doi:10.1088/1757-899X/944/1/012023

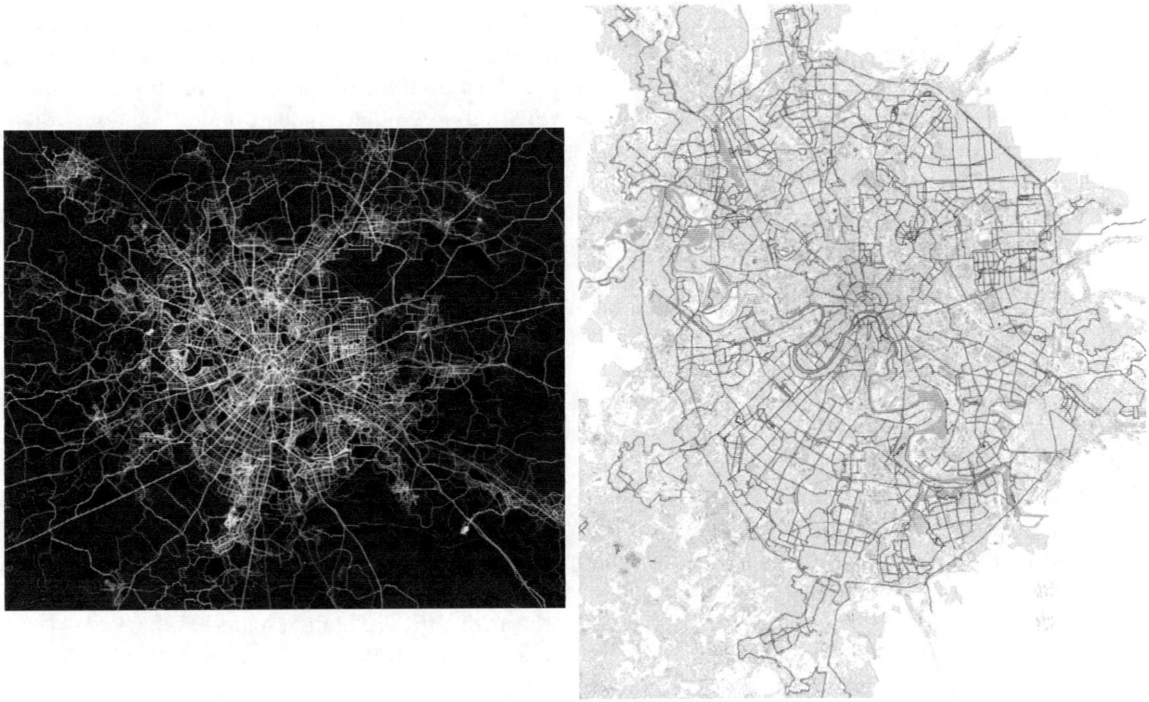

Figure 1. Strava heat map of Moscow cyclists' activity and BranchCycle Development Scheme developed by Moscow General Planning Research and Project Institute

The main **features** of cycling network include [6]:

- Connectivity – the provision of the ability to get to the destination via a **continuous** cycle transport network - the cycle and pedestrian network should be a continuous structure, connecting residential areas with destination objects. At the same time, the connection between residential areas and destination objects should be laid along the shortest routes.

- **Directness** –this parameter can be determined through the overrun coefficient (the indicator should be close to 1);

- Possibility of **free** transfer to other types of transport;

- **Safety** – in order to ensure this parameter, the following must be observed:

- Reduction of conflict points – the intersections of the traffic current of cyclists with cars and pedestrians by separating cycle lanes and dividing flows. Mixed movement of cyclists and cars on the roadway is extremely unsafe due to the difference in speeds of at least 2 times. The reason is the high speed of vehicles with a "non-penalty" margin of 20 km / h, as well as the large width of traffic lanes, which allows car drivers to move comfortably at a speed of 60-80 km / h. Combining movement on pedestrian paths also entails the creation of emergency situations due to the difference in speed with a pedestrian;

- limiting the speed of movement at intersections with other road users;

- mutual visibility of road users, the visibility distance of a cyclist should be at least 10-30 m.

Only if these two basic features of the cycling network are ensured, it is possible to develop a convenient and attractive cycling infrastructure in the city.

In many foreign cities, due attention is paid to the development of cycling infrastructure. "The current leaders of the use of cycling are Holland, Denmark and Germany" [3]. Each of these countries has developed a special program for the development of cycling, which is different in its goals and objectives. Cycling rates in various cities are as follows: Amsterdam - 50%, Copenhagen - 33%, Munich - 17%, Berlin - 13%, London - 2%.

IOP Conf. Series: Materials Science and Engineering **944** (2020) 012023 doi:10.1088/1757-899X/944/1/012023

The formation of bicycle infrastructure in **Stockholm** began with the adoption of a program to develop the city bicycle network, providing the parallel movement of cars, cyclists and pedestrians. The city has a rental system called CityBikes. Bicycles are often used for excursions. This type of transport is popular among all segments of the population. Cyclists are regulated by the police. The popularity of this type of transport is justified by statistics: more than 60% of Stockholm residents use a bicycle for regular trips, every fifth Swede rides a bicycle daily to work, university or school, 18% use a bicycle during winter season. Cycle paths in the city are either separated by markings on the roadway, or laid along the sidewalk. Each house has a dedicated bicycle parking space.

The number of cyclists in **Japan** is constantly growing. In Japan, bicycles are mainly used for short distances. The principles for the development of Japan's bicycle infrastructure include the development of attractive routes, the creation of a comfortable and safe environment, information support, and the creation of an accessible environment.

The improvement of road safety is due to the development of cycling and the rapid bus system. Due to this, the death rate on the roads of the Colombian capital, **Bogotá**, has decreased by more than 60 percent.

In **Brussels**, the speed limit for cars in most of the city is 30 km / h. This limitation reduces road accidents and creates favorable conditions for the movement of cyclists and pedestrians.

The main goal of the development of cycling in **London** is to increase the physical activity of residents, improve public health and use this means of transport for short-term trips and transportation of small loads. Safe cycling routes for short-term and transit trips are being created in the city.

Cycling infrastructure must be adapted to any weather conditions, as in the Finnish city of **Oulu**, despite low temperatures, cycling is actively developing.

There are the examples of successful development of cycling infrastructure of city design in Russia. **Almetyevsk** is the first Russian city to create a full-fledged bicycle road network, taking into account the main parameters as connectivity and safety at the design stage. The city cycling infrastructure is built mainly separately from the roadway and sidewalks. It is planned to create a safe bicycle road network with a length of 250 km with the possibility of functioning at night and during winter season. The city has already equipped more than 600 bike parking lots.

According to the calculation data, a value was determined that characterizes the quality of the network development in the city, which can be the length of the bicycle road network per 1 inhabitant: London - 2.1 m / person; Oulu - 4.3 m / person; Malmo - 1.6 m / person; Almetyevsk - 1.6 m / person (after project completion).

Today in **Moscow** due attention is also paid to the development of bicycle infrastructure. Unlike foreign examples, the bicycle network of the capital is just beginning to form. 230 km of bike paths have been built: 90 km on the road network and 140 km in parks. Since 2015, Moscow cyclists have been allowed to use dedicated traffic lanes for public transport. Thus, the bicycle network is 773 km. Despite the impressive number of the cycle road network, it does not meet the requirements as it is discontinuous and needs to be carefully improved. The cycling infrastructure includes 1867 parking lots for 11656 spaces [5].

Today the total length of the city road network is 6064 km. Thus, the share of bicycle paths in Moscow road network is 3.8%, taking into account the allocated lanes for public transport - 12.8%, per one inhabitant - 0.06 m of bicycle paths.

The main purposes of the development of Moscow cycling network are to reduce the transport and environmental load, increase the mobility and physical activity of citizens [3]. For a number of reasons, Moscow is unlikely to develop as Amsterdam, where bicycles are the main means of transport, but with the proper development of bicycle infrastructure and the creation of the correct image of cyclists in the eyes of city residents, a bicycle can become one of the main alternative means of transport to a car, along with public transport. The residents of Moscow consider a car as an indicator of status and wealth. In order to change the attitude of the residents of Moscow to alternative means of transport, including bicycles, as not prestigious and unfashionable, it is necessary to conduct public and social campaigns.

IOP Conf. Series: Materials Science and Engineering **944** (2020) 012023 doi:10.1088/1757-899X/944/1/012023

The indisputable advantage of cycling is the ability to lay a convenient route along the shortest distance and the lack of attachment to the public transport schedule and traffic jams. According to the analysis, the time for cycling along the selected routes is often less than the time for driving a car during the rush hours of the road network. Moreover, in the areas with an existing cycle road network, the time for cycling is comparable to the time for driving a car, even during off-rush traffic hours.

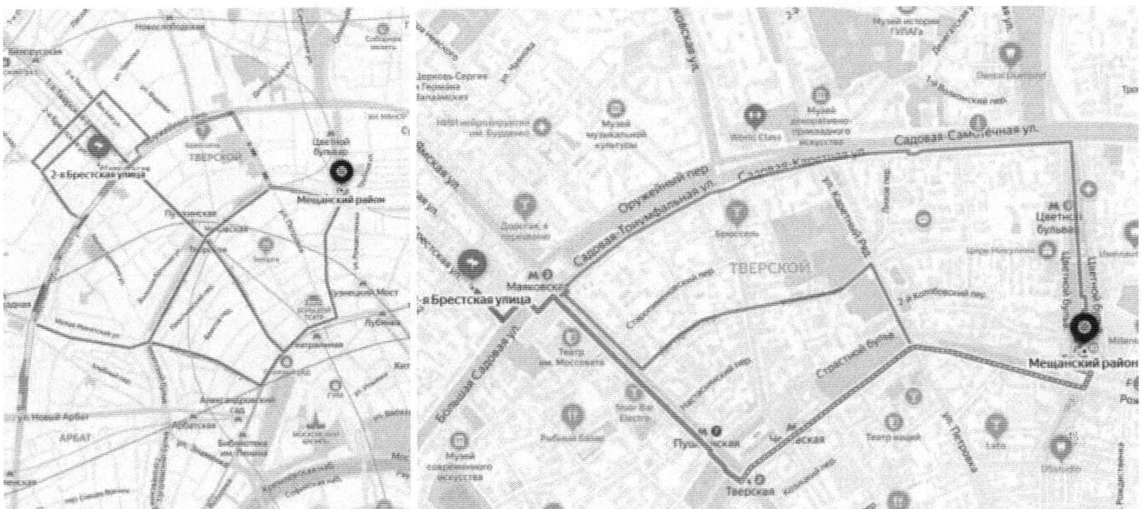

Figure 2. Comparative options for the routes of car driving and cycling – the 2nd Brestskaya Street - Meshchansky district

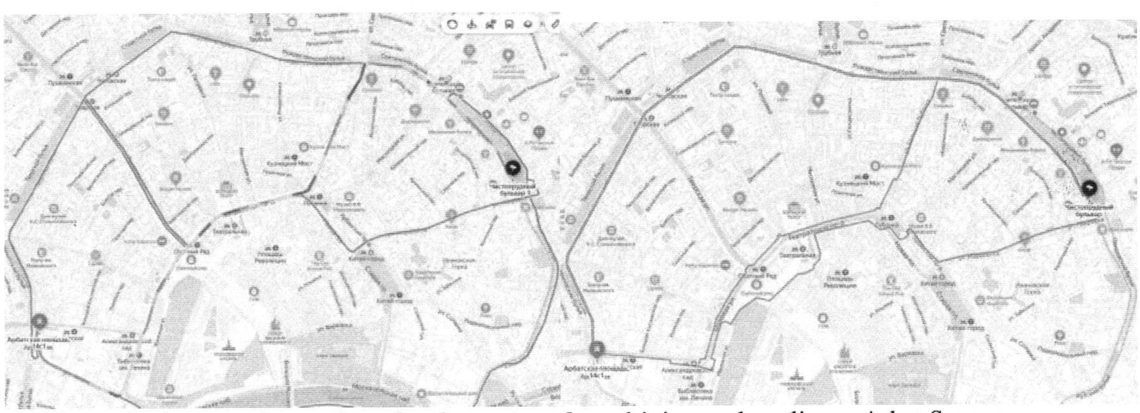

Figure 3. Comparative options for the routes of car driving and cycling – Arbat Square - Chistoprudny Boulevard

Route: the 2nd Brestskaya Street - Meshchansky district (4.3 km) is shown in Figure 2: travel time by car (without traffic jams): 14 minutes (6.4 km); Cycling time: 11 minutes (2.6 km).

Route: Arbatskaya Square 14, building 1 - Chistoprudny Boulevard (4.3 km) is shown in Figure 3: travel time by car (without traffic jams) - 16 minutes; cycling time - 17 minutes.

Accordingly, during rush hour driving by car will take no less time than cycling. In addition, cycling, due to its maneuverability, allows reducing the distance between objects.

In comparison with a car, a bicycle has a number of advantages: a daily bicycle run of 2.5 km saves 180 liters of gasoline and reduces carbon dioxide emissions by 495 kg [8]; 1 parking space places 10 bicycles.

IOP Conf. Series: Materials Science and Engineering **944** (2020) 012023 doi:10.1088/1757-899X/944/1/012023

Figure 4. Scheme of the existing Moscow cycling network [11]

IOP Conf. Series: Materials Science and Engineering **944** (2020) 012023 doi:10.1088/1757-899X/944/1/012023

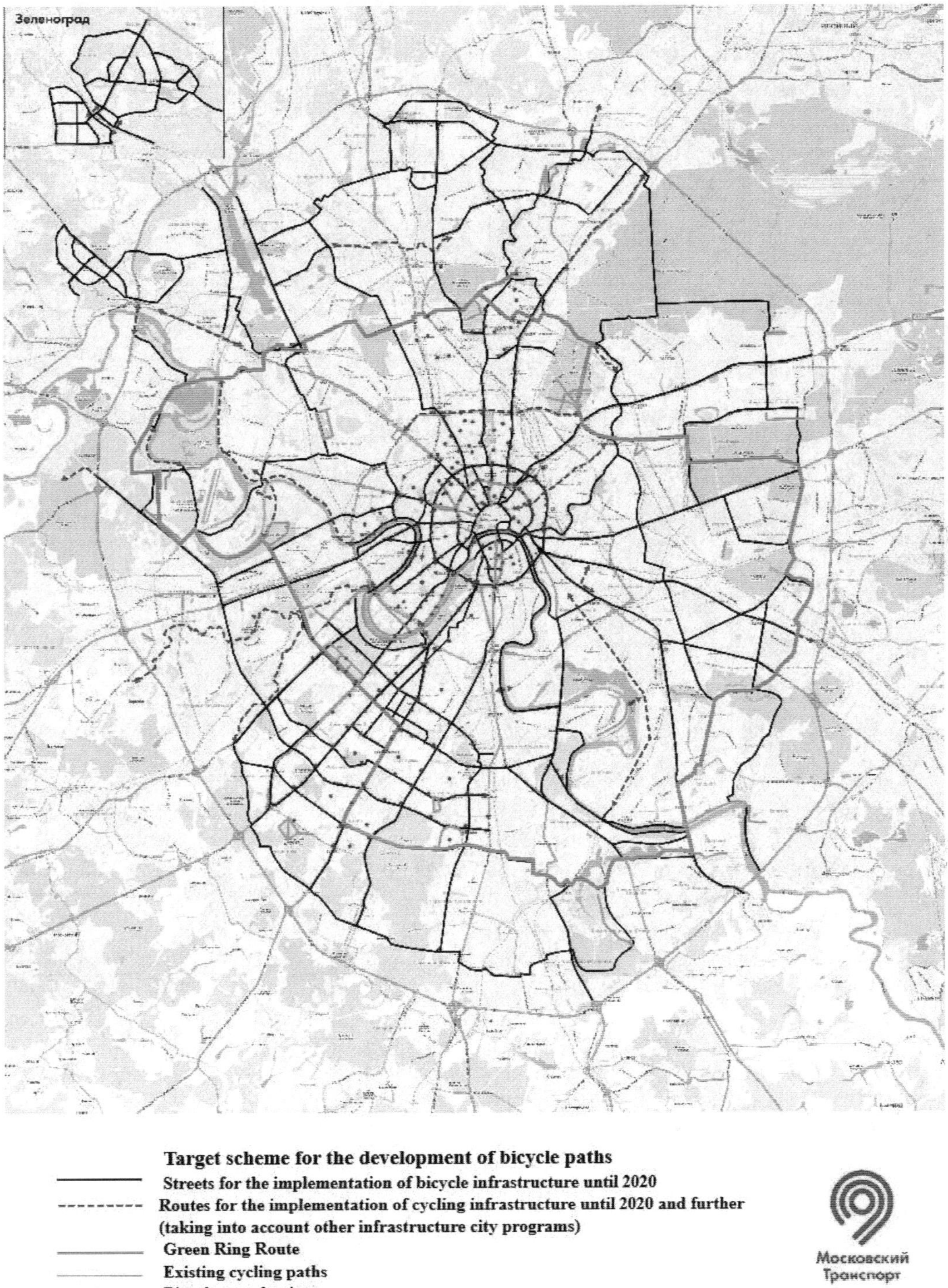

Figure 5. Scheme of the existing Moscow cycling network

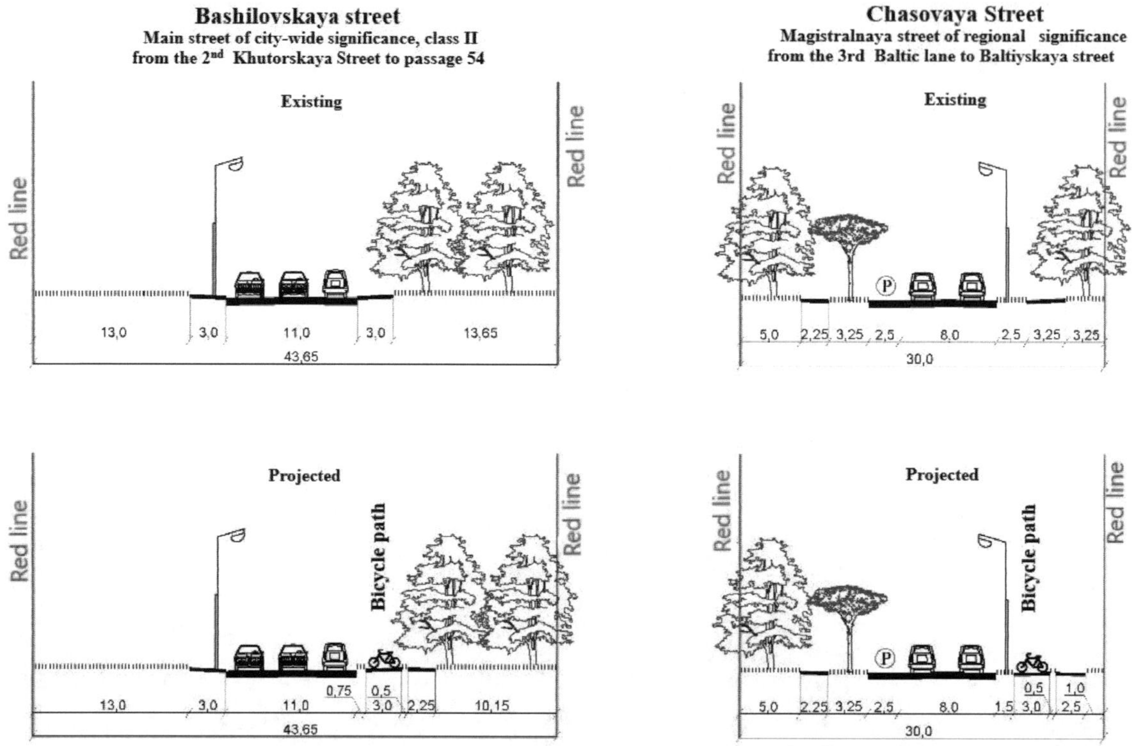

Figure 6. Examples of the development of bicycle road network within the red lines of the street and road network of Moscow

Analyzing the existing position of the network of bicycle paths, there is a lack of structure and connectivity of the districts (Figure 4). Today, the share of leisure trips in Moscow is higher than business trips, due to the development of cycling infrastructure in city parks along the embankments. Despite this, during warm season, bicycles are actively used by couriers throughout the city, as well as by office workers and residents in the central planning zone of the city.

In Moscow, it is possible to arrange bicycle paths within the red lines of the road network along the existing streets, which have room for development, thus meeting the demand of Moscow residents for cycling. Through the example of Bashilovskaya and Chasovaya streets, it can be seen that the organization of cycle lanes is possible by reducing the width of the greening strip, the pedestrian area or narrowing the strip of the mainline (Figure 6). Along many Moscow streets there is a reserve for organizing bicycle lanes separately from the mainline and sidewalks [9].

The development of cycling infrastructure has a positive impact on other spheres of human life: **social, transport, environmental.**

Thus, the development of the city cycling network will meet the basic principles of sustainable development of the city and the trends for zero death rate - Vision Zero and will accelerate the achievement of the goals set within the framework of national projects Safe and high-quality highways "Road safety strategy for 2018-2024" and the national project "Safe and Quality Highways" [10].

3. Conclusion

Cycling is an integral part of the city transport service system, which is a necessary factor to increase the mobility of citizens and the share of personal and business trips on environmentally friendly transport. It leads to a balanced transport mobility of all population and also provides the accessibility and connectivity of territories, which is important for Moscow, so as in the metropolis there are many natural and man-made obstacles.

IOP Conf. Series: Materials Science and Engineering **944** (2020) 012023 doi:10.1088/1757-899X/944/1/012023

For effective work of Moscow cycling network the following is necessary: the development of cycling infrastructure, which primarily ensures the connectivity of the cycling network, providing safe conditions for travel, popularization of cycling - regular monitoring of the attitude of citizens to the level of development of cycling infrastructure is necessary, a policy for the development of cycling, the development of a public transport system, the organization of free transfers to other types of transport, the development of a system of intercepting parking lots for cyclists.

Thus, providing residents of megalopolises with high-quality, high-speed public transport, as well as a developed cycling and pedestrian infrastructure, will be the reason for citizens to prefer environmentally friendly and sustainable means of transport rather than personal car.

References

[1] Order of the Government of the Russian Federation dated January 8, 2018 No. 1-r "On approval of the road safety strategy in the Russian Federation for 2018-2024", available online: http://www.consultant.ru/document/cons_doc_LAW_288413/

[2] Site of the Russian national project "Safe high-quality roads", available online: https://bkdrf.ru/

[3] Zavyalova N B and Saginova O V 2018 Comfortable Urban Environment: Bicycles In A Big City *Human capital and vocational education* **1 (25)** pp 13–20

[4] Evseeva A I 2017 Monitoring of bicycle traffic in a city *Public administration. Electronic bulletin* **64** pp 82–108

[5] The official portal of the Moscow Mayor and Moscow Government, available online: https://www.mos.ru/

[6] Dufour D 2010 *PRESTO Cycling Policy Guide: Promoting Cycling for Everyone as a Daily Transport Mode* (Netherlands)

[7] Kotova T P, Hamadeeva Z A and Khisamutdinova A F 2017 Bicycle Tourism In The Republic Of Bashkortostan: The History, Current State, And Development Prospects *Modern problems of service and tourism* **3** pp 105–119

[8] Kozikova A A 2016 Cycling in big cities in Russia and abroad *Young scientist* **29 (133)** pp 93–95, available online: https://moluch.ru/archive/133/37220/

[9] Nemchinov D M 2017 Cycling in urban environment *Urban Planning* **6 (52)** pp 39–56

[10] Zhbanova C A and Vdovichenko V V 2019 Road Safety Ensuring: Challenges And Solutions *Road safety management: state, problems, ways of improvement* **1 (2)** pp 171–177

[11] The official site SUE Genplan Institute of Moscow, available online: https://genplan-mos.ru/

Experience of reorganization of depressed industrial areas

M V Perkova[1], **M Yu Drebezgova**[1]

[1] Belgorod State Technological University named after V.G. Shukhov, Russia, 308012, Belgorod, Kostukov Street, 46

E-mail: perkova.margo@mail.ru

Abstract. Historic industrial sites and their territories present the integral part of urban environment. Depressive industrial territories in cities need reorganization and transformation, since they do not correspond to the modern requirements for the development of territory and destroy the image of cities. The paper studies the prerequisites and problems of the reorganization of industrial territories, as well as the examples of successful reorganization of industrial facilities depending on their location in urban structure: in the historical center of the city, in industrial zones in city outskirts, reorganization of quarries and dumps distantly located from city structures. Three types of reorganization of depressed industrial territories were identified, depending on the degree of preservation of industrial functions: reorganization with preservation of industrial functions, partial re-functionalization and complete re-functionalization. The applicability of the ways to reorganize various industrial sites against the identified types of depressed industrial areas is evaluated.

1. Introduction

Today the topic of reorganization and transformation of industrial zones becomes more and more acute. Historically industrial zones generally formed in the central part of cities do not meet modern development requirements and destroy the image of cities as an integral architectural and urban planning formation, limit the possibilities for the development of the territory, reduce economic and urban planning value of lands and aggravate the ecological situation in cities. However, the depressing industrial zones of cities present a potential resource for the development of urban areas, construction of housing, business and shopping centers and recreational areas. This is a significant land reserve, which can make it possible to correct the deficit in building areas in the central part of cities.

Historic industrial sites and their territories present an integral part of urban environment. There is no doubt that it is necessary to understand the meaning and content of the industrial territories subject to reorganization, to assess their transformation over time in the implementation of federal and regional programs and strategies [1]. Almost all industrial enterprises of the 18th - early 20th centuries are monuments of architecture, and the problem of their preservation and subsequent operation is one of the most urgent.

Today, there are urban contradictions in the development of such territories [2]. In this regard, it is necessary to determine the existing conflicts between participants in urban planning activities (land use, social and functional activity, transport, regulatory, property) and search for a compromise in solving problems of renovation of the existing buildings [3].

IOP Conf. Series: Materials Science and Engineering **944** (2020) 012024 doi:10.1088/1757-899X/944/1/012024

2. Materials and methods

The theoretical and methodological basis of the study is presented by conflictological approach, which allows studying the cultural and historical process, the period of industrialization and the processes taking place on the territory of industrial enterprises at the present stage of development of society. In addition it makes it possible to identify the interests and needs of participants in urban planning activities and to find a compromise that allows not only creating a comfortable living environment but also preserving nature. The research is based on a comprehensive study of the process of adaptation of industrial zones in the context of changes in urban planning system and socio-cultural priorities.

3. Reorganization of depressed industrial areas

At the end of the 19th and the beginning of the 20th centuries, in big cities around the world, the main city-forming factors were industrial and transport enterprises, where citizens worked and lived in a close proximity (from 70 to 80%). As a result of the constant increase in production capacity, industrial areas increased and residential areas appeared within the production zones. Such buildings were typical for densely built-up cities of Russia and Europe.

During the research, it was found that the first problems of irrational use of industrial areas arose in 1970, when technology was intensively developing. Then there was a decline in industrial production: the shutdown of factories, their disrepair or re-equipment into non-industrial premises. In 1990, the profitability of traditional industrial production fell and the demand for non-industrial use of sites and premises in the centers of big cities increased. Coastal industrial areas also suffer from mismanagement, as industries often developed along rivers. With the development of cities, the requirements for ecological situation increased and a conflict between industrial and recreational zones arose [4]. The need to adapt industrial facilities and integrate them into the modern urban and social environment appears.

Thus, the non-compliance of industrial enterprises with modern requirements leads to the need to reorganize industrial territories [5].

The prerequisites for the reorganization of industrial territories are as follows:

• *Architectural planning* (insufficient density and intensity of use of industrial areas; low architectural and aesthetic quality of industrial buildings; convenient location of industrial areas in the city structure; shortage of residential and public spaces, especially in the central districts of cities);

• *Administrative and legal* (non-compliance with modern urban planning requirements, a fairly high level of environmental impact);

• *Management prerequisites* (the tendency to reduce large industrial facilities with the development of small and medium-sized enterprises, new forms of management and economic models);

• *Technological prerequisites* (development of new high-tech enterprises in smaller industrial areas and spatial production models - mixed use zones, industrial parks and technology parks [1].

For city-forming industrial territories, the best land plots were usually allocated from the city fund, which were within transport accessibility to cities or settlements with developed infrastructure. However, during the reorganization of these territories, a number of problems arise, such as:

• Lack of clear government programs and administrative procedures for working with industrial sites (decisions are usually made on individual basis).

• Higher economic costs compared to new construction.

• The presence of several owners with different plans for its reorganization.

• Deterioration of engineering networks.

• Environmental problems [6].

Foreign and Russian experience in the adaptation of degrading industrial areas in the historical environment allowed determining the main potential of industrial zones in order to adapt to modern urban conditions, as well as diverse and interrelated problems of adaptation - environmental, security and rehabilitation, aesthetic, design, construction, economic, environmental and organizational ones [7]. There are various ways of renovation and reorientation of industrial areas to meet modern requirements. The results of such design solutions are public spaces in the form of commercial

IOP Conf. Series: Materials Science and Engineering **944** (2020) 012024 doi:10.1088/1757-899X/944/1/012024

premises, creative clusters aimed at the development of the territory, as well as residential complexes and apartments with increased level of comfort and life support.

In the course of the research the authors consider the examples of successful practices of the reorganization of industrial facilities [8] depending on their location in urban structure: in the historical center of the city, in industrial zones in city outskirts, reorganization of quarries and dumps distantly located from city structures (Figure1).

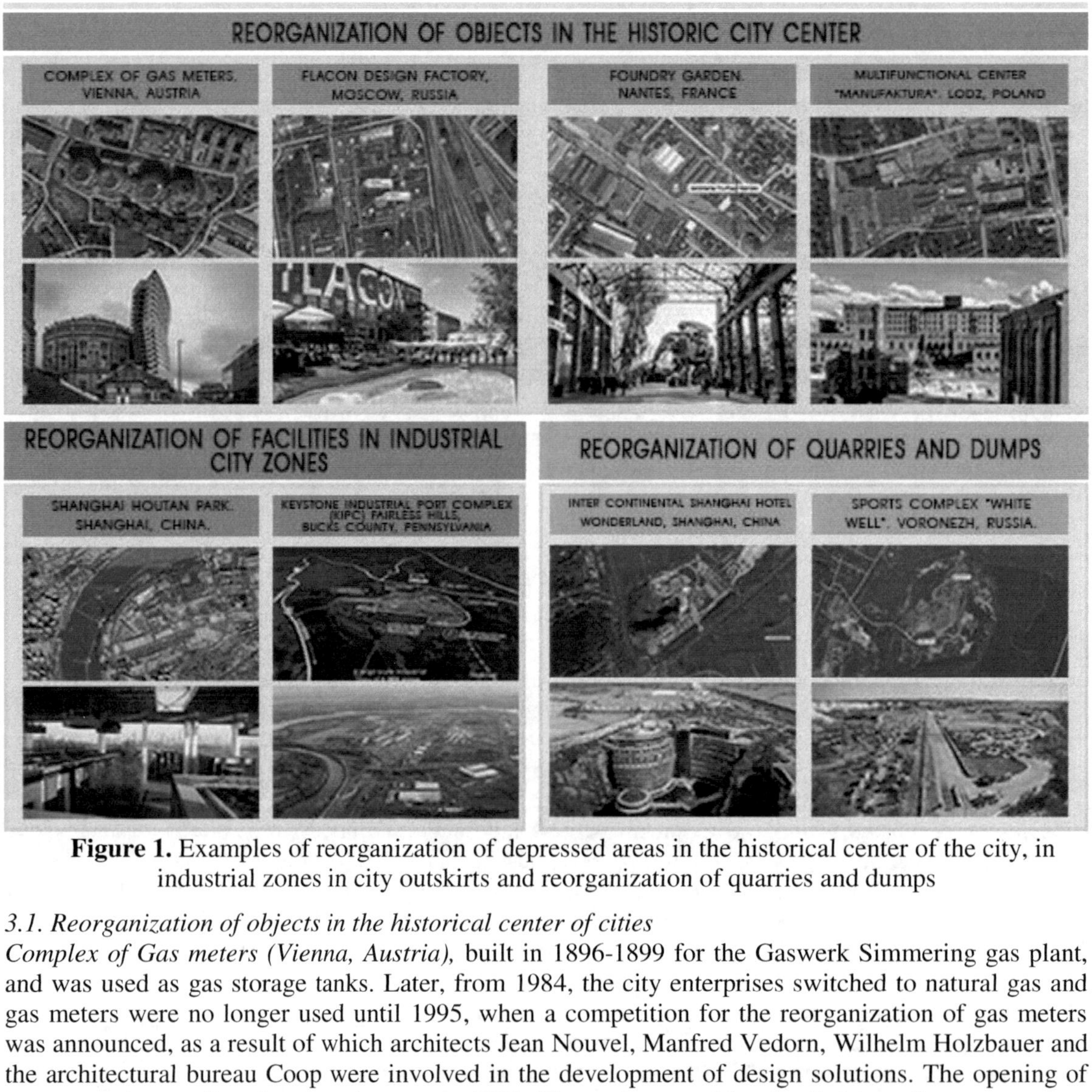

Figure 1. Examples of reorganization of depressed areas in the historical center of the city, in industrial zones in city outskirts and reorganization of quarries and dumps

3.1. Reorganization of objects in the historical center of cities

Complex of Gas meters (Vienna, Austria), built in 1896-1899 for the Gaswerk Simmering gas plant, and was used as gas storage tanks. Later, from 1984, the city enterprises switched to natural gas and gas meters were no longer used until 1995, when a competition for the reorganization of gas meters was announced, as a result of which architects Jean Nouvel, Manfred Vedorn, Wilhelm Holzbauer and the architectural bureau Coop were involved in the development of design solutions. The opening of gas meters in new functional zoning was in 2001. The project provides the division of gas meters into a residential area, office and retail and entertainment areas while preserving the historical exterior of the building. The complex of gas meters also includes a student dormitory (70 rooms for 250 students), a cinema, a concert hall (for 2,000-3,000 people) and an archive [9].

Design factory FLACON (Moscow, Russia), is a trade, exhibition and office complex with a total area of 25 thousand m^2 in the Butyrsky district of Moscow on the territory of the former Kalinin Crystal Factory. In 2005 it was bought out, but only in 2009 the replacement of the infrastructure of buildings and utilities began along with the removal of obsolete production equipment and the cleaning of the plant territory. In 2010, the complex was completely reoriented. Various advertising agencies and design studios and craft workshops are located on its territory [10]. For various events, the complex has three specialized sites. The first one is The Cube. It occupies the space of one of the workshops and includes an exhibition and conference hall. The second site is focused on cultural and educational events and located in a separate building of the complex. The third area is used for small events. On the territory of the FLACON plant, there are also two coworking spaces: Flacon Coworking and Start Hub (positioned as a "business accelerator" or "business incubator").

Foundry Garden (Nantes, France) is a 350 hectare complex located on the island in the center of Nantes, on the site of the former Fonderies Atlantique, which produces propellers for sea liners. The reorganization project was based on the principles of maximizing the use of existing buildings without demolition, as well as preserving the social environment, historical and geographical heritage. The main objectives of the project were to create a covered public space ("roof garden") for daily operation and show of preexisting industrial activities. The roof and surrounding buildings protect the garden from adverse weather conditions, resulting in a "greenhouse effect" that allows exotic plants to be grown. They are irrigated with rainwater collected in tanks (2x50 m^3) and distributed through various irrigation networks. Moisture and freshness is maintained by misty spraying of plants [11].

Multifunctional center «Manufaktura» (Lodz, Poland) - the center, which opened in 2006, is a shopping and entertainment complex that has preserved the historical atmosphere of this place and the original industrial architecture (project of the Lodz architect Hilary Majewski, 1872), with a total area of 27 hectares. It is located in the central part of Lodz. There are a restaurant complex, parking lots, a cultural and entertainment center, including a multiplex cinema, a bowling alley, a climbing wall, a fitness club and a skate park on its territory [12]. The project of the complex was developed by the British firm Virgile & Stone (London) together with the French architects from Sud Architectes (Lyon).

3.2. Reorganization of objects in industrial zones in city outskirts

Shanghai Houtan Park (Shanghai, China) is located in the abandoned industrial area along the banks of the Huangpu River in China, on a 14 hectare, 1.7 km long and 30 - 80 m wide. The goal of the project was to restore the coastline and clean up the polluted river. Previously, the park placed a steel mill, and in recent years it has been used as a garbage dump. The Shanghai Houtan Park project was designed to accommodate a large flow of visitors during Expo 2010, with the option of converting into a public park after it. The central part of the park is decorated with cascades and terraces with a variety of plants capable of absorbing water polluting elements, saturating the air with oxygen. They are designed to compensate the apparent elevation difference between the city and the river. The surviving structures of the former steel mill have been converted into hanging gardens and viewing decks [13].

Keystone Industrial Port Complex (KIPC) (Fairless Hills, Bucks county, Pennsylvania) is a "green industrial complex" located on the territories of a by-product coke, steel and chemical plant, power plant and others. Since these industries were highly polluting, in 1993, a joint agreement was concluded between the US EPA and the US Steel Fairless Works to clean up the areas with the purpose of developing new industries and creating jobs. The reorganization of the site consisted in the processing and reuse of coal dust, scrap metal, land reclamation, the production of electricity from landfill gas, the development of renewable energy production, that is, the sustainable development of the US EPA enterprise in order to improve the environmental and economic performance in relation to the consumption of natural resources and the use of energy of man-made raw materials [14].

3.3. Reorganization of quarries and dumps distantly located from city center

Inter Continental Shanghai Wonderland Hotel (Shanghai, China) is located 30 km from the center of Shanghai on the site of an abandoned quarry, close to the Sheshan Mountain National Forest Park and the Chengshan Mountain Botanical Garden. The hotel building consists of three volumes, made of glass and metal, which are tightly adjacent to the wall of the quarry. The 18-storey complex with 336 rooms includes public areas and conference rooms, as well as an underwater restaurant under the

IOP Conf. Series: Materials Science and Engineering **944** (2020) 012024 doi:10.1088/1757-899X/944/1/012024

artificial lake at the bottom of a quarry and the artificial waterfall on a sheer wall. Its central part is a glass tower resembling a waterfall with an elevator and an upper observation deck. The exploited roof has mini-gardens and sports grounds.

Sport Complex «White Well» (Voronezh, Russia) is the complex of motor sports located 12 km from the city of Voronezh in a chalk quarry. It has existed since 2003. On the territory of the sports complex, car tracks are designed. There are beach volleyball courts. Fishing sites are organized on the chalk lake, and rock climbing is expected on the steep walls of the chalk quarry [15]

As a result of the study of world practices for the reorganization of industrial zones, three types of reorganization of depressed industrial areas were determined, depending on the degree of preservation of the industrial function: reorganization with the preservation of the industrial function, partial re-functionalization and complete re-functionalization (Figure 2).

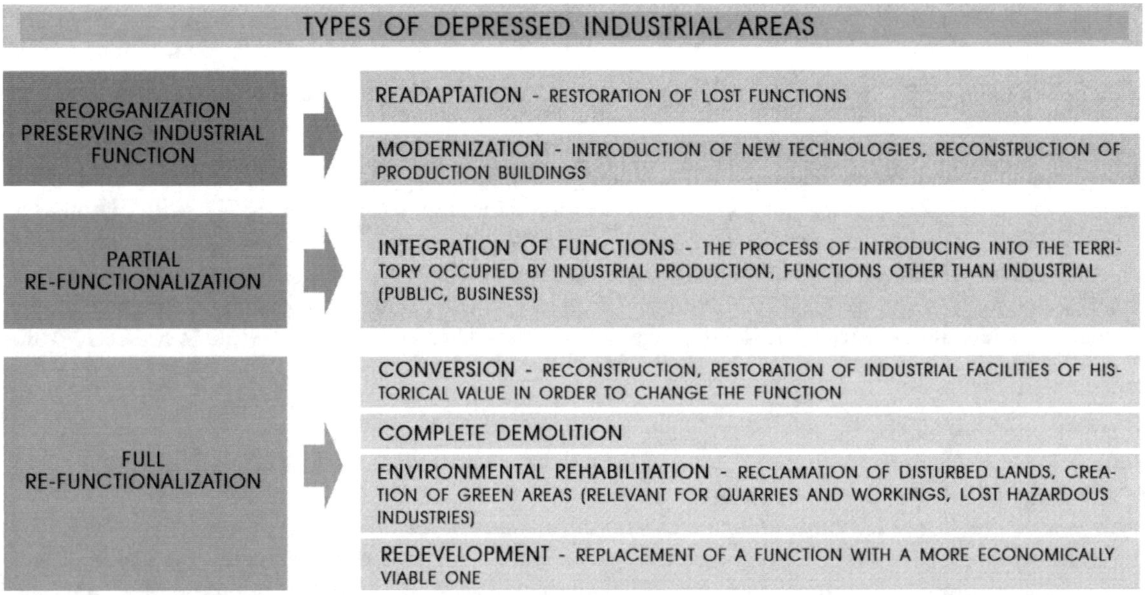

Figure 2. Types of depressed industrial areas developed by M.Yu. Drebezgova

4. Conclusion

1. The authors considered the examples of successful practices of reorganization of industrial facilities depending on their location in urban structure: in the historical center of the city, reorganization of facilities in industrial zones in city outskirts, reorganization of quarries and dumps distantly located from the city structure.

2. Three types of reorganization of depressed industrial areas were identified, depending on the degree of preservation of industrial functions: reorganization with preservation of industrial functions, partial re-functionalization and full re-functionalization.

3. The applicability of the ways of the reorganization of industrial sites against the identified types of depressed industrial areas was assessed:

- for the territories in the historical center, the most applicable way is full or partial re-functionalization;

- for the sites in industrial zones in city outskirts the most applicable way is partial re-functionalization (while maintaining the existing industrial function and replacing it with a new industrial function);

- for the reorganization of quarries and dumps distantly located the city structure, the most applicable way is complete re-functionalization in the form of environmental rehabilitation with the possible arrangement of recreational areas, leisure facilities and sites.

5. Acknowledgements
This work was realized in the framework of the Program of flagship university development on the base of the Belgorod State Technological University named after V G Shukhov, using equipment of High Technology Center at BSTU named after V. G. Shukhov.

References
[1] Shikovets A V 2018 Preconditions for the renovation of industrial territories in the neighboring countries *Science - education, production, economy: materials of the 16th International Scientific and Technical Conference* (Minsk: BNTU) **2**
[2] Perkova M V, Vitens A G and Baklazhenko E V 2018 Classification of urban planning conflicts *Bulletin of BSTU named after V G Shukhov* **12** pp 83–90
[3] Perkova M V 2018 Methods for identifying and resolving urban planning conflicts and contradictions in development on the example of small towns in the Belgorod region *Bulletin of BSTU named after V G Shukhov 9* pp 74–84
[4] Perkova M V and Krushelnitskaya E I 2016 Recreation and tourism organization issues in modern urbanization conditions *International Journal of Pharmacy & Technology* **8 (4)** pp 24919–24929.
[5] Topchiy D V 2008 Reconstruction and re-profiling of industrial buildings *Publishing house of the association of building universities* pp 3–6
[6] Chadovich A A 2013 Preservation or demolition? Compromise! *Architecture and Modern Information Technologies* (Moscow: MARHI) **1 (22)** pp 34–47
[7] Perkova M V, Blagovidova N G and Tribuntseva K M 2015 Features of Design of Ecovillages in Depressed Areas in the City *Research Journal of Applied Sciences* **(10)** pp 608-619
[8] Drozhzhin R A 2015 Renovation of industrial territories *Bulletin of the Siberian State Industrial University* **1 (11)** pp 84–86
[9] Sasi Yu 2009 Life in the Gasometer Vienna: The History of the City / per. with him. E. Krivtsova (Moscow: Eksmo) pp 210–225
[10] Grishin N 2013 Creative meters *Secret of the firm* **3 (328)** pp 34–39
[11] ADH & Hervé Abbadie. Foundries' Garden [Electronic resource] access mode: http://www.landezine.com/index.php/2009/09/foundries-garden/ (date of access 05.06.2019)
[12] Kondratieva S How to turn a closed plant into the liveliest place in the city [Electronic resource] strelkamag.com. Access mode: https://strelka.com/ru/magazine/2017/01/31/lodz-case (date of access 06.06.2019)]
[13] Houtan Park. Turenscape Landscape Architecture [Electronic resource] landezine.com - access mode: http://www.landezine.com/index.php/2011/02/shanghai-houtan-park-by-turenscape/ (date of access 06/02/2019)]
[14] Keystone Industrial Port Complex [Electronic resource] Western Pennsylvania Brownfields Center - access mode: http://www.cmu.edu/steinbrenner/brownfield (access date 01.06.2019)].
[15] About the sports complex [Electronic resource] Sports complex "White well" sportkompleksbk.ru - access mode: http://sportkompleksbk.ru/history.html (date of access 06.06.2019)]

Design of thermal-resistant special clothes in three-dimensional environment

I A Petrosova[1], N I Konstantinova[2], E G Andreeva[1], E S Bokova[1] and G M Kovalenko[1]

[1] Kosygin State University of Russia, Moscow, Russia
[2] FGA FSRI MES of Russia, Balashiha, Russia

E-mail: 76802@mail.ru, konstantinova_n@inbox.ru, elenwise@mail.ru

Abstract. The article is devoted to fundamental and applied research in the field of fire safety of textile materials. The purpose of this work is the design and creation of special-purpose clothing, involving the effects of electricity and / or elevated temperatures, by: scientifically-based selection of fabrics and accessories with protective properties. Special design solutions were developed to protect the most likely sites of injury. The authors completed the studies to assess the quality of planting samples of heat-resistant products confirmed that the magnitude of the increments, calculated taking into account the dynamic movements and properties of heat-resistant materials, provide high ergonomic characteristics and ease of use of clothing when performing professional duties in extreme conditions involving the effects of electricity and elevated temperatures. At present, the process of 3D representation of the external basic shape of the product has been studied to the greatest extent since simple silhouette shapes of the clothing can be mathematically described. The surface of the outer clothing can be described by silhouette projections in the frontal and sagittal planes, horizontal sections, as well as informative points located on the contours of the sections characterizing the silhouette. The quantitative characteristics of the shape of the designed product in three dimensions can be magnitudes of projection increases as the gap between the skin and the clothing. In addition, such parameters of horizontal section contours as the spatial gap and its angular measure facilitate the process of describing behavior of the material in the finished product relative to the surface of the person's figure. According to the above mentioned aspect, it is obvious that the three-dimensional scanning method can be a tool not only for studying the behavior of heat-resistant materials in special clothes to protect against an electric arc, but also it allows creating an actual database of properties of heat-resistant materials.

1. Introduction

Heat-resistant protective clothing is used. It is designed for military, firefighters, utility and industrial fields [1], emergency services [2] and first responders in order to protect against intense heat exposure during performance. The use of heat-resistant clothing is aimed at preventing burn injuries, heat stress and reducing the risk of death due to thermal effects on the human body [3, 4].

One of the reasons hindering the development and use of three-dimensional (3D) automated clothing design systems in the clothing industry is the lack of a reliable database of the properties and behavior of various materials. This issue is particularly relevant for the design of special-purpose clothing, operated under extreme conditions and involving the effects of electricity and / or elevated temperatures, etc. Such clothing, in addition to solving functional tasks –it ensures work safety in case of electric arc and ignition hazard, it also should provide comfort wear and do not cause discomfort when making necessary movements in the working poses.

Thus, design and creation of special-purpose clothing require setting and solving a set of tasks, which include not only a scientifically-based choice of fabrics and accessories with protective properties, but also the development of special design solutions that protect the most vulnerable areas, weight reduction and ease of movement performed by a worker.

The solution of such multicomponent problem is promoted by the use of modern three-dimensional tools for assessing the quality of designed heat-resistant special-purpose garments in three-dimensional environment (Fig. 1).

Three-dimensional environment which includes: creating a database (DB) of heat-resistant material properties; formation of requirements for the composition of materials in design of heat-resistant clothing; the creation of DB for dynamic effects of dimensional signs, which can change when performing characteristic movements by people at risk of an electric arc; recalculation of constructive additions taking into account the properties of the materials used and the dynamic effects of dimensional signs; construction of ergonomic heat-resistant clothing; quality assessment of design solutions in three-dimensional environment.

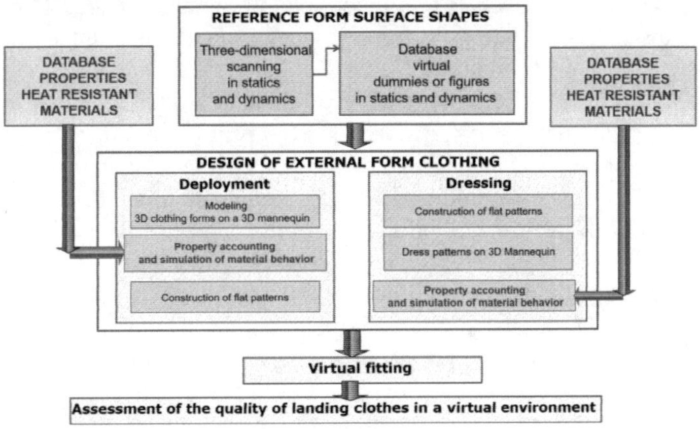

Figure 1. Conceptual model of the formation and evaluation of the quality of clothing fit, taking into account the properties of materials used

2. Literature review

Power line technicians [5–7] who need protection from electric arc exposure and fire with appropriate arc protective clothing [8] are most subject to dangerous situations such as risk of severe burns and electrical contact injuries. An electric arc is a flash of electric current between conductors through air, which emits thermal energy, depending on the strength of the current, duration and length of the arc [7], and which is able to ignite conventional materials and cause burns. Standards define "thermal hazard as the heat energy sufficient to cause burn injury to human tissue subjected to a momentary electric arc" [9].

Work on power lines can be executed both under voltage and in de-energized conditions, when a human can be electrocuted, or a thermal flash from an electric arc can cause burns or serious injury. Such situations are sudden, incidental and invisibly dangerous [10]. Therefore, while working with electricity, it is especially important to observe proper safety rules and use appropriate personal protective equipment (PPE) [10–12]. Flame-retardant clothing helps workers avoid accidents and injuries during electric-arc accidents [13]. Careful application of established safety standards for protection from electric shock and heat exposure, including the use of appropriate personal protective clothing (PPC), increases the safety of staff working near electrically conductive circuits [14]. As the culture of labor protection and safety is developing, the demand for high-performance heat-resistant clothing has increased [4].

The reliability of the virtual model of the designed product largely depends on the prediction of the behavior of the materials on the figures of different constitution. Most modern CAD systems clothing contains a module simulating the behavior of the fabric in the product and considers some material properties, both visual (color, texture, ornament) and mechanical (stretching on the base and weft, flexibility, surface density, thickness, maximum stretching, etc.). There is a close relationship between the distribution and the size of the air gaps in clothing and the mechanical properties of various fabrics, established by 3D scanning, subsequent 3D visualization of clothing models and computer processing of graphic data. Air gap space in clothing is positively correlated with bending stiffness, the length of the garment and the density of the base (stiffness-bending, length and warp density) [15]. In many 3D design systems, a certain database of materials is already laid. Any user can add any information to it. For correct 3D modeling of fabrics that accept an ambiguous spatial configuration in clothes, it is possible to use the links between the parameters of the embroidered and scanned sample of the finished product and its 3D model obtained by virtual design, on the basis of which the database of material behavior in space is formed [16].

To assess thermal protection of clothing 3D scanning method is used. This method measures the distribution of thickness air gaps between the protective clothing and the skin [17, 18]. The distribution of the thickness air gaps affects the temperature resistance of clothes, since they participate in heat exchange under clothes and help to achieve thermal balance. Increasing the thickness of the air gap between the protective clothing and the skin makes it possible to reduce the effect of heat flow on human skin during both dry and moist thermal effects. The study results of the distribution of air gaps in heat-resistant clothing using a three-dimensional scanner showed that the thermal protection of clothing depends on the nature of uneven distribution of air gap sizes, which is influenced by the silhouette and style of clothing, the contour of the human body [18] and the degree of thermal shrinkage of the material [17].

At present, the process of 3D representation of the external basic shape of the product has been studied to the greatest extent [19] since simple silhouette shapes of the clothing can be mathematically described [20]. The surface of the outer clothing can be described by silhouette projections in the frontal and sagittal planes, horizontal sections, as well as informative points located on the contours of the sections characterizing the silhouette. The quantitative characteristics of the shape of the designed product in three dimensions can be magnitudes of projection increases as the gap between the skin and the clothing. In addition, such parameters of horizontal section contours as the spatial gap and its angular measure facilitate the process of describing behavior of the material in the finished product relative to the surface of the person's figure.

Based on the above mentioned aspects, it is obvious that the three-dimensional scanning method can be a tool not only for studying the behavior of heat-resistant materials in special clothes to protect against an electric arc, but also it allows creating an actual database of properties of heat-resistant materials.

3. Materials

The authors studied two groups of materials:

The first group consists of fabrics based on cotton, viscose, cellulose and wool yarns, including those based on Zirpro® fire-resistant wool, Lenzing FR® cellulose, which have moderate or high heat resistance and fire resistance;

The second group of fabrics is based on para-aramid, meta-aramid and polysulfonamide yarns P-140, Nomex®.

4. Methods

Research methods are based on the analysis and synthesis of theoretical and practical material, grouping and comparison, and a systematic approach. The theoretical foundations of metrology, image processing and computer vision methods are used in the work. In the course of the study, methods of systematization and classification, expert assessments, engineering methods for obtaining unwinds of clothing parts for typical and individual figures were used.

For obtaining three-dimensional models, methods of three-dimensional scanning of a human figure were used, and for solving visualization problems and operations on three-dimensional images, free software libraries of Graphics Magick and VTK and a number of specialized CAD software were used.

The flammability of the tissues was determined according to ISO 15025, the heat-shielding efficiency with an open flame – according to ISO 9151, the heat-shielding efficiency with heat radiation – according to ISO 6942.

5. Methodology

Thermal stability of materials and material packages are investigated in this study and the results that formed the basis of a database of their properties in a three-dimensional clothing design system are obtained.

Speaking about the requirements for heat-resistant materials, we must first proceed from the specific conditions of usage. The specifics of work in energy industry imposes requirements for the protective suit of power engineering operators, the main features are the following: preservation of thermo-fire-resistant properties throughout the entire service life, convenience and lightness, environmental friendliness, hygroscopicity and breathability of used fabrics and materials. Special fire-resistance requirements and test methods for both the suit's top material and the clothing package as a whole are established by technical regulations and a number of regulatory documents [21–32].

The standards define the requirements for clothing materials that protect from an electric arc, including thermal stability, flame resistance for 2 seconds after ignition, the fabric should not melt, drip, peel and ignite, get fit more than 10% within 5 minutes at 260 °C [9]. Thermal resistance is a primary factor in protecting a human body from flames and extremely high temperatures [33]. In addition, heat-resistant materials which protect against thermal hazards of an electric arc must have constant thermal properties, providing protection against the incident energy of an electric arc effect of at least 20 cal/cm^2 (protection level 3 [34–36]), also they ensure heat protection efficiency energy of at least 20 cal/cm^2, and must be resistant to the effects of an open flame and don’t melt. It is important to preserve the heat resistance of clothing after 5 and 50 washings.

Heat-resistant materials are characterized by diversity, effectiveness of protection and suitability for workwear [37], can be woven and non-woven constructions [38]. Today, the market for arc-rated textiles is one of the fastest growing in the world for protective clothing [39], it is regulated by IEC, ASTM, ISO and other national standards [9, 34–36, 21–32]. It should be noted that when choosing heat-resistant clothing it is the consumer who is responsible for determining the appropriate level of protection, based on the actual conditions of use and load [40].

Fire resistance is provided either by incorporating fibers with inherent fire resistance into clothing, or by applying flame retardant treatment to fibers that are usually flammable [3]. Heat and flame resistant textiles can be divided into two groups according to the main components of the fibers.

The first group of textiles consists of cotton, viscose, cellulosic, wool and certain man made fiber-containing fabrics, including flame resistant wool Zirpro®, cellulose Lenzing FR®, which have moderation to high heat and flame resistance and are available at a reasonable price [41–44]. Currently, flame resistant cotton is popular [2]. The operational limitations of these materials are associated with the heating of the fibers and the need for treatment with flame retardants [43].

The second group includes aromatic and carbonized fibers, which are distinguished by mechanical stability, high cost and low environmental friendliness due to difficulty to recycle [43]. The following materials should be distinguished: para-aramid and meta-aramid materials, P-140, Nomex®, polysulfonamide, which are characterized by strong thermal shrinkage and reduction of heat resistance after multiple exposures [2, 8, 41, 45]. The combination of para-aramid Kevlar® fiber with polybenzimidazole [2] is characterized by good thermal stability. CarbonX® carbon fabric for heat and fire-resistant clothing does not burn, char, shrink or decompose when exposed to flame, arc flash or high heat [1]. Waterproof materials such as Neoprene / Nomex and PVC / Nomex / Kevlar [46] are used for flame resistant clothing. Waterproof heat-resistant clothing made of Hypalon® rubber-coated material and of batt-type material has a relatively long protection time, depending on the level of heat flow [47].

The combinations of flame retarded char-forming fibers and intumescents leads to an increase in thermal and fire resistance. Intumescent combinations are especially effective when based on more cost-effective and environmentally acceptable raw materials [43].

Interpolymer complexes (IPC) based on polycarboxylic acids and non-ionic polymers, stabilized by hydrogen bonds are of great interest. [48] Modifying compositions based on IPC of polyacrylic acid with non-ionic polymers (polyvinyl alcohol, polyethylene oxide, polyacrylamide) are presented in this study, the conditions for the formation of polymer-polymeric complexes in the aquatic environment are considered and a method of impregnating fabrics like canvas with stoichiometric and non-stoichiometric composition is proposed. The possibility of regulating the weight gain of the IPC in the structure of the tissue and the effect of the IPC on increasing the oxygen index of materials, reducing their inflammability and ability to maintain independent combustion are shown.

It is not recommended to use such thermoplastic fibers as polyester, polypropylene and polyamide), which, despite flame retardants functions during the polymerization and / or extrusion process, melt and / or form holes when exposure to flame, therefore, they do not meet the requirements for fabrics for protective clothing [43].

The composition of fibers has a greater effect on relative humidity of the microclimate underneath the PPC and the total clothing insulation [41]. With prolonged thermal exposure, thermal energy can accumulate inside multilayer protective clothing, which leads to a decrease in level of protection expected [49]. To improve thermal insulation, we can increase the thickness of a fabric or a package of materials [2, 47, 50], but with an increase in fabric weight physical load increases, the level of physiological stress of workers increases [51, 52] and, as a result, their performance [41].

The innovations of heat-resistant materials are associated both with their high functionality and performance in protection against flame and heat as “smart” textile materials [38] develop. The Teflon® treated fabrics in addition to heat resistance, have stain and water repellency properties and a low level of energy absorption [53]. The development of systems of intellectual thermal protection based on the use of thermal sensors and programs with algorithms of intellectual behavior is known [54]. In case of flame, a sensor installed in protective clothing can mechanically activate protective components, for example, enable the quick release of a liquid flame retardant through a vinyl tube sewn into clothing to create a protective barrier around the human’s face and neck when unexpected heat exposure [5].

When choosing the fabric for power engineers, along with the effectiveness of its fire protection, the surface density of the material, hygroscopicity and breathability, strength were considered. Table 1 shows the results of studies for heat-resistant materials used in some models of summer and winter clothing of electric welders.

As it follows from the data in the table, when choosing an effective fire-resistant fabric for the top of the clothing, it is possible to model any type of clothing for power engineers, depending on the environmental conditions that meet the standardized requirements. As a rule, only the index of thermal diffusivity changes (group B1–B5) since its measurement is directly related to the conditions of heat transfer through a package of materials and is characterized by its thermal-physical characteristics.

In addition to the requirements for the main material of the clothing, there are requirements for auxiliary materials, such as threads and accessories. The requirements for sewing threads used to manufacture protective clothing against the thermal effects of an electric arc should eliminate their melting and provide resistance to high temperatures and chemical cleaning.

Regarding the accessories, external metal parts are excluded from the clothing set. If it is impossible to avoid such accessories as zippers, buttons, etc., then they should be covered both on the external and internal sides with heat-resistant material. In addition, the design of the clothing should provide the ability to quickly remove it after the impact of an electric arc due to preserving the functions of accessories.

Table 1. Study of heat-resistant properties of materials and material packages

Name of the material (package of materials), their characteristics ISO 15025, ISO 9151, ISO 6942	The value of parameters according to the classification		
	Flammability ISO 15025	Heatproof effect with open flame ISO 9151	Heatproof effect with heat radiation ISO 6942
Nomex fabric, $\rho = 220$ g/m^2, heat-resistant knitted fabric, $\rho = 220$g/m^2	Group A	B2	C2
Nomex fabric, $\rho = 220$ g/m^2 , heat-resistant fabric, $\rho = 135$ g/m^2 insulation, $\rho = 150$ g/m^2 windproof fabric, $\rho = 150$ g/m^2 insulation, $\rho = 150$ g/m^2 heat-resistant fabric, $\rho = 135$ g/m^2	Group A	B5	C2
Fabric of fire-resistant mixed fibers, $\rho = 300$ g/m^2 Heat-resistant knitted fabric, $\rho = 220$ g/m^2	Group A	B2	C2
Fabric from fire-resistant mixed fibers, $\rho = 300$ g/m^2 2 layers – modacrylic fiber, quilted with cotton lining, fabric made of fire-resistant mixed fibers, $\rho = 400$ g/m^2	Group A	B4	C2

The next stage of research is the development of design features of clothing for performing work under the conditions of high-temperature flow and high-voltage impacts.

The methods for designing materials and clothing are important for obtaining optimum thermal protective performance of clothing and ensuring the physiological comfort of workers [4, 55, 56]. Design criteria include the level of thermal protection of clothing, the functionality of the design, the requirements for physiological and psychological comfort, the efficiency of production and maintenance [45].

The temperature of the human body during work under hot conditions is influenced by arc-flash and fire-resistant AFR clothing [6, 47, 60]. Sets of protective heat-resistant clothing have a single-layer or multi-layer construction consisting of a combination of fire-resistant outer layers with inner layers of natural or artificial fibers [6]. Under electric arc conditions, flame-resistant rainwear is effective [46].

A small but significant impact on the temperature resistance of clothes is exerted by the completeness, fitting and fasteners [45]. Better thermal protection provided free silhouette than close-fitting; tight fitting adjustable cuffs on sleeves and legs were more effective than zippers; and a stand-up collar better protected the neck than a turn-down collar. In practice, thermal protection sets consisting of a one-piece jumpsuit and two-piece trousers and jackets are used oftener. According to some researchers, the smallest thermal protection was provided by two-piece set of jackets and pants, one-piece jumpsuit is slightly better and the best protection is by a two-piece overall, mainly due to the effect of layering [45]. According to other researchers, on the contrary, a two-piece set of fire-resistant material contributes to greater heat removal [56]. A variety of recommendations on the completeness of heat-resistant clothing predetermined further research in the field of consumer preferences.

It is known that most of the electricity companies use protective heat-resistant clothing in the form of a set consisting of a jacket and trousers or a jacket and semi-overalls for the employees. Indeed, as shown in the survey, aimed at studying consumer requirements for protective heat-resistant clothing [57], 153 electricians aged 20–45 gave the greatest preference to the set consisting of a jacket and semi-overalls, which provide an opportunity to work in a more comfortable way than trousers or overalls.

The first place consumers gave to the quality of heat-resistant materials, which do not sustain combustion and do not melt after exposure to a thermal arc.

Experts chose patch pockets on the front of the heat-resistant jacket and on the sleeve, which are located as follows: two pockets are at chest level, two pockets are at waist level, one pocket is at the top of the left sleeve.

The most comfortable and providing the greatest protection of the neck, electricians consider the collar "stand", and the most convenient fittings for taking off the set are a zipper and Velcro tape.

Heat-resistant clothing should not only protect from dangerous temperature conditions, but also be comfortable, fit, provide mobility of the employee. Therefore, in addition to the regulatory requirements for thermal protection, it is important to design dynamic garment (5). It is known that the convenience of operating protective clothing during the work in enclosed spaces and in open areas provide ergonomic performance. The design of ergonomic clothing takes into account the thickness of materials and the dynamic effects of sizes [58].

Anthropometric studies of male electricians, who were attended by 103 people, aged 20 to 45, their work was under observation during the shift, as well as photo and video recording of characteristic movements and postures of electricians are done. A list of characteristic movements was drawn up, among which are the following: climbing a power line – (Fig. 2, a), working on a pole (Fig. 2, b), working with an electric shield while standing (Fig. 2, c) and in a half-squat.

a b c

Figure 2. The characteristic movements of electricians: a – climb the pole, b – work on the pole, c – work with an electrical shield

A list of sizes, which can increase and decrease during characteristic movements, is made. The scheme of measurements in dynamics for the characteristic movement "Work at an electric board while standing" is shown in Fig. 3, a. With the help of 3D scanning, measurement of dimensional signs of men's figures in statics (Fig. 3, b) and dynamics (Fig. 3, c) was performed. Subsequent studies were performed both for naked body and in clothing made of heat-resistant materials.

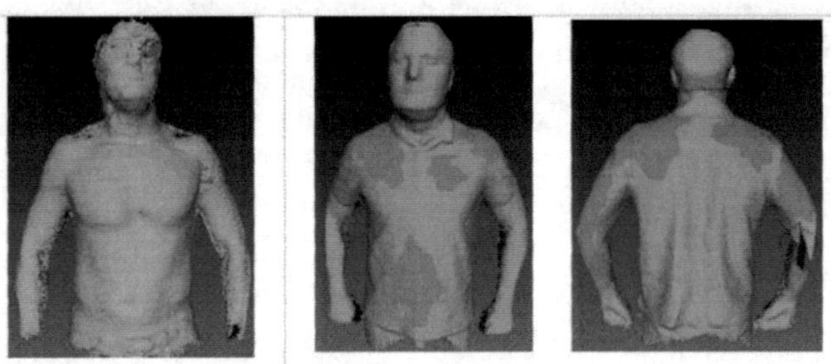

a

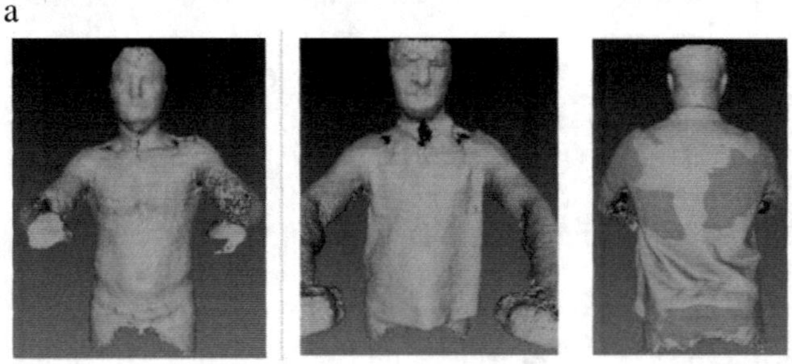

b

Figure 3. Study of a figure using three-dimensional scanning: a – research in statics, b – research in dynamics

The measurements were carried out in accordance with the methodology of anthropometric research which are given in recommendations on the standardization of the adult and child population. Maximum and minimum dynamic effects are calculated by the formula (1).

$$d_i = \frac{X_i^{(d)} - X_i^{(s)}}{X_i^{(s)}} \times 100\% \qquad (1)$$

where d is the dynamic effect of the i figure (i = 1, 2, ... n), cm; $X_i^{(d)}$ – the value of the dimensional sign in the dynamics of the i-figure, cm; $X_i^{(s)}$ the value of the dimensional sign in statics of the i-figure, cm.

The detected values of changes in dimensional signs in dynamics were used to determine the movements, taking into account the properties of the material used in the design of heat-resistant protective clothing. The calculation of increases in movement is made according to the formula (2)

$$IM = \frac{(di - \varepsilon)}{100\%} \times Nc \qquad (2)$$

where IM – increase in movement, cm; di is the dynamic effect, %; Nc is the size of the dimensional sign in statics, cm; ε is the value of conditionally residual deformation, depending on the group of tissue, %. Heat-resistant materials belong to groups of fabrics with a value of conditionally residual deformation up to 0.5 % [64, 65].

Regarding consumer preferences, requirements for heat-resistant clothing and research results, models of a jacket and semi-overalls for a male electrician were developed. A sketch of the model and the look of the finished garment are shown in Fig. 4. The clothing is transferred for testing.

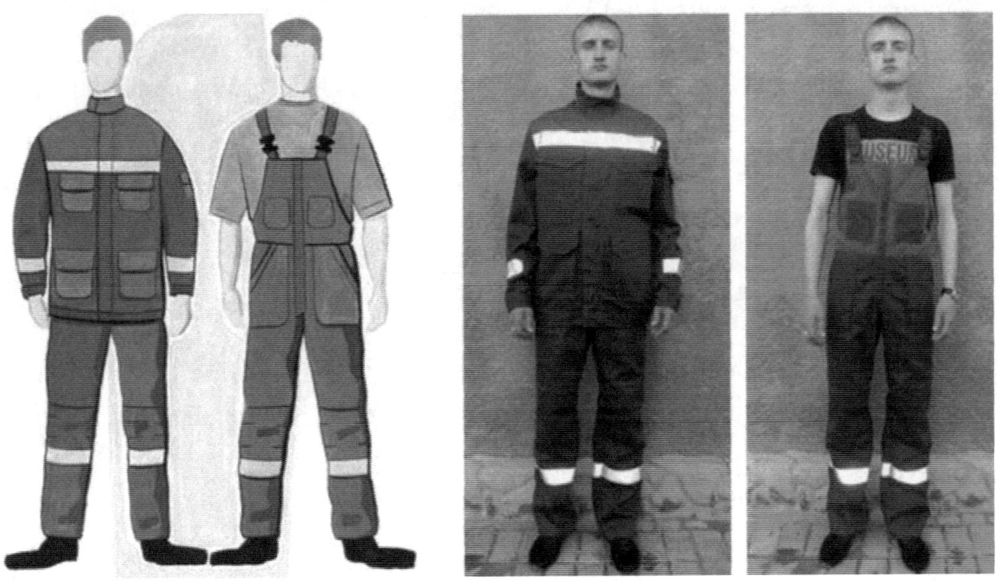

Figure 4. Sketch of the finished heat-resistant clothing

In the final part of the work, the assessment of the quality of the of heat-resistant clothing samples and the assessment of the projection gaps between the inner layer of the package of heat-resistant materials and the human body using three-dimensional scanning are made. The calculation of the free fitting in the product was done using the SizeReader program developed at the Department of RSU named after A.N. Kosygin with the support of the Ministry of Industry and Trade of the Russian Federation [17, 18]. An increment has been established for the armhole section, both in statics (Fig. 5, a) and in dynamics (Fig. 5, b). The increase in the area of the armhole was determined as the difference in the lengths of arm sections of the diameter.

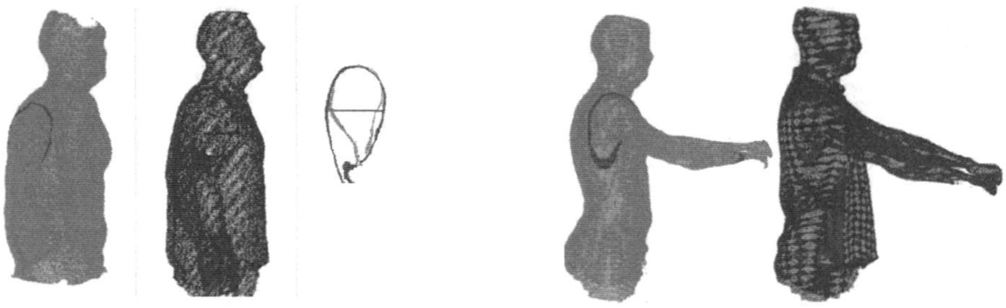

a b

Figure 5. Estimation of projection gaps using three-dimensional scanning: a- in statics, b- in dynamics

6. Conclusion

Large increments contribute to excessive accumulation of material on the construction site, which leads to laying of folds and deterioration of the appearance of the product, small – limit the amplitude of movements. The studies to assess the quality of heat-resistant products confirmed that the increase, which was calculated with the dynamic movements and properties of heat-resistant materials provide high ergonomic characteristics and ease the use of clothing when performing professional duties under extreme conditions involving the effects of electricity and elevated temperatures.

Acknowledgments

The reported study was funded by The Kosygin State University of Russia and FGA FSRI MES of Russia, Balashiha.

References

[1] Young J 2008 Markets: Safety and protective. Taking the heat: Innovations in fire-retardant fibers and fabric products are keeping people and structures safer *Spec. Fabrics Rev.* **93(10)** 97–100

[2] Wang M and Li J 2016 Thermal protection retention of fire protective clothing after repeated flash fire exposure *J. of Industr. Textiles* **46(3)** 737–55

[3] Linnane D M, Fogarty A and Hunt A 2017 The effects of flame resistant protective clothing on heat exchange and thermal strain *J. of Sci. and Med. in Sport* **20(2)** 73–4

[4] Song G, Mandal S and Rossi R 2016 *Thermal Protective Clothing for Firefighters* (Cambridge, UK: Woodhead Publ.) 242 p

[5] Mendoza E, Cooper J P, Evangelista J W, Auerbach M and Arnas A Ö 2012 On Demand Thermal Protection (ODTP): A New Approach for Designing Garments Exposed to Flash Flame Incidents pp 2111–2120, *Int. Mechanical Engineering Congress and Exposition* (Houston, TX, US: ASME)

[6] Neal T E, Bingham A H and Doughty R L 1997 Protective clothing guidelines for electric arc exposure *IEEE Transact. on Industry Applicat.* **33(4)** 1041–54

[7] Wulf K 2009 Flame-Resistant Clothing Its Role In Protecting Electric Line Workers From Arc Flash Burns *Professional Safety* **54(6)** 685–6

[8] Doughty R L, Neal T E, Dear T A and Bingham A N 1999 Testing update on protective clothing and equipment for electric arc exposure *IEEE Industry Applicat. Magazine* **5(1)** 37–49

[9] ASTM F1959/ F1959M – 14e1 *Standard Test Method for Determining the Arc Rating of Materials for Clothing* 2014

[10] Blume S W 2016 Personal Protection (Safety) In *Electric Power System Basics for the Nonelectrical Professional* (Hoboken, New Jersey, US: Wiley-IEEE Press) pp 209–23

[11] ISO 11999-1:2015 *PPE for firefighters, Test methods and requirements for PPE used by firefighters who are at risk of exposure to high levels of heat and/or flame while fighting fires occurring in structures* Part 1: General

[12] ISO 11999-3:2015 *PPE for firefighters, Test methods and requirements for PPE used by firefighters who are at risk of exposure to high levels of heat and/or flame while fighting fires occurring in structures* Part 3: Clothing
[13] Mäkinen H and Mustonen S 2003 Features of electric arc accidents and use of protective clothing in Finland *Safety Sci.* **41(9)** 791–801
[14] McClung B and Mohla D C 2005 Electrical design – refined for safety *Industrial and Commercial Power Systems Technical Conf.* (Hoboken, New Jersey, US: IEEE)
[15] Yu M, Wang Y, Wang Y and Li J 2013 Correlation between clothing air gap space and fabric mechanical properties *J. of the Textile Instit.* **104(1(1))** 67–77
[16] Kenkare N, Lamar T A M, Pandurangan P and Eischen J 2008 Enhancing accuracy of drape simulation Part I: Investigation of drape variability via 3D scanning *J. of the Textile Instit.* **99(3(6))** 211–8
[17] Petrosova I A, Andreeva E G and Guseva M A 2019 The system of selection and sale of ready-to-wear clothes in a virtual environment pp 1–5 *Int. Sci. and Technol. Conf. "EastConf"* (Vladivostok) DOI: 10.1109/EastConf.2019.8725390
[18] Petrosova I A, Shantseva O A and Andreeva E G 2017 Assessment of a right size ready-to-wear clothes to the consumer's figure in the three-dimensional system *Izv. Vysshikh Uchebnykh Zavedenii, Ser. Teknol. Tekstil. Promyshl* **5(371)** 139–42
[19] Gazhur A A and Lukiyanchuk I N 2018 Trends of development of the sphere of services in Russia *Proc. of the Voronezh State Univer. of Engineer. Technol.* **80(3)** 444–50 Retrieved from: https://doi.org/10.20914/2310-1202-2018-3-444-450
[20] Gurina M A and Rumyanceva J V 2019 Comprehensive tools and methodologies quality improvement in the management competitiveness trade enterprises *Proc. of the Voronezh State Univer. of Engineer. Technol.* **81(2)** 320–35 Retrieved from: https://doi.org/10.20914/2310-1202-2019-2-320-33520.
[21] ISO 6942:2002 *Protective clothing, Protection against heat and fire, Method of test: Evaluation of materials and material assemblies when exposed to a source of radiant heat*
[22] ISO 11612:2015 *Protective clothing, Clothing to protect against heat and flame, Minimum performance requirements*
[23] ISO 12127-1:2015 *Clothing for protection against heat and flame, Determination of contact heat transmission through protective clothing or constituent materials Part 1: Contact heat produced by heating cylinder*
[24] ISO 12127-2:2007 *Clothing for protection against heat and flame, Determination of contact heat transmission through protective clothing or constituent materials Part 2: Test method using contact heat produced by dropping small cylinders*
[25] ISO 13506-1:2017 *Protective clothing against heat and flame Part 1: Test method for complete garments, Measurement of transferred energy using an instrumented manikin*
[26] ISO 13506-2:2017 *Protective clothing against heat and flame Part 2: Skin burn injury prediction, Calculation requirements and test cases*
[27] ISO 14116:2015 *Protective clothing, Protection against flame, Limited flame spread materials, material assemblies and clothing*
[28] ISO 15025:2016 *Protective clothing, Protection against flame, Method of test for limited flame spread*
[29] ISO 16073:2011 *Wildland firefighting personal protective equipment, Requirements and test methods*
[30] ISO 17492:2003 *Clothing for protection against heat and flame, Determination of heat transmission on exposure to both flame and radiant heat*
[31] ISO 9151:2016 *Protective clothing against heat and flame, Determination of heat transmission on exposure to flame*
[32] ISO/TR 2801:2007 *Clothing for protection against heat and flame, General recommendations for selection, care and use of protective clothing*

[33] Mandal S, Song G, Ackerman M et al 2013 Characterization of textile fabrics under various thermal exposures *Textile Res. J.* **83(10)** 1005–19
[34] IEC 61482-1-1:2009 *Live working, Protective clothing against the thermal hazards of an electric arc, Part 1-1: Test methods, Method 1: Determination of the arc rating (ATPV or EBT50) of flame resistant materials for clothing*
[35] IEC 61482-1-2:2014 *Live working, Protective clothing against the thermal hazards of an electric arc. Part 1-2: Test methods, Method 2: Determination of arc protection class of material and clothing by using a constrained and directed arc*
[36] IEC 61482-2:2018 *Live working. Protective clothing against the thermal hazards of an electric arc, Part 2: Requirements*
[37] Brewster E P and Barker R L 1983 A Summary of Research on Heat Resistant Fabrics for Protective Clothing *Amer. Industr. Hygiene Associat. J.* **44(2)** 123–30
[38] Shishoo R 2002 Recent developments in materials for use in protective clothing *Int. J. of Cloth. Sci. and Technol.* **14(3/4)** 201–15
[39] Hoagland E H 2013 Flame resistant textiles for electric arc flash hazards. Chapter In *Handbook of Fire Resistant Textiles* (Cambridge, UK: Woodhead Publ.) pp 549–80
[40] Doughty R L, Neal T E, Laverty G M and Hoagland H 2002 Minimizing burn injury: electric-arc hazard assessment and personnel protection *IEEE Industry Applicat. Magazine* **8(3)** 18–25
[41] Carballo-Leyenda B, Villa J G, López-Satué J and Rodríguez-Marroyo J A 2017 Impact of Different Personal Protective Clothing on Wildland Firefighters' Physiological Strain *Frontiers in Physiol.* **8** 185–7
[42] Day M, Cooney J D and Suprunchuk T 1988 Durability of Firefighters' Protective Clothing to Heat and Light *Textile Res. J.* **58** 141–7
[43] Horrocks A R 1996 Developments in flame retardants for heat and fire resistant textiles – The role of char formation and intumescence *Polymer Degradat. and Stability* **54(2-3)** 143–54
[44] Mehta P N 1980 Engineered Wool Industrial Protective Clothing *Textile Res. J.* **50(3)** 185–93
[45] Tan Y B, Crown E M and Capjack L 1998 Design and evaluation of thermal protective flightsuits. Part I: The design process and prototype development *Cloth. and Textiles Res. J.* **16(1)** 47–55
[46] Hoagland H and Morrow B 2000 Using rainwear as switching jackets: a reasonable solution for electric arc exposure *IEEE Transact. on Industry Applicat.* **36(5)** 1241–6
[47] Stull J O 2000 Comparative thermal insulative performance of reinforced knee areas of firefighter protective clothing In *Performance of Protective Clothing: Issues and Priorities for the 21st Century* vol 7 (Austin, TX: ASTM) pp 312–28
[48] Bokova E S, Kovalenko G M, Woźniak B et al 2016 Interpolymer complexes as modifying compounds for reducing cotton blended fabric flammability *Fibres and Textiles in Eastern Europe* **24(6(120))** 157–60
[49] Song G, Cao W and Gholamreza F 2011 Analyzing stored thermal energy and thermal protective performance of clothing *Textile Res. J.* **81(11)** 1124–38
[50] Dale J D, Crown E M, Ackerman M Y et al 1992 Instrumented Mannequin Evaluation of Thermal Protective Clothing ed McBriarty J P, Henry N W In *Symposium on Performance of Protective Clothing. Philadelphia* (PA, USA: Amer. Society for Test. and Mater. (ASTM)) pp 717–33
[51] Wen S, Petersen S, McQueen R and Batcheller J 2015 Modelling the physiological strain and physical burden of chemical protective coveralls *Ergonomics* **58(12)** 2016–31
[52] Yoo S and Barker R L 2005 Comfort Properties of Heat-Resistant Protective Workwear in Varying Conditions of Physical Activity and Environment Part I Thermophysical and Sensorial Properties of Fabrics *Textile Res. J.* **75(7)** 523–30
[53] Devarajan S, McQueen R H and Wen S 2017 Can common finishing treatments used in chef jacket fabrics improve protection against scald injury *Fashion and Textiles* **4(19)** 1–15

[54] Rogale S F, Rogale D, Dragčević Z et al 2007 Technical systems in intelligent clothing with active thermal protection *Int. J. of Cloth. Sci. and Technol.* **19(3/4)** 222–33

[55] Song G, Barker R L, Hamouda H et al 2004 Modeling the Thermal Protective Performance of Heat Resistant Garments in Flash Fire Exposures *Textile Res. J.* **74(12)** 1033–40

[56] Poirier M P, Meade R D, McGinn R et al 2015 The Influence of Arc-Flash and Fire-Resistant Clothing on Thermoregulation during Exercise in the Heat *J. of Occupat. and Environ. Hygiene* **12(9)** 654–67

[57] *Anketolog, Online questionnaire to determine the requirements of consumers to heat-resistant clothing* Retrieved from: https://anketolog.ru/survey/manage/preview/153883, 2018 (accessed 07 November 2018)

[58] Martynova A I, Andreeva E G and Maksutova M T 2011 Development of a method for assessing the dynamic characteristics of the design of women's shoulder clothing *Cloth. Industry* **1** 47–8

[59] Dunaevskaya T N, Koblyakova E B, Ivleva G S and Ievleva R V 2005 *Conf. Fundamentals of applied anthropology and biomechanics* ed E B Koblyakova (Moscow: MSUDT) 280 p

[60] L'vov E Eh, Zubkova N S and Konstantinova N I 2010 Justification of criteria for selection of fabrics of the upper layer of special protective clothing of workers of metallurgical enterprises *Fire safety* **1** 90–6

IOP Conf. Series: Materials Science and Engineering **944** (2020) 012026 doi:10.1088/1757-899X/944/1/012026

Three-dimensional scanning of a figure as the basis for mass customization of industrial clothing collections

I A Petrosova, E G Andreeva, R S Romanovsky, A A Kopwlov and M A Rodionova

Kosygin State University of Russia, Moscow, Russia

E-mail: 76802@mail.ru, elenwise@mail.ru, romansk88@mail.ru

Abstract. The subject of research is clothing development according to individual design chosen by a consumer. The application of the concept of mass customization has become one of the most high-demanded areas of development in light industry. Manufacturers invest heavily in the development of software products, which can help the consumer to choose the right-sized and well-fitting clothes from industrial collections on manufacturers' websites. While constructing clothing designs on an individual order, and then while manufacturing process under conditions of industrial flow production, exact body dimensions are extremely important. The world light industry is actively using digital technologies in order to measure shapes and determine body features (made-to-measure systems). Such systems make it possible to better take into account the individual characteristics of customers, which allow designing high-quality clothing, characterized by a good fit on a figure of a particular consumer. There is a developed system for obtaining a digital image of a shape, which based on the use of depth sensors. An experimental study of the anthropometric matching of clothes with a figure in a virtual environment was carried out by comparing scanned three-dimensional models of figures with and without clothes. The permissible values of projection gaps were revealed, within them the product visually corresponds to a figure. An approach where there is an information which consists of customers' preferences, dimensional characteristics, shape and body type in specially created databases, which are used when it is necessary to help to create clothing products that are best suited to the consumer in size and design, is proposed.

1. Introduction

In modern society under the conditions of progressive digital automated transformation of production, the consumer has become a direct participant of light industry products' design. Introducing digital technologies and systems of three-dimensional scanning of a figure and virtual modeling of clothing designs, manufacturers more and more prefer to use mass customization. Now a buyer can purchase a unique and individualized thing that will meet his or her personal requirements, whether it is the color scheme of the product, its design features or technological processing. Today, a number of the largest companies such as Zara, H&M, Adidas, American Eagle, Children`s Place, Gloria Jeans and others are successfully practicing the principle of mass customization, which helps them optimize design processes and, accordingly, increase their income.

The following prerequisites of using mass customization in the fashion industry can be distinguished:

- increased competition between companies which produce clothes and shoes;
- growth of customer requests;
- market transparency, caused primarily by the growth of online sales;

- short product life cycles and rapid obsolescence of stocks;
- high market volatility;
- low forecasting efficiency;
- seasonality;
- wide range of products [1].

According to the majority of scientists, application of mass customization allows the manufacturer to retain the ability to save costs due to the economy of scale, and at the same time, products are becoming more attractive to customers. Due to this fact, the main task of customization is not just the creation of an optimal product according to the consumer view, but the offer to a customer an individual product at the price level as for mass-produced one, which is certainly a very competitive advantage in the face of high competition in the fashion industry.

2. Literature review

The topic of mass customization and the adaptation of its principles of work in the computer-aided design of garments is more and more frequently becoming an object of study for both domestic and foreign scientists, which confirms its relevance and significance in the development of production potential. By "mass customization" (mass customization of garments) is meant the manufacturing of products under industrial conditions with the possibility of adapting or modifying them in accordance with the requirements of consumers, in other words, it is a set of approaches of mass production of clothing and individual sewing of products at the same time. To implement this production concept, the "customer relationship management system (CRM)" is used. It acts as a software application designed to automate interaction with consumers in order to increase sales level and strengthen communication with the circle of clients by accumulating information about them and their purchases for future analysis and making of a product release plan and sales strategy [2].

In particular, the authors M. Khajeh, P. Payvandy, S.J. Derakhshan have developed a system for designing new clothing models using the principle of mass customization while choosing fabric samples for future models [3].

The authors N.A. Kazakova, O.V. Ivanova, E.A. Khammatova in their studies analyze and propose their own algorithm for creating new models using neoprene with author's textures, which can be performed online [4]. The application of the concept of mass customization has become one of the most high-demanded areas of light industry development. Manufacturers invest heavily in the development of software products, with the help of them the consumer can choose clothes which are appropriate in size and fit well from mass-oriented collections on manufacturers' websites [5].

While constructing clothing designs on an individual order, and then during manufacturing process under conditions of industrial flow production, exact body dimensions are extremely important. The world light industry is actively using digital technologies in order to measure shapes and determine body features (made-to-measure systems). Such systems make it possible to better take into account the individual characteristics of customers, which allow designing high-quality clothing, characterized by a good fit on the body figure of a particular consumer [6]. 3D scanning technologies are still relatively new and expensive technologies, and only few retail clothing stores can afford to install scanning fitting rooms. At their flagship store in central London, Levi's has offered their customers to use this technology to choose a pair of jeans that fit them perfectly. Bodi.me Company which supplies three-dimensional scanning fitting rooms collaborates with brands such as Topshop, Forever 21, Next and some luxury brands (Ralph Lauren, Emporio Armani and Boss). The software creates a 3D avatar of a client, compares the measurements taken with the exact parameters of the clothes in order to choose the optimal size.

New technologies open up great possibilities for individual tailoring (bespoke) [7], providing the opportunity to integrate the principles of customization and the conditions of industrial apparel industry. The manufacturer collects maximum information online to create a virtual model of the client's body. The system provides storage of data on customer preferences, size characteristics, shape and type of body in specially created databases that are used as needed to help to create pieces of clothing that are best suited to the consumer. Thus, a system of long-term cooperation is created. As necessary the data can be corrected if a client has lost / got some weight, but in general the process of creating clothes takes tens of times less time than before. A client can interactively model his or her clothes from a number of elements / blocks on the website after a certain period of time they obtain the ready-made product fitted to a figure [8].

Amazon acquired Body Labs Company in 2017 in order to improve the Prime Wardrobe service, which allows a wide range of goods to be sent to a client for fitting, including shoes and accessories, and after that to be returned if pieces do not suit him or her. For greater Amazon customer satisfaction and, consequently, a reduction of returns' rate, development of new and more targeted technology can be of great benefit [9].

The analysis of modern software products proves that in the field of retail sales of clothes, modern technologies for attracting customers are developing, they aim to make choices of the most suitable products easier, as well as to manufacture products according to individual size characteristics obtained with the help of using three-dimensional scanning. Existing systems to one degree or another allow visualizing the assortment of the store and identifying the pieces most suitable for a consumer, while the reliability of determining the conformity of the actual size of clothes with the consumer's figure still remains low.

3. Methodology

In this regard, the automation of the process of determining the size characteristics of figures, and subsequently the selection of appropriate clothes from the industrial mass collection or the order a customized product is really actual task.

The perspective solutions in this area are seen in the following sequence: 1) development of scanning systems that are distinguished by simple use and low cost 2) comparison of size features of a three-dimensional avatar and a scanned model of a client's body 3) determination of the parameters of the ready-made clothing and their comparison with the dimensional characteristics of the figure of client who makes the order 4) ordering a customized product on the manufacturer's website.

To implement the described approach, the following research was conducted:

- Development of a system for obtaining a digital image of a figure, determination of size features using a three-dimensional model of a figure;
- Development of men's shirts designs' drawings with different options of widening of fitting freedom, and manufacturing of clothing models;
- Identification of the main dimensional features of the figure and the associated structural parameters of clothing that determine a comfortable fit on the figure;
- Development of a customer interaction interface while ordering a customized product.

Nowadays, there are a large number of different difficulty levels and different market price segments installations for obtaining a figure's digital image. Based on the analysis of existing systems, the use of depth sensors is identified as the most promising direction, providing high accuracy and high speed of obtaining initial information about the surface shape of objects. There has been developed a three-dimensional scanning system, which is based on the Microsoft Kinect depth sensor [10]. The information and technical structure of the module of obtaining a digital image of a figure consists of a material part built on serial depth sensors and, on the other hand, of proficient software that allows using the equipment as a module. Two versions of the module for obtaining a digital image of a figure are proposed. The first option involves rotating the figure around its own axis while the scanning process takes 60 seconds (figure 1).

IOP Conf. Series: Materials Science and Engineering **944** (2020) 012026 doi:10.1088/1757-899X/944/1/012026

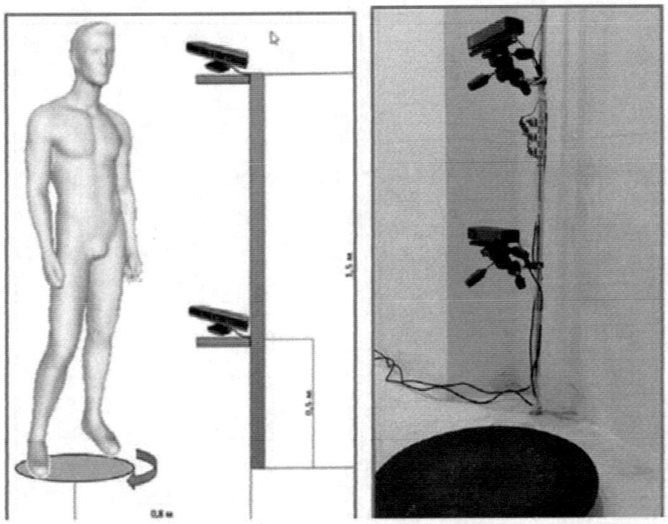

Figure 1. Diagram and appearance of a system for obtaining a digital image of a figure with two sensors

It is suitable for use as a portable complex, measurement error is 3 mm. The creation of digital image of human body to full height occurs when two scanned three-dimensional models are combined together – the upper supporting surface and the lower supporting surface.

During scanning, a person stands straight without any tension and clamps in the body, heels are put together and toes are separated. The position of the legs should be stable, calm and comfortable for the person being measured. Hair is collected revealing the back of the neck. The second option of the system consists of six sensors; a person stands motionless during scanning process (figure 2). Obtaining a digital image of the human body to full height occurs when six separate scanned sections of the surface of the human body are combined together.

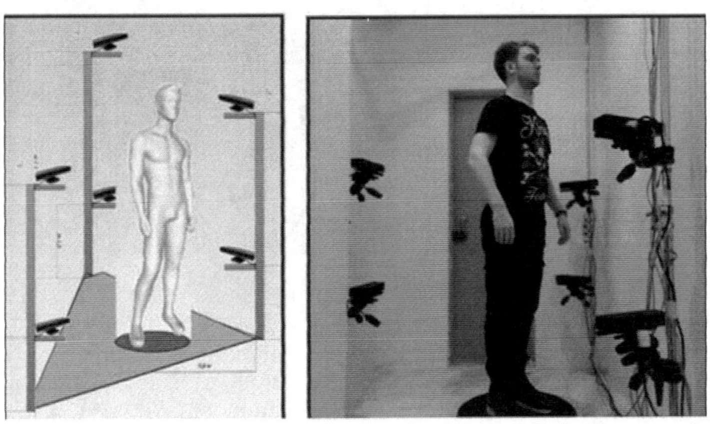

Figure 2. Diagram and appearance of a system for obtaining a digital image of a figure with six sensors

The visualization technology of the scanned object, proposed by the specialists of the Department of Art Modeling, Design and Technology of Garment of the RSU named after A.N. Kosygin, allows getting reliable anthropometric characteristics (figure 3): projection, circumferential, arc measurements; to perform an assessment of the quality of fitting; to identify the presence of defects; to assess the conformity of the ready-made product and a sketch [11].

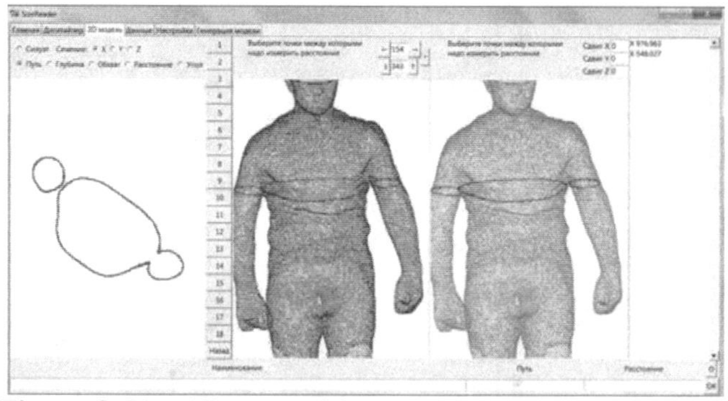

Figure 3. Identification of dimensional features of a figure

The assessment of the anthropometric compliance of designer's clothing is based on a list of parameters that require monitoring. To determine this list, a survey was conducted with 110 people as experts, including light industry quality control experts and retailers. Chest girth and neck circle are determined as the leading dimensional parameters of men's shirts. There were designed and manufactured 17 models of men's shirts. Nine models were made with a change in widening of chest girth. Eight models were made with a change in widening of neck circle.

One of the main tasks in assessing ready-made clothing models is the visual correspondence of a figure with a product and the anthropometric correspondence of sizes of clothes with a figure. At the next stage, we compared the scanned three-dimensional models of the figure wore shirts with undressed models in a virtual environment. To do this, initially a model of the item was put on a motionless human figure (figure 4, c) and was scanned. Then the product was being removed and re-scanning of a person wore underwear (Fig. 4 a, b) was performed. After that we combined virtual images of the figure wore the investigated clothing and without it.

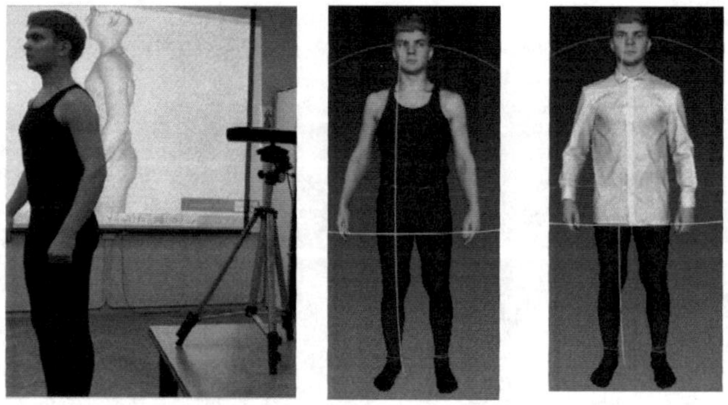

a b c

Figure 4. The process of virtual assessment of a clothing model: a – scanning of a person's figure; b – a virtual model of a human figure; c – a virtual model wore a shirt.

At the next stage of the experimental research the projection gaps between the figure and the product were studied using the Size Reader program [12], which allows two models to be loaded simultaneously in .ply format and displayed in two windows at the same time. When pointing to the desired area, the program draws a section and shows a comparison of two sections in the third window by geometric overlapping (Figure 5). For research and taking of measurements in a virtual environment, sections of the chest and hips were selected as the largest in size. It was revealed that the optimal value of widening of chest girth is a range from 2 to 9 cm. The most suitable models due to the neck area for the consumer were products with widening of 3 cm to the total neck circle, then with widening of 2 cm and widening of 1 cm. According to the results of study, the choice of widening of the neck circle in the range of 1–4 cm allows ensuring visual conformity of a product to a figure of a person in the neck area.

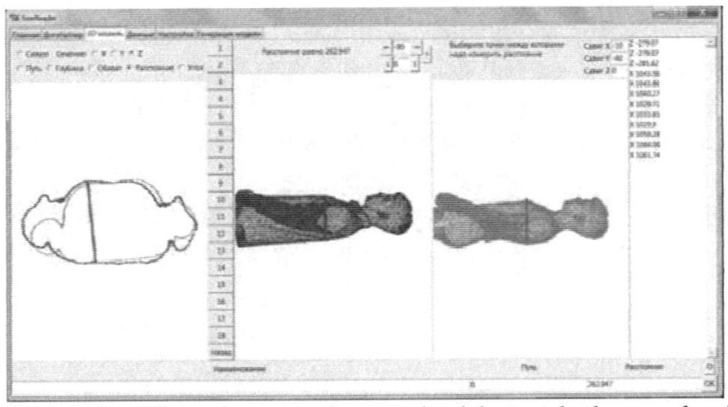

Figure 5. Comparison of 3D models wore clothes and without clothes on horizontal sections of the figure at a chest level

At the next stage, a consumer is implemented for the automated design of cust interaction interface omized garments. The system collects information about customers' preferences, size characteristics, shape and type of body in specially created databases, which are used when it is necessary to help to create clothing products that are best suited to the consumer in size and design, is proposed.

Before starting product design, the user is invited to watch a brief video review how to work with the program, which describes each stage of creating a new model. On the start page of the program it is proposed to determine the product being designed, for this we need to select one of the categories "Women / Men / Children" and indicate the required one from the list that appears. After the selection, the user goes to the design section, where it is proposed to select one or more design solutions in the categories "Front", "Back", "Sleeve", "Collar" (figure 6). In the lists of subcategory options of the design section, one of them is indicated automatically. The user can change the default options or add the necessary ones to the existing. The selected option(s) of the "Silhouette" subcategory of the "Front" category automatically generates the same option(s) of the "Back" category. Similarly, subcategories of the design section are formed that are structurally and technologically connected at the front and back of the product (for example, "Belt", "Product Length", "Decomposition", "Decorative Elements", etc.).

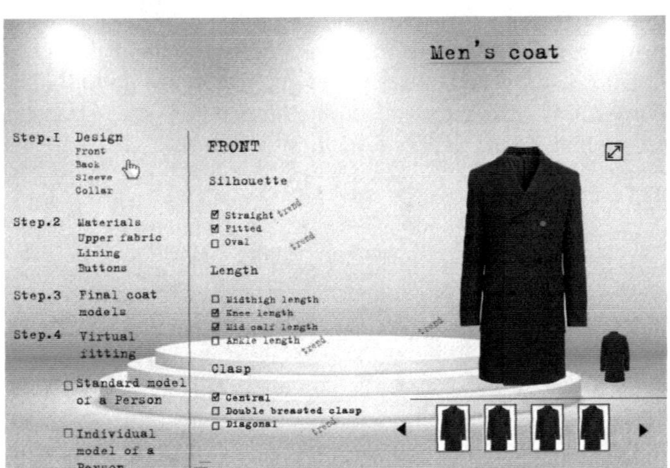

Figure 6. Design section of the program

At the next stage of designing, the user determines necessary materials for the product model in the relevant categories of "Upper fabric", "Lining fabric", "Accessories / Buttons". The principle of operation in the "Materials" section is similar to the "Design" section. In each category of the "Materials" section, it is necessary to select the type of fabric / material in the "Select" field, located

IOP Conf. Series: Materials Science and Engineering **944** (2020) 012026 doi:10.1088/1757-899X/944/1/012026

under the name of the category of the section (natural fabrics, synthetic fabrics, mixed fabrics, etc.). Subcategory options are indicated by default. We can leave, remove or add others from the proposed list. By specifying fabrics / materials for one of the categories, it is possible to automatically exclude some samples for other categories (types of fabrics / materials, their properties can "argue" with each other during technological processing of product units).

In all sections of the program, there is a demonstration of graphic images of the designed product model with selected options and / or default options for each subcategory of the available design sections.

The selected number of options in each subcategory automatically generates the corresponding number of sketches in the lower part of the demonstration field, and also in the "Final Models" section.

The signs (the name of the subcategory of the structural element of the garment) are denoted by Sn, and the solution's options are denoted by S^n. While choosing the necessary options, the selected features are combined, thus obtaining various conceptual sketches of the designed models ($S^1 1 * S^1 2 * S^1 3 * S^1 n$ .... $S^n 1 * S^n 2 * S^n 3 * S^n n$).

Based on the matrix method, the program creates all possible variants of sketches of the designed product, combining structural elements of clothing selected by a user.

The user can see all the options for the graphic image of the designed product in the "Final Models" section in accordance with previously selected options for the subcategories of the "Design" and "Materials" sections.

Graphic images of the models are divided into the "Classic" and "Creative" groups. The "Classic" group presents models in the classical style, but the "Creative" group contains other styles. The process of dividing sketches into groups occurs depending on the structural elements of the product, which form the common style of the item.

Each group can be minimized and maximized by clicking on the appropriate buttons next to the name of the group. There are scroll bars at the left side of each block in case all drafts could not be fitted on the screen. To mark the desired sketch, we have to click on a cell located in the upper right corner of the sketch (when we click on the cell a checkmark will be displayed). To select all the options, use the "Mark all" button in the lower right corner of the block (the checkmark will appear on all options). In the right area of the program section, selected models are shown with the ability to enlarge images, as well as see information about the model (type of fabric, fabric composition, article, etc. at the discretion of the manufacturer). In the lower right corner of the program section there are buttons "FITTING" and "ORDER". The client has the opportunity to go to the "ORDER" section after demonstrating the model without trying on and order the product after choosing the size, or use the virtual trying on by pressing the "FITTING" button (figure 7).

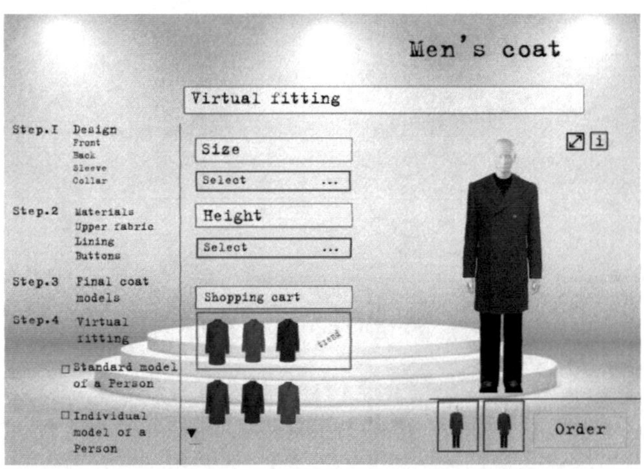

Figure 7. Virtual trying on of the designed model

Marked sketches in the "Final coat models" section form the "Basket" block of the "Virtual fitting" section. The section has two subsections, "Standard Model of a Person" and "Individual Model of a Person". When choosing the subsection "Standard Model of a Person", we should indicate the values of the parameters "Size" and "Height", selecting the one from the proposed list (press the "Ellipsis" button in the "Select" field). When choosing the "Individual Model of a Person" subsection, we should indicate the values for the parameters "Height", "Chest Girth", "Waist Girth", "Hip Girth", after that a model is being selected in order to demonstrate the designed product. It is also possible to download a ready avatar made with the help of appropriate devices and applications.

4. Conclusion

The developed systems of the three-dimensional scanning make it possible to obtain reliable anthropometric characteristics: projection, girth, and arc measurements; to assess the conformity of the ready-made product and the dimensional characteristics of the consumer's figure. The conducted studies allow substantiating the rational values of permissible gaps at given horizontal levels in order to correctly determine the correspondence of different clothing models with the sizes of each figure in the three-dimensional virtual environment. The proposed approach provides the consumer with the opportunity to try on various types of products on his or her three-dimensional avatar so that the reading unit of the program independently selects the most suitable clothing model from the offered list, based on the correspondence of the design parameters of the product and the dimensional characteristics of a particular person. Thus, a buyer has the opportunity to create a unique customized product, taking into account his or her consumer requirements for the desired model.

Acknowledgments

The reported study was funded by RFBR, project number 19-37-90089

References

[1] Gurina M A and Rumyanceva J V 2019 Comprehensive tools and methodologies quality improvement in the management competitiveness trade enterprises *Proc. of the Voronezh State Univer. of Engineer. Technol.* **81(2)** 320–35 Retrieved from: https://doi.org/10.20914/2310-1202-2019-2-320-335

[2] Gazhur A A and Lukiyanchuk I N 2018 Trends of development of the sphere of services in Russia *Proc. of the Voronezh State Univer. of Engineer. Technol.* **80(3)** 444–50 Retrieved from: https://doi.org/10.20914/2310-1202-2018-3-444-450

[3] Khajeh M, Payvandy P and Derakhshan S J April 2016 Fashion set design with an emphasis on fabric composition using the interactive genetic algorithm *Fashion and Textile* **3(8)** 1–16

[4] Kazakova N A, Ivanova O V and Khammatova E A 2017 Use of author's structure in the customization of garments *Proc. of Technol. Univer.* **21** 70–2

[5] Xu B and Sreenivasan S V 1999 A 3-D body imaging and measurement system for apparel customization *J. of the Textile Instit.* **90(2(1))** 104–20

[6] Song H K and Ashdown S P 2012 Development of automated custom-made pants driven by body shape *Cloth. and Textiles Res. J.* **30(4)** 315–29

[7] Park L I 2016 *FXmirror 3D Virtual Fitting Solution* Retrieved from: http://seoulspace.com/2016/06/29/fxmirror-3d-virtual-fitting-solution. (Viewed: 29.07.2019)

[8] *The acustom experience* Retrieved from: https://acustom.com/pages/ theacustomexperience (Viewed: 29.07.2019)

[9] Kokoszka P 2018 *Is 3D body scanning the future of fashion?* Retrieved from: https://www.verdict.co.uk/3d-body-scanning-fashion-future/ (Viewed: 29.07.2019)

[10] *Website developer.microsoft.com* (2020) Retrieved from: https://developer.microsoft.com/en-us/windows/kinect/ (Viewed: 01.01.2020)

[11] Petrosova I A, Andreeva E G and Guseva M A 2019 The system of selection and sale of ready-to-wear clothes in a virtual environment pp 1–5 *Int. Sci. and Technol. Conf. "EastConf"* (Vladivostok) DOI: 10.1109/EastConf.2019.8725390
[12] Petrosova I A, Shantseva O A and Andreeva E G 2017 Assessment of a right size ready-to-wear clothes to the consumer's figure in the three-dimensional system *Izv. Vysshikh Uchebnykh Zavedenii, Ser. Teknol. Tekstil. Promyshl.* **5(371)** 139–42

IOP Conf. Series: Materials Science and Engineering **944** (2020) 012027 doi:10.1088/1757-899X/944/1/012027

Motion design in 1990-2010 and its entertainment function: social and technological aspects

O V Petrukhina

Saint Petersburg Stieglitz State Academy of Art and Design, 13, Solyanoj pereulok, Saint Petersburg, 191028, Russia

E-mail: oks.petruhina@yandex.ru

Abstract. The increased importance of applied animation associated with the active introduction of products of motion design into national design practice requires due attention and research. In order to understand and use correctly the techniques and principles of modern design activities related to dynamic images, it is necessary to study its origins and conduct a historical analysis of expressive methods and means. The paper studies one of the stages in the development of motion design, covering the 1990-2010s and describes the types and techniques of applied animation used in advertising.

1. Introduction

The development of the basic principles of motion design, created at the turn of the century, in particular, the reliance on new technologies, led to the fact that one of the most prominent tools of the motion graphics at the turn of the 1990s was the widespread use of the computer graphics capabilities. It manifested in the creation of such forms of motion graphics as commercials, captions, educational, political and social videos, as well as elements created by the motion design capabilities for Web industry.

The researcher Ya. B. Ioskevich studying the art culture of the Internet space writes that motion design "seeks to use the specifics and new aesthetic possibilities offered by the Web" [1]. N. G. Krivulya in the research "Animated poster: origins, specificity and originality" writes that a characteristic feature of the works of applied animation, created in the period 1980s - 2000s was the "uneven use of digital technologies in various subspecies of this genre". Their implementation turned out to be the most productive in the field of advertising posters, while realism tendencies increased in the political poster and digital technologies were practically not involved. This trend continued until the early 2000s [2].

2. Evolution of motion design (1990-2010)

The vector of evolution development of motion graphics since the early 1990s is directed by market requirements and commercial feasibility, which determine the conditionality of certain expressive means.

Starting from this period, the features of formation are more and more determined by the tendency to use the most simple and laconic solutions from the point of view of graphics: local colors, stylization and simplification of forms, minimal use of light and shade, etc. Since the late 1990s during the creation of applied product, motion graphics increasingly refers to the methods and means used at the stage of its formation. With the advent of digital technologies in the art of motion graphics, the trend of attraction, artistic expressiveness is increasing, simulacra images are actively used.

The strategy in the creation of political animation is changing. Since the beginning of the 2000s in political animation, the heroic-patriotic style begins to prevail. A candidate is idealized with hypertrophy, and the general intonation of the video focuses on the image of the "defender" and "friend" of voters. As an example of such videos, we can cite the video of "Boris Yeltsin's electoral company" and "SLON" political party.

Another strategy used in pre-election advertising was in the comparison of a candidate and an opponent. A candidate was opposed to his opponent, giving negative traits to the opponent's image, which, together with the ironic context, changed the general tone of work from "official to caricature" ("She has no choice, but you have!"). The graphics of political commercial videos "were characterized by the techniques of popular print, comic, carnival and collage components"[3].

Late 1990s - early 2000s are characterized by the widespread introduction of three-dimensional graphics capabilities in motion graphics tools, which leaves its unconditional imprint on the stylistic component.

At the same time, the beginning of the 2000s is associated with a new stage in the development of motion graphics. In this period, there was a significant change in aesthetics and a renewal of the strategies used by animators when communicating with audience. Many works appear that use the author's underground style, which makes it possible to bring confidential personal intonations into communication.

At the present stage of development of digital technologies, the possibilities of computer graphics are considered to be unlimited and "the imagery of animation art can give rise not only to surprise and entertainment, but also to make a person think, understand and learn complex social, philosophical or aesthetic aspects of society by means of art" [4]. Animated graphics synthesize the principles and methods of many types of art, visual communication and technology.

Nowadays, we should talk about an already established type of project activity - motion design, which combines graphics, sound, photo and video information, animation, and combines the capabilities of two-dimensional and three-dimensional graphics.

In the modern world, animation has a special place in the creation of computer games, advertising (including banner advertisement), music videos and TV airtime design. A multimedia product increases the efficiency of information perception that is why it is one of the most promising and popular areas. N.S. Kuznetsova, studying new developments in the field of human-computer interaction, in which there is a simultaneous and constantly improving influence on the organs of human perception, draws attention to the implementation yet fundamentally different "game" principles of information presentation [5].

The scope of motion graphics, which combines graphic design, photography, typography, sound effects, animation and storytelling, is constantly expanding. In products created by motion graphics, 2D or 3D objects come to life using 2D and 3D graphics - new images of game and advertising characters appear. With the help of the visual channel of perception, information is structured and systematized. Applied animation graphics have large figurative and expressive means, which, unlike static graphic objects, acquire new qualities in the process of dynamization of movement. The main segments of the application of motion graphics are the film industry, television, marketing, advertising (social, political, commercial), education, entertainment, presentations, etc.

The process of "animation" captures all the new, previously untouched spaces of adjacent areas. One of these spaces is the Internet.

According to T. Yu. Kitaevskaya a website is a system of visual elements aimed at the provision of a holistic perception of both the entire object (website) and its individual elements (pages, objects) [6].

A positive emotional reaction of a user is incorporated into the design of the site at the stage of its design. A designer should not only provide such a client's reaction to the Internet site, but also contribute to the emergence of a connection with the "feelings of individual consumer about a design product" - the advertised object [7].

Motion design is used in the creation of online advertising, interface elements and navigation. It allows simplifying user interaction between the objects of a web page and user actions when it appears on the download page. At the junction of two or more subject areas, new meanings and spheres of interaction appear.

If the product of motion graphics does not have a detailed plot and initially simply illustrates certain ideas, then in conjunction with the web interface it creates and broadcasts a previously nonexistent virtual model of an object or animation phenomenon. As an example, combining game and informational functions of two subject areas of design (motion and web), we can cite the replacement of static screensavers dedicated to various events and anniversaries on the main page of Google by Google Doodles, the development of which is directly related to the use of motion graphics capabilities. These capabilities simplify search functionality and transform reality into motion graphics for videos, splash screens, commercials etc.

The power of 2D digital graphics is usually used for web animation. 2-D shape animation transforms both the simplest geometric shapes, transforming them into each other as well as rather complex shapes.

a b

Figure 1. Shape 2-D animation: a – transformation of the simplest objects; b – character shape transformation.

In products created by motion graphics in the 2000s the emphasis was on art and technological innovation and eclecticism. Artists tried to create a capacious memorable image, in some cases deliberately using a minimum of means, while in others, on the contrary, they filled the screen space with a visual kaleidoscope that combined the capabilities of video, photo, hand-drawn and three-dimensional graphics. The common points for all animated applied products of this period were the unconditional predominance of elements created by means of computer graphics, the absence of a pronounced narrative in the video plots (emphasis on action). Information was presented through dynamic and expressive scene changes, where the only, often, connecting semantic element was a character (mascot) or the advertised object. The visual range of the video was often supplemented by the introduction of dynamic font elements, which significantly enriched the visual range.

N. G. Krivulya, formulating the distinctive features of modern works of motion graphics, draws attention to the predominance of "semantic reduction and clip nature in the presentation of verbal and visual information" [8]. It is also necessary to note the emphasis on strengthening the entertainment and game components of an animated commercial during the studied period. Over time the presentation of the material increasingly begins to express tendencies towards the strengthening of active and rather aggressive intonations. In addition, the convergence of the subject areas of design leads to the emergence of new products created at the junction of animation and other types of design.

Motion graphics, developing over time, has changed its attitude towards meaningful and depictive presentation of information. Roland Barthes writes about the need to include advertising "among the main mental food products, which are for us literature, theatre, cinema, sports, press, fashion" [9]. Thus, advertising according to Barthes is one of the most effective means of the formation of the spiritual experience of a person. In this regard, one of the examples is the transformation of meanings

IOP Conf. Series: Materials Science and Engineering **944** (2020) 012027 doi:10.1088/1757-899X/944/1/012027

in social (anti-alcohol) advertising: in the films of the 1920s, the problem of alcoholism was seen as a consequence of illiteracy and remnants of the dark past, in the videos of the 1950s - 1960s the weakness of the drinking person is exposed and his rehabilitation is assigned to the family and work collective, and in the 1980s the accusatory slogan gives way to a narrative form, later acquiring a harsh, gloomy-ironic sound ("Reanimation", "Blue devils" projects).

One of the most effective leverages on a viewer (a consumer of advertising) is a dynamic advertising character (mascot). The task of such a character consists in the volumetric and "consistent implementation of the concept, style, specific design solutions for the visualization of the trademark, corporate logo of a company and brand" [10]. In general, animated according to M. Baitler, should contribute to the fact that "the attention of any person who falls into the area of perception of this notification" was filled with the subject of such notification [11]. For motion graphics, this task is simplified by the fact that viewers "... since childhood have retained a completely supportive attitude towards cartoon characters" and are subconsciously friendly towards what is happening [12].

A well-known example of a commercial that used a child's friend in communication with a user is the products of Prostokvashino brand. The commercials for Prostokvashino use the two-dimensional stylistics originally used for cartoons. The possibilities of 2D graphics are also demonstrated in commercials for Alpen Gold, Pit Breweries, Sorti dishwashing products.

The example of the use of 3D graphics is the commercials of the companies M&M's, Nesquik, Invite +, Picnic, Heinz, Michelin, etc. 3D animation of the early 2000s focuses on the creation of an animation object or scene in which animation takes place in specialized computer 3D programs, when "it is necessary to create geometric models of all its constituent objects, configure light sources and cameras, create and assign materials to objects for which it is necessary to select or draw textures" [13]. The example of this is the anthropomorphic advertising characters used by M&M's. These characters are multi-colored candies that enter into active communication with a viewer from the screen.

It is possible to say that the specifics of motion graphics at the turn of the 2000s is the active use of 3D animation for the creation of commercial advertising products, where technologies become an integral part of the information flowing from the screen.

The key features of computer animation in 3D space are sets of textures, objects, light sources, bones, etc. These components create complex solutions for visualization of special effects, modeling and rendering using the capabilities of functional programs (3Ds Max, Maya, Cinema4D, etc.), which have a wide base of tools and additional plugging. 3D animation, digitally creating an object, provides it with a "skeleton" that allows moving the model. The capabilities of a computer allow interpolating between frames, thereby creating motion, where visual transformations are controlled by the matrices of rotation, shift, scaling and the morph effect, which smoothly transforms the "flow" of one object into another.

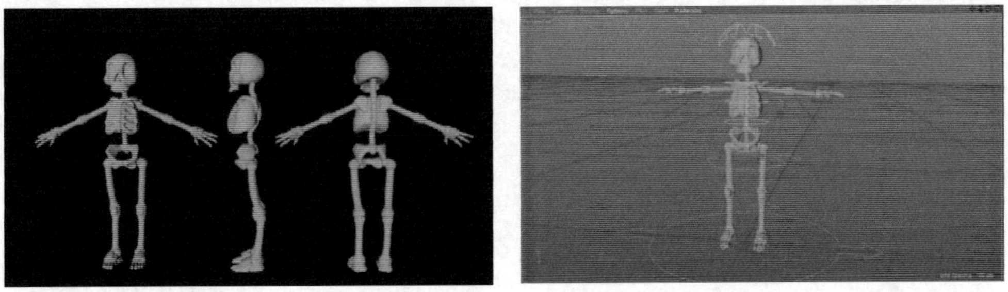

a b

Figure 2. Computer 3D animation: a – creation of a character; b –animation of a character.

Procedural animation is used only in 3D animation to create effects and movements that describe how parameters change over time. It is calculated by a computer using Java-Script and Action-Script with the help of which the movements of animated objects are programmed, creating simulated particle systems (water, smoke, fire), hair and fur dynamics, character animation and their movements.

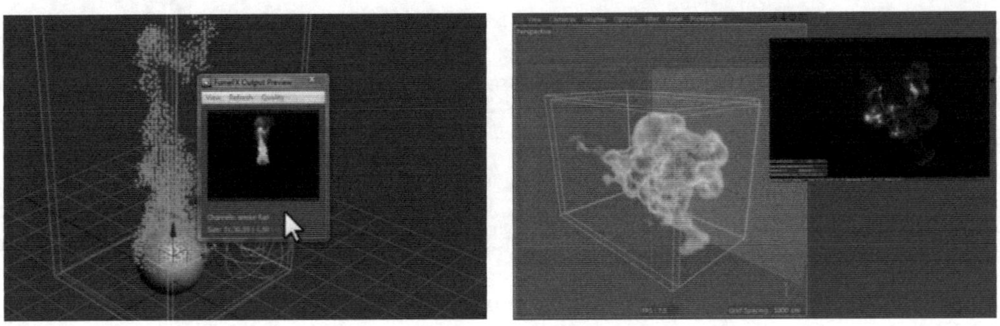

a b

Figure 3. Procedural 3-D animation: a – fire animation; b – smoke animation.

Commercials use all the capabilities of computer technology (2D, 3D animation and their combination), but 3D graphics are preferred. In social and instructional videos, the possibilities of 2D animation are predominantly used, as a cheaper option in terms of cost.

One of the most popular social advertising projects that used the possibilities of both 2D AND 3D graphics is the project "Reanimation", to the creation of which well-known Internet artists and comics artists were involved (Khikhus, A. Pronkin, A. Velitov, A. Shelyakina, L. Zilonova, P. Muntyan, M. Polyakov, E. Sidoruk, K. Komardin and others). The musical accompaniment of the project was provided by the Russian rock movement (musicians of the groups "Splin", "Auction", "Crematorium", "Sky is here", "Purgen", "Kasta", etc.). The figurative solution of the project is an eclectic, maximally sophisticated aesthetics of graffiti, comics, anime, abundantly mixed with black humor, where "the informal, controversial language, used by the creators and developers of this project, baffles not only critics, but also the representatives of blogosphere, raising doubts about the adequacy of the means used" [14].

The project uses cut-out animation, 2D and 3D digital animation, frame-by-frame and stop-motion animation.

IOP Conf. Series: Materials Science and Engineering **944** (2020) 012027 doi:10.1088/1757-899X/944/1/012027

Addressing the issues of technology for the creation of an animated work, it is possible to state that frame-by-frame technology is used in 2D and 3D animation (drawn and claymation-animated films), giving a dynamic nature to information. Frame-by-frame animation sequentially demonstrates a series of frames and is used to make changes to the time series. This method is close to traditional hand-drawn animation, only the role of a phase-shifter is performed by a computer: the hand- drawn frames are programmatically combined into a video sequence, creating complex phases of movement. Frame-by-frame animation is widely used in animation films, advertisements, games, educational programs.

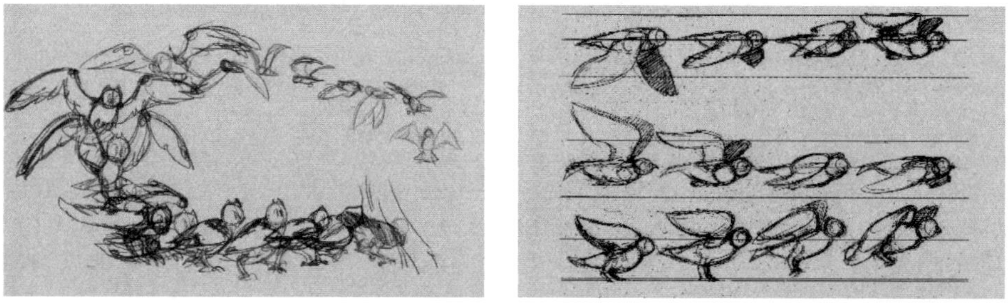

a b

Figure 4. Stop-motion animation: a – complex motion; b – linear motion.

Digital technologies changed constructivist epistemology. They allowed the manifestation of creative individuality, becoming a new aspect of modern culture. They are the next stage in the evolution of animation technologies, a new way of influencing a viewer. Digital animation is a highly efficient technology for the development of media content that allows creating animated advertisements using 2D graphics, 3D modeling, stop-motion animation, computer-aided cutout and the works that combine various technologies, the defining factor of which is the original concept that increases the efficiency of the animated work.

Motion design is becoming more and more popular as a kind of symbiosis of graphic design and dynamic animated graphics. For the visualization of statistical information, creation of a nonexistent environment for an advertising object, integration of a branded character, hybrid graphics are used to in order to increase the effectiveness of advertising products. During the animation the following programs are popular: Adobe Animation, Ulead Animation, Ulead GIF Animator 5, CoffeeCup Animation Software, CoffeeCup GIF Animator, CoffeeCup Firestarter, Firestarter, Reallusion Animation Software, Effect3D Studio, 3D Max, Toonboom Studio etc.

3. Conclusion

New technologies had great influence on the form of motion design, significantly expanding the scope of such products. Motion design addresses us from on-screen street panels and banners. The works of motion design are becoming a means of global communication, going beyond geographic boundaries, acquiring interactivity, performing the function of activating attention, entertainment and information. These works (videos, screensavers, animated interface elements, etc.) open the facets of reality, not copying it, but increasing the number of visual forms of perception and knowledge of the surrounding world, creating other relationships with meaning, time, history and becoming an obligatory component of the media and advertising.

Being integrated into all the spheres of life of modern society, motion design indirectly affects the formation of public consciousness, even if this is happening through the culture of consumption nowadays.

IOP Conf. Series: Materials Science and Engineering **944** (2020) 012027 doi:10.1088/1757-899X/944/1/012027

In the conditions of highly increased information flows, motion design, again, as at the beginning of its appearance, is forced to use deliberate limitation of visual means in order to remain noticeable against the background of general “digital noise”. The means developed by first motion designers, avant-garde artists at the dawn of motion design: the contrast of scales, the capacity of verbal formulations, contrasts of form, font, the use of graphics and collage techniques, the use of color contrast, the symbolism of the hero's image encoded in a graphic sign (modern advertising hero), the use of editing techniques - are becoming a relevant tool in which modern designers use in order to create applied animation. More recently, the established toolkit has been supplemented by the possibilities provided by the limitless color range of computer graphics.

References:

[1] Ioskevich Ya B 2006 *Internet as a new medium of artistic culture* (Sant Petersburg: Federal agency of culture and cinematography of the Russian Federation [and others]) RIIII 166
[2] Krivulya N G 2016 Animation poster: origins, specificity and originality *VGIK Bulletin* (Moscow: VGIK) **2 (28)** pp 8–23
[3] Krivulya N G 2016 Genre formation of animation posters in the second half of the twentieth century *Bulletin of VGIK* **4 (30)** p 28–42
[4] Khitruk F S 2007 Profession - animator *FS Khitruk: in 2 volumes* Vol 2 (Moscow: Gayatri) p 126–132
[5] Kuznetsova N S 2010 Web design as a tool for modeling the virtual space of modern culture: PhD thesis on cultural studies (Saransk)
[6] Kitaevskaya T Yu 2014 Alternative styles in web design *Bulletin of Tomsk State University* **2** pp 569–570.
[7] Glazychev V L 2006 Design as it is (Moscow: Europe) p 106
[8] Krivulya N G 2016 Genre formation of animation posters in the second half of the twentieth century *Bulletin of VGIK* **4 (30)** p 8–42
[9] Barthes R 2003 Advertising message *Fashion system. Articles on the semiotics of culture* (Moscow: Publishing house of Sabashnikov) p 415
[10] Design concept of the project. Development of the design concept [Electronic resource] / P.I.X. – Access mode: http://pixmedia.ru/dizayn-koncepciya-proekta-razrabotka-koncepcii-dizayna%5D.A
[11] Baitler M 1926 Advertising and film advertising (Moscow: Kinopechat) **112** p 8–9
[12] Vasiliev M 2000 Animation in advertising *Advertising technologies* **5**
[13] Marov M N 2005 *3ds max. Real animation and virtual reality* (Saint Petersburg: Peter)
[14] Krivulya N G 2016 Genre formation of animation posters in the second half of the twentieth century *Bulletin of VGIK* **4 (30)** p 28–42

IOP Conf. Series: Materials Science and Engineering **944** (2020) 012028 doi:10.1088/1757-899X/944/1/012028

The development of protected natural areas and adjacent territories

O E Roshchupkina[1], M V Perkova[1], E V Baklazenko[1], A G Vaytens[2]

[1] Belgorod State Technological University named after V.G. Shukhov, 46 Kostyukov str., Belgorod, 308012, Russia
[2] Saint Petersburg State University of Architecture and Civil Engineering, 4 Vtoraya Krasnoarmeiskaya str., St. Petersburg, 190005, Russia

E-mail: olyarossof@gmail.com

Abstract. In Russia, the most common form of preservation of specially protected natural areas is formation of nature reserves. Due to nature protection regulations, which limit activities within the boundaries of the reserve, and due to the difference between the natural boundaries and those of the protected natural areas, the authors propose the strategy for urban development of the territories adjacent to the protected natural areas. The study identifies the main types of the areas adjacent to the reserves and provides their classification. It is proposed to form three zones: strictly preserved, buffer and protected natural zones. Strategies for urban planning development of the territories adjacent to the protected natural areas and measures have been developed for three sites of the Belogorye reserve: Forest on Vorskla, Yamskaya Steppe, Stenki Izgorya. This will ensure preservation of unique natural complexes; strengthen the natural framework of the region and stabilize the ecological situation; intensify research and educational activities in the reserves; enhance the socio-economic development of the territories adjacent to the reserve and the development of the region's tourist and recreational infrastructure.

1. Introduction

Various natural complexes and/or objects with their unique ecosystems make the basis of the protected natural areas. Their formation at the level of administrative-territorial division depends on numerous factors: on the historically established natural and landscape-geographical features, the specifics of public administration, on current spatial boundaries and characteristics of the territories of the administrative units. Social and economic development of the adjacent territories, which develop to meet the needs and requirements of the population, is of great importance for the status of the protected natural areas. As a result, most of the territories adjacent to the protected natural areas are to some extent anthropogenic. The processes in anthropogenic territories that are key to the regions can refer to agricultural, agro-industrial, mining complexes, etc. Their direct and indirect impact on the environment can be extremely negative for the natural framework and lead to the loss of the unique natural and biological originality of the protected natural areas. Landscape design will ensure competent and safe nature managemen to eliminate current problems of the protected natural areas and their sustainable development. The tourist and recreational potential of the territory is interrelated with the natural, historical, cultural and transport framework of the territory [4]. Landscape design of the the territories adjacent to the protected natural areas involves:

IOP Conf. Series: Materials Science and Engineering **944** (2020) 012028 doi:10.1088/1757-899X/944/1/012028

– identification of typological features of these territories, and development of (proposals for) their classification.

– identification of the most favorable territories for development and establishment of regulations for their use with subsequent eleboration of development projects.

– functional zoning of the territory to identify the optimal ratio of the type of nature management and the environmental regulations.

2. Materials and methods

The theoretical and methodological basis of the study is a conflictological approach used to study the resource potential of the territory, identify interests and needs of the participants of urban planning activities in the territories adjacent to the reserve, and to find a compromise in solving issues of land use optimization in order to preserve and strengthen the natural framework.The study also employs a comprehensive analysis of the process of urban development of territories with regard to the socio-economic function of various types of landscape.

3. Classification of the territories adjacent to the protected natural areas

It is proposed to classify the lands adjacent to the protected natural areas in accordance with the interstate standard [1], in accordance with the basis for the development of principles of landscape classification by A.G. Isachenko [3], with the classification of modifications of primary and conditionally primary facies developed by V.I. Prokaev [9]. These classifications are based on the degree of impact of human economic activity on the territory, the depth and degree of anthropogenic transformation. This allows most full application of these classifications to the territories adjacent to the protected natural areas, which are largely anthropogenic. When studying the territories adjacent to the protected natural areas, it is proposed to correlate their classification with urban planning conflicts developed by M.V. Perkova [6, 7]. Thus, the lands adjacent to the protected areas can be classified as follows (Table 1).

1. With regard to the main types of landscape, depending on the socio-economic function:

- agricultural;
- forestry;
- water management;
- industrial;
- landscapes of settlements;
- recreational;
- reserved;
- not currently used [1].

2. Depending on the location in relation to the protected natural areas.

- Land use.
- Material.
- Linear.

3. Depending on the types of urban planning conflicts.

- Land use;
- Property;
- Social and functional;
- Transport.

4. Depending on the degree of transformation.

- Indigenous (unchanged and conditionally unchanged);
- Derivatives (modified to varying degrees);
- Anthropogenic (completely modified).

Indigenous landscapes include modern territorial units that did not undergo or were indirectly subjected to conomic activity [2]. Specially protected natural areas are indigenous landscapes.

Derived landscapes refer to territories with changes within their boundaries related to the features of their use. The features characteristic of these landscapes have retained their priority, and changes occuring can be temporary and reversible unless the interference of human activity.

Anthropogenic landscapes are among the most deeply changed territories. These include man-made parks, naturally formed residential and historical industrial territories, landscapes with strong religious, and artistic and cultural associations [5].

5. *By economic value.*

- Reasonable;
- Excessive;
- Depressive.

Reasonable territories include areas of agricultural, industrial and other anthropogenic activities with the results needed to maintain a high level of economic, social, economic, cultural and other functions.

Surplus territories include areas with irrational economic activities within their boundaries with outcomes that have an adverse affect on the state and development of the protected areas.

Depressive territories include areas where no or insufficient activities performed, which deteriorated their functional and aesthetic qualities. For example, natural waterlogging of floodplain areas; degradation of landscapes in need of restoration; soil erosion due to no measures to protect them undertaken, etc.

4. Potential for development of the territories adjacent to protected natural areas

The study of different types of territories adjacent to the sites of the Belogorye reserve revealed different levels of their resource potential (high, medium or low). Thus, the lands with rural settlements, their transport links and areas needed for sustainable socio-economic development of the territorial unit were recognized as unsuitable for the formation of buffer territories of the protected natural areas within their boundaries. At the same time, water and/or floodplain and agricultural territories are of the highest potential for development, which makes them a priority basis for formation of a buffer zone of the protected natural areas.

For a number of natural reasons (a decrease in livestock, abnormal weather conditions, etc.), 30–40 percent of these lands are not used. At the same time, the land of the protected natural areas make up 49.4 million hectares, as a result of which many territories adjacent to the protected natural areas are agricultural land. The presence of floodplain landscapes is characteristic of these territories. The river network with adjacent territories has a great potential for development as buffer zones. A coastal strip, a coastal protective strip and a water protection zone restrict the formation of any infrastructure, which takes place with regard to the established mode of economic activity.

Thus, agricultural lands and floodplain territories adjacent to the protected natural areas need measures to prevent environmental degradation. Partial withdrawal of degraded territories from agricultural lands and renaturation of their landscapes can expand the boundaries of the protected natural areas to natural boundaries.

5. Transformation of the territories adjacent to protected natural areas

Based on the functional zoning model of the territories adjacent to the protected natural areas proposed by M.V. Perkova [8], strategies for the development of three sites in the Belogorye reserve have been developed. The urban development strategies need allocation of three main zones: strictly preserved, protected, and buffer zones. *A strictly preserved natural zone* is the natural core, which implies limited human presence due to the specifics of nature conservation activities. Protection and buffer zones are formed along the perimeter of the protected natural areas. *A protected natural zone* includes recreational activities within its boundaries, which will meet the requirements of special conditions for land use. At the same time, the territories adjacent to the protected natural areas belong to the protected natural zone. It is proposed to form *a buffer zone* as territories of the protected landscape and territories with objects of tourist and recreational infrastructure – territories for scientific research, treatment-and-prophylactic and recreational territories.

IOP Conf. Series: Materials Science and Engineering **944** (2020) 012028 doi:10.1088/1757-899X/944/1/012028

M.V. Perkova and O.E. Roshchupkina have developed a strategy for transforming the territories adjacent to the protected natural areas in three areas of the Belogorye reserve in Belgorod region. At the same time, the fundamental goal is to form a single interconnected network of territories of the Belogorye reserve through a continuous system of connecting elements (ecological corridors, protected natural areas, buffer zones, riverine territories, etc.) and to ensure its development as a large federal museum-reserve, corresponding to the status of a particularly valuable heritage site of the peoples of the Russian Federation.

The urban development strategy for the territories adjacent to the reserve includes:

1. Expansion of the boundaries of the reserve areas to natural boundaries by changing the status of currently unused state territories.
2. Formation of conservation and buffer zones to maintain a more stable ecological status (including the renaturation of landscapes and reclamation of technogenic landscapes).
3. Formation of ecological corridors to provide relationship between the cores of the natural framework and to increase biodiversity.
4. Organization of visit centers for implementation of educational, cognitive, teaching and research functions of the reserve, as well as the development of tourist and recreational infrastructure in the territories not currently used by the reserve (Figs 1–3).

The urban development measures for the Les na Vorskla area (Fig. 1) include:

1. Organization of tourist, recreational and educational space on the basis of an arboretum adjacent to the reserve.
2. Construction of a wooden nature trail in the forest and a structure covered with perforated metal over the ravine.
3. Design and construction of a new building for a nature museum with a greenhouse in place of the old building that is to be demolished.
4. Reconstruction and adaptation of the cultural heritage site of regional significance *The house where the geobotanist, academician V.N. Sukachev worked. 1934–1941.*

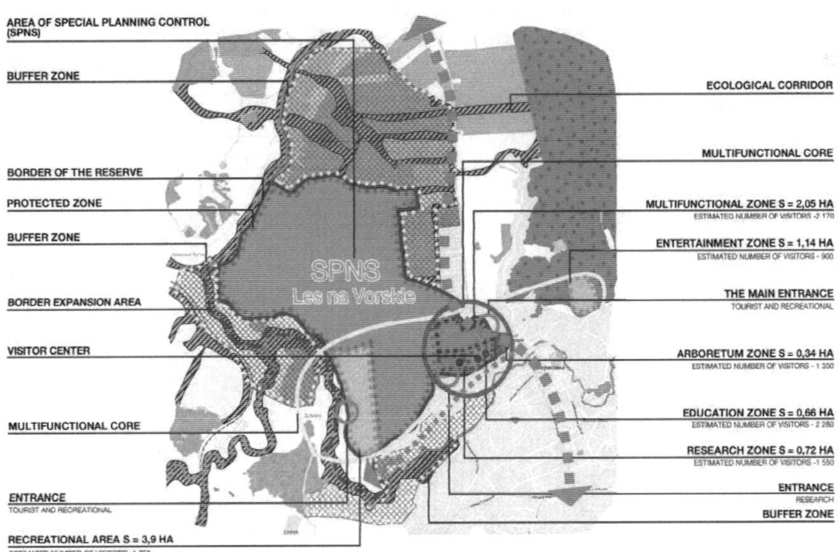

Figure 1. The strategy for transformation of the territories adjacent to the protected natural areas (Forest on Vorskla, Belogorye reserve). Developed by M.V. Perkova and O.E. Roshchupkina

The urban development of the Yamskaya Steppe site (Fig. 2) includes:

1. Organization of ecotrails for walking and horseback riding in the reserve with unpaved surface.
2. Construction of a facility over the paleosoil layer in the territory of the reserve.
3. Breeding of ungulates for preservation of identical steppe landscapes in Belgorod region (saiga,

kulan, Przewalski's horse) in the territories adjacent to the reserve.

4. Creation of the tourist and recreational infrastructure in the territories adjacent to the reserve (an asphalt road and public transport stops).

5. Strengthening of the dust protection in the Lebedinsky mining complex: treatment of the tailing dumps with chalk suspension in order to fix the surface and form stable relief – artificially created chalk mountains with identical vegetation.

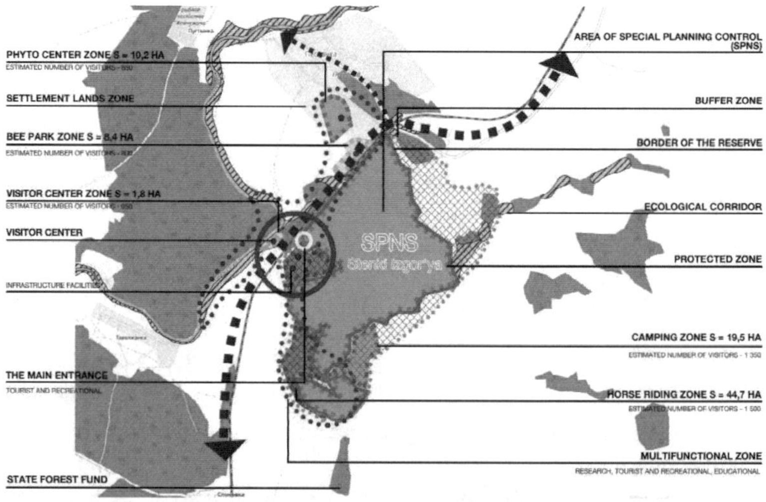

Figure 2. The strategy of transformation of the territories adjacent to the protected natural areas (Yamskaya steppe, Belogorye reserve). Developed by M.V. Perkova and O.E. Roshchupkina.

The urban development of the Stenki Izgorya site (Fig. 3) includes:

1. Construction of a nature trail in the territory of the reserve.
2. Design and construction of a hippotherapy center in the territories adjacent to the reserve.
3. Organization of a bee park in the agricultural lands adjacent to the reserve.
4. Design and construction of a phytocenter in the territories adjacent to the reserve.

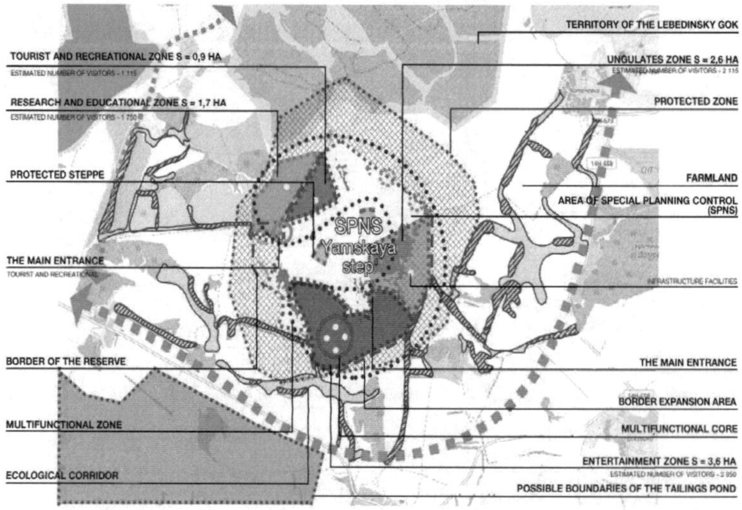

Figure 3. The strategy for transformation of the territories adjacent to the protected natural areas (Stenki Izgorya, Belogorye reserve). Developed by M.V. Perkova and O.E. Roshchupkina.

6. Conclusion

1. The classification of the territories adjacent to the protected natural areas by groups has been proposed: by the types of landscape, depending on the socio-economic function; by the type of location; by the type of urban planning conflicts; by the degree of transformation, by the economic value. This classification will take into account various interests of those involved in urban planning activities in relation to the considered area and contribute to realization of its resource potential through optimization of land use.

2. The strategy has been developed for transformation of the territories adjacent to the protected natural areas, which can be the basis for optimization of land use. It will preserve unique natural complexes; strengthen the natural framework of the region; contribute to research and educational activities in the reserve; develop the tourist and recreational infrastructure in the territories adjacent to the reserve.

3. Development of the tourist and recreational infrastructure will make it possible to create an investment-attractive environment for formation of tourist and recreational clusters of educational tourism in the territories adjacent to the protected natural areas. It will develop unique recreation and tourism forms (hippotherapy center, phyto-center, bee park) to identify the region's environment and help attract internal and external tourist flows.

7. Acknowledgements

The work was carried out as part of the Program of flagship university development at Belgorod State Technological University named after V G Shukhov, using equipment of High Technology Center at BSTU named after V G Shukhov.

References

[1] GOST 17.8.1.02-88 Nature protection. Landscapes. Classification [Electronic resource]. Information on http://base.garant.ru/5369968/ (In Russian)

[2] Gurevskikh O Y 2019 Otsenka antropogennykh modifikatsiy prirodnykh kompleksov pri landshaftnom proyektirovanii regional'nykh sistem OOPT (na primere Sverdlovskoy oblasti) *Problems of steppe science* **15** 82–86 (In Russian)

[3] Isachenko A G 1991 Landshaftovedeniye i fiziko-geograficheskoye rayonirovaniye *Vyshchaya Shkola Publishing House* 365 (In Russian)

[4] Ladik E I 2017 Regional features of formation and development prospects of the planning structure of recreation and tourism areas in the central black earth region on an example of the Belgorod region *Bulletin of BSTU named after V.G. Shukhov* **2** 111–119 (In Russian)

[5] Leonova V A 2015 Cultural landscape – concepts, history and problematics *Forestry bulletin* **19(5)** 83–87 (In Russian)

[6] Perkova M V, Baklazhenko E V and Vaytens A G 2018 Urban Conflicts of the Belgorod Regional Settlement System and its Elements *IOP Conference Series: Materials Science and Engineering* **463 (3)** 032014

[7] Perkova M V and Krushelnitskaya E I 2016 Recreation and Tourism Organization Issues in Modern Urbanization Conditions International Journal of Pharmacy & Technology **8** 24919-24929

[8] Perkova M V, Roshupkina O E, Kolesnikova L I and Vaytens A G 2019 The urban conflicts identification and resolution on the Belogorie reserve territory *IOP Conf. Series: Materials Science and Engineering* **698** 033006

[9] Prokayev V I 1980 Ob uchete antropogennoy differentsiatsii sushi pri fiziko-geograficheskom rayonirovanii *Geography and Natural Resources* **2** 24–30 (In Russian)

IOP Conf. Series: Materials Science and Engineering **944** (2020) 012029 doi:10.1088/1757-899X/944/1/012029

Possibilities to integrate wearable biomonitoring sensors into adaptive lighting systems

S V Roslyakova[1], D A Chirimisina[1], Y S Lyubakova[1]

[1] ITMO University, Kronverkskiy prospekt, 49A, St. Petersburg, 197101, Russia

E-mail: svetlana.roslyakova@itmo.ru

Abstract. There are two trends in modern lighting: the first one is lighting intelligence and adaptability, and the other is an orientation on the biological effects of light on the human body. This article analyzes the possibilities for further development of adaptive lighting systems from human biomonitoring data integration point of view. A review of wearable medical devices, which in the future could work as sensors of an adaptive lighting system, is presented.

1. Introduction

Modern lighting is no longer limited only by its functionality, such as combating darkness and prolonging daylight hours. Lighting has become an important part of a smart home, features adaptability, and can work with a person's state. The concept of «adaptive lighting» is usually defined as an intelligent system, where lighting dynamically adapts to the presence of people or current environmental conditions through data transmitted by different sensors [1]. Lighting systems do not include modules that work with objectively obtained data about a person. At the same time, due to the development of distance medicine, the first devices that work directly with a person in real-time have already appeared. This paper analyzes the possibilities for further development of adaptive lighting systems, taking into account the development trends of wearable medical devices as sources of biomonitoring data about a person.

The reason for such discourse may be the trend to develop medical devices that are becoming smaller, more convenient, and portable, and provide the opportunity to obtain reliable data outside large laboratories. According to analysts, the market of wearable medical devices is destined to grow from USD 6,231.7 million in 2017 to USD 23,310.9 million by the year 2025 [2], while smart lighting is expected to grow from USD 7.93 billion in 2018 to USD 20.98 billion by 2023 [3]. There is also a notable trend to create a human-oriented environment.

The eye is a sensory organ that reacts to light and provides vision. However, at the beginning of the 21st-century studies showed that parts of the visual system (eyes and nerve channels) are used not only to provide visual functions. The so-called non-image-forming effects of light were discovered. The non-visual effect of light also impacts our mood, concentration, vigor, alertness, mental and physical performance [4]. In other words, 3 types of light exposure on the human body are distinguished: circadian, cognitive, and emotional.

Lighting control systems were much more expensive and complicated just 15-20 years ago and are fast developing. Previously, there were a large number of restrictions in lightning systems related to the number of lighting devices, their compatibility, and control units. A big number of luminaires is no

IOP Conf. Series: Materials Science and Engineering **944** (2020) 012029 doi:10.1088/1757-899X/944/1/012029

longer a problem since there are solutions for lighting control of entire cities now [5]. The indoor lighting control system usually works by receiving a signal from illuminance sensors. The rapid development of smartphones has allowed the replacement of sensors with mobile phones, which almost everyone has nowadays [6]. The number of publications about lighting control systems is growing; for example, in the journal "Energy and Buildings" the share of such publications is about 33%. This fact indicates a huge interest of professionals in various spheres in the development and integration of lighting control systems [7]. Lighting is not just light at this stage of technology development. Lighting is a monitoring tool of the system as a whole, which is a metering system for electricity consumption. This is a way of influencing human biological rhythms [8] and tracking customers` position in a store as well [9].

The majority of modern adaptive lighting systems are based on external environmental parameters. The systems that react to the movements of a person, their presence, as well as climate change, day and night cycles are quite common. Lighting systems, which are positioned as biologically effective, are widely represented on the market. However, lighting settings that affect circadian rhythms (color temperature and luminous flux of luminaires) are changed per the programmed script [10], [11], or are adjusted by the user [12]. Adaptability in these systems is primarily associated with the use of presence/motion sensors, as well as photosensors. The adaptability, in this case, is aimed to reduce energy consumption.

As a transitional stage from external to internal parameters, there are systems that collect data from the external environment to influence the human condition. For example, an adaptive system can analyze the color of the sky to determine the correlated color temperature of natural light and bring artificial lighting closer to natural [13]. Thus, harmonization of the interaction of human circadian rhythms and artificial lighting is achieved. It is worth noting that such adaptive systems are still at the level of scientific development and have not been introduced into mass production.

There are also prototypes of adaptive lighting systems that work directly with human data received from wearable devices and sensors. In the Swedish Healthy Home project, the system collects the user's lighting history with the Daysimeter wearable device and supplements it with the necessary quantity and quality of artificial light ensuring the circadian effect. Achieved by the data integration system directly controls the lighting system or gives the users information about what kind of light they need [14].

These days scientists are considering options for integrating biomonitoring devices into adaptive lighting systems. For example, the ALADIN project aims to create a lighting system that will help older people live independently at home for a longer time and improve their quality of life. The prototype should help people achieve a state of mental activity or relaxation in specific situations by adjusting lighting [15]. Shifting the focus from external to internal parameters, a great potential is hidden in the possibilities of monitoring the human condition in real-time. The trends in the development of medical devices are wearability, miniaturization, as well as a combination of several sensors to read physiological and psychophysiological data in one device.

However, there are no examples of wearable sensors as devices that collect, process, and transmit information about a person on the smart homes market, especially on the market of smart lighting. At the same time, a large number of patents in human biomonitoring indicate the relevance and prospects of the combination of wearable monitoring devices and a controlled lighting system in one device. Patents describe lighting systems working with wearable sensors, that determine body temperature, pulse, user location, combined with external environmental factors, such as temperature and light [16]. Some biodynamic systems are described with a heartbeat sensor to control light devices when the user falls asleep and wakes up [17]. There are patents to adjust the color temperature of the source due to wearable sensors monitoring human condition [18].

To promote human health and well-being, the principles of how lighting works and its dynamic quality must include knowledge and data on the effects of light throughout the day, and light interaction with biological systems [14]. An intelligent lighting system based on biomonitoring can cope with this better than manual control, since it can perform complex human condition tracking

IOP Conf. Series: Materials Science and Engineering **944** (2020) 012029 doi:10.1088/1757-899X/944/1/012029

around the clock in real-time, and the control algorithms are based on scientifically proven patterns of light effect on the human body. Shortly (Figure 1) one small sensor, a finger patch (already developed by scientists), can become a "receptor" of the lighting system for a home or an office and provide information about a person's psychophysiological state: level of arousal and stress, circadian rhythm, concentration and performance, emotional and cognitive fatigue. At the same time, the adaptive system monitors a person through surveillance cameras and can identify complex and nuanced emotional and cognitive states when analyzing their behavioral patterns. This real-time data has not only the potential to determine the psychophysiological state of a person but also to correct or maintain it through light.

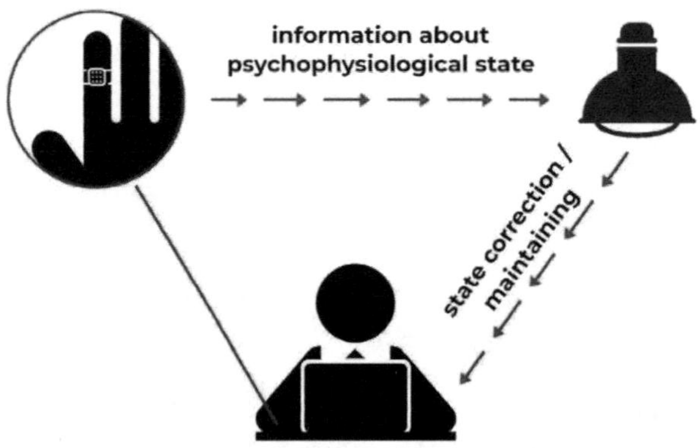

Figure 1. The lighting system monitors and reacts to a human's psychophysiological state via the "receptor" finger patch.

2. Materials and methods

The criteria to analyze wearable biomonitoring devices to select gadgets are presented:
1. Functionality. Devices should read only the psychophysiological parameters affected by light. Psychophysiological parameters, which can change when exposed to light, are:

- hormone level;
- brain signal;
- heart rate;
- electrodermal activity.

2. Portability of sensors.
3. Independence from medical centers. Mandatory data analysis and decoding by a medical institution can create a large delay in the reaction of the light system.

3. Results

Based on market analysis wearable sensors are divided into 3 groups: devices to study circadian rhythms, cognitive and emotional states of a person. The development of technologies in each group is described. The existing wearable sensors presented in the analysis can be integrated into lighting systems to collect and analyze biomonitoring data in these groups.

3.1. Circadian rhythms

The objective methods to analyze the state of a person undergo significant changes. Until recently, to analyze a person's hormonal balance when studying circadian rhythms, it was necessary to conduct research and process data only in specialized medical centers. Now electrochemical patches (portable wearable patches measuring hormone levels) can determine the effect of various light characteristics on the sleep-wake cycle of a person.

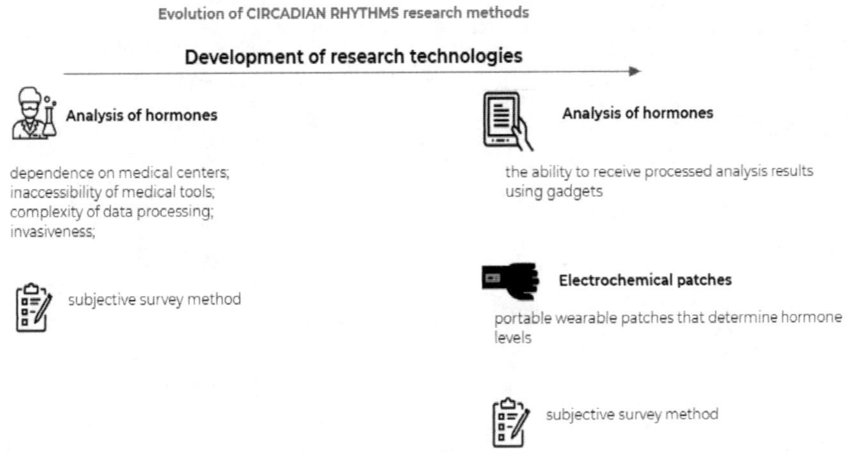

Figure 2. Evolution of circadian rhythms research methods.

For example, lab tests home kits are already available for purchase from EveryWell [19]. This is the equivalent of laboratory medical research. According to the producers, it ensures the reliability of medical data, independence from medical centers, and remote decryption. This product cannot work in an adaptive system, but its appearance demonstrates a series of changes that have affected the market for health self-monitoring.

Portable sensors that can read human condition and collect these data during the day have already been developed. An elastic wearable patch to measure cortisol level in sweat has been created at Stanford University [20]. This sensor ensures non-invasive and continuous monitoring of various biomarkers and independence from medical centers.

Eccrine Systems has developed prototypes of Sweatronics technology that facilitates the collection and analysis of sweat using wearable devices [21]. A big advantage is the continuity of data collection during sleep-wake periods, which increases the reliability of the data.

These fully autonomous sensors can become one of the main types of devices that collect and analyze the effect of light on a human body during the day and, in turn, change lighting conditions following the state of a user.

3.2. Cognitive state

Currently, there are many devices from completely different industries to research a person's level of attention and concentration under certain environmental conditions. Studying the effect of specific light characteristics on human cognitive activity is becoming easier and more widespread. Today a researcher does not depend on huge non-portable machines, can conduct tests in the natural environment, and provide subjects with a wider field of action.

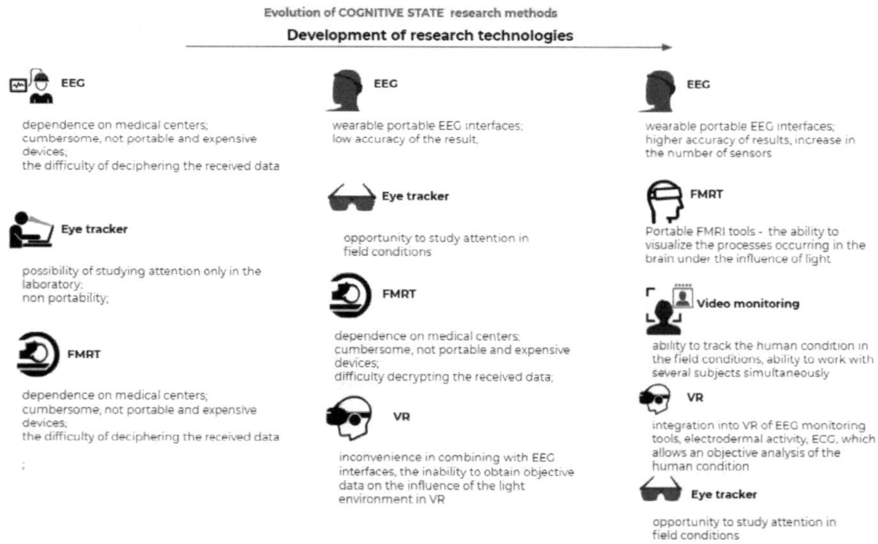

Figure 3. Evolution of cognitive state research methods.

OBELAB has created the NIRSIT, portable functional magnetic resonance imaging (fMRI) [22]. It visualizes the processes in the brain, tracks the progress and level of depression, checks the signs of emotional reactions to specific stimulations, and monitors the learning process. This tool can be used to determine the effects of light on psychological processes such as learning, memory, depression, and anxiety. This device is not unique on the market right now. OPENWATER has made the same wearable device disguised as a regular hat [23].

Another perspective tool is a surveillance camera that allows to evaluate the behavior of people on the street. Affectiva is a platform for reading the cognitive and emotional biomarkers of the human face in real-time. It can serve as a tool to determine the influence of light characteristics on the cognitive functions of a person and allows working with several subjects in parallel [24]. The authors of BrainWaveBank emphasized the possibility of measuring the cognitive states of a person based on the health study of older people [25].

3.3. Emotional state

Another set of studies of interest to light designers is the emotional impact of lighting solution or its specific characteristics on the user. Modern science helps the researchers allowing studying the human condition easily and conveniently. Sensors combine several types of data collection functions (for example, measuring the electrodermal activity of skin and pulse) and are completely portable.

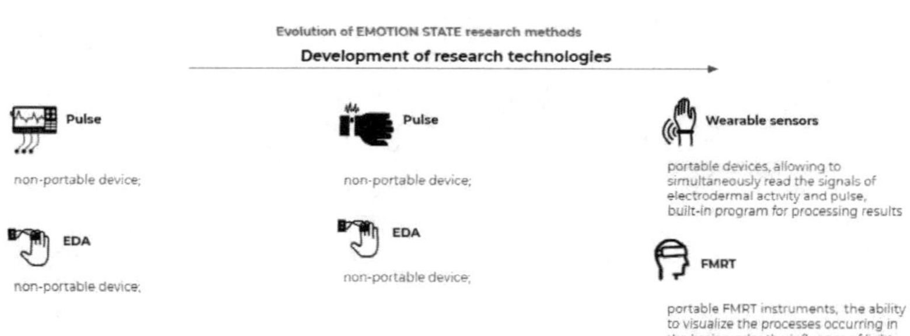

Figure 4. Evolution of emotion state research methods.

Portable electroencephalography (EEG) or fMRI devices like the aforementioned NIRSIT and OPENWATER hat can become one of the tools for obtaining data on a person's emotional state. OPENWATER provides an opportunity to study the influence of light on such psychological processes as depression and anxiety.

Feel is a gadget bracelet, integrates several biomonitoring sensors. It measures skin electrodermal activity and heart rate, which give objective data of a person's emotional state [26]. The Spire device takes into account the respiration rate, heart rate, sleep quality, and displays a person's stress level, which can be applied directly to adaptive systems.

Market analysis of portable biomonitoring devices shows that modern medical technologies provide comprehensive monitoring of a person's state in real-time. Their integration as sensors of adaptive lighting systems allows a more accurate effect on the human body, based on scientifically proven laws, and result monitoring.

Table 1. Types of portable medical devices.

State	Indicator	Type of device	Example
Circadian rhythm	hormone level	patch	Stanford's sensor
		bracelet	Sweatronics
Cognitive and emotional statement	brain activity (fMRI)	headset	NIRSIT by OBELAB, OPENWATER
	brain activity (EEG)	headset	BrainWaveBank
	facial expressions	platform for computing vision	Affectiva
Emotional statement	electrodermal activity of the skin heart rate	bracelet	Feel

4. Discussion

Being aware of the problems and threats associated with artificial intelligence, artificial intelligence sphere continues evolving, and we are entering the sphere of ubiquitous sensors. The design of such systems requires an understanding of both functional and ethical issues. Specific sensor properties impose certain restrictions on them. For example, they are hyper-personalized: if a system adapts to the psychophysiological state of several people, it will have to average the results that refute the existence of an adaptive system. Hence, the area of application is limited to indoor space.

Medical portable devices may collect various personal data. It is important to note that all medical personal data must be protected from theft and alteration, therefore data protection is a crucial point in the development of such systems. Software and data transfer systems must protect personal data and portable medical devices from hacker attacks. Hacking a diabetic's smart contact lenses to get erroneous readings can lead to their owner either not receiving warning signals or overreacting to the exaggerated glucose readings. This can not only have serious consequences but also be fatal [27]. There are several difficulties for fully running adaptive lighting systems that work with a person's medical data:

1. Portable medical devices often need to send data to a smartphone or control unit for the correct operation of adaptive lighting systems. To do this, a significant increase in the battery capacity of the device or a constant device charging is required.

2. Portable devices may have difficulty in positioning a person in an apartment. Considering the smart home concept, where the lighting varies depending on a person's medical indicators, the positioning will be easier to organize using special cameras than with portable devices.
3. There may be a delay in transmitting a large amount of information between a device, a smartphone/control unit, or a lamp since the mathematical capabilities of the device are limited. Then, the sense of adaptability is lost in principle, and the device with a long delay will not be in demand.
4. Another challenge is the construction of a "psychological" relationship between a person and the system. The possibility of choice (or its illusion) is a natural need in human life. Depriving a user of this opportunity, transferring to the algorithm the right to dictate living conditions, there is a risk of putting a person in a subordinate state, which will cause an internal (and then external) protest. The system, first of all, should not be a dictator for a person, but an adviser.

To avoid negative emotional manifestations, a person must be aware of the causal relationship between their psychophysiological state and the recommended parameters of the light environment and be able to follow or not follow the recommendations received.

5. Conclusions

The development of technology leads to a constant change in the relationship between a man and a machine. Adaptive lighting systems might be created with the integration of sensor types analyzed in this paper. Such systems have a lot of opportunities and challenges. Adaptive lighting systems with human health monitoring will develop further despite all technical and ethical problems. A balanced and skillful approach to their development with the integration of medical sensors makes lighting an efficient assistant of the person in the near future.

References

[1] Pihlajaniemi H 2016 Designing and experiencing adaptive lighting : case studies with adaptation, interaction, and participation (academic dissertation) University of Oulu
[2] Wearable Medical Devices Market to 2025 Global Analysis and Forecasts Device Type, Application and Geography. Available from: https://www.researchandmarkets.com/reports/4669217/wearable-medical-devices-market-to-2025-global/.
[3] Smart Lighting Market by Offering (Hardware: Lights & Luminaires, Lighting Controls; Software, and Services), Communication Technology (Wired and Wireless), Installation Type, End-use Application, and Geography - Global Forecast to 2025. Available from: https://www.marketsandmarkets.com/Market-Reports/smart-lighting-market-985.html/.
[4] Figueiro M, Nagare R and Price L 2018 Non-visual effects of light: How to use light to promote circadian entrainment and elicit alertness *Light. Res. Technol.* **50(1)** pp 38–62
[5] Wojnicki I et al. 2017 Application of distributed graph transformations to automated generation of control patterns for intelligent lighting systems *J. Comput. Sci.* **23** pp 20–30
[6] Tang S et al. 2017 Development of a prototype smart home intelligent lighting control architecture using sensors onboard a mobile computing system *Energy Build.* **138** 368–376
[7] Wagiman K R et al. 2020 Lighting system control techniques in commercial buildings: Current trends and future directions *Journal of Building Engineering* **31** 101342 (*Preprint* gr-qc/101342)
[8] Brainard G C and Hanifin J P 2005 Photons, clocks, and consciousness *Journal of Biological Rhythms.* **20(4)** pp 314–325
[9] Indoor positioning for superior customer service at Carrefour. Available from: https://www.lighting.philips.com/main/cases/cases/food-and-large-retailers/carrefour-lille/.
[10] Human Centric Lighting – ESYLUX. Available from: https://www.esylux.com/innovations/human-centric-lighting/.
[11] Imagine. Helvar. Available from: https://helvar.com/ru/решения/imagine/.
[12] Illustris Power. Helvar. Available from: https://helvar.com/ru/решения/illustris-power/.

[13] Catalbas C 2017 Environment Adaptive Lighting Systems for Smart Homes *Adv. Sci. Technol. Res. J.* **11(3)** pp 172–178
[14] Piippo K and Rea M 2014 Report on the Development of the Swedish Healthy Home. Available from: https://docplayer.net/9667728-Report-on-the-development-of-the-swedish-healthy-home.html/
[15] Hazzam A et al. 2011 Implementing ambient assisting technologies in elder-care: Results of a pilot study *Synesis A J. Sci. Technol. Ethics, Policy.* **2(1)** pp 27–38
[16] CN109640483A - A kind of indoor feedback lighting control system based on intelligent wearable device - Google Patents. Available from: https://patents.google.com/patent/CN109640483A/en?q=lighting+physiological+parameters+wearable+sensors&oq=lighting+physiological+parameters+using+wearable+sensors&page=2
[17] CN104411036B - A kind of method and Intelligent worn device controlling luminaire based on physiological feature - Google Patents. Available from: https://patents.google.com/patent/CN104411036B/en?q=lighting+physiological+parameters+wearable+sensors&oq=lighting+physiological+parameters+using+wearable+sensors&page=2
[18] TWM541548U - Smart lighting system - Google Patents. Available from: https://patents.google.com/patent/TWM541548U/en?q=lighting+physiological+parameters+wearable+sensors&oq=lighting+physiological+parameters+using+wearable+sensors&page=1
[19] Everlywell: Home Health Testing Made Easy - Results You Can Understand. Available from: https://www.everlywell.com/
[20] A wearable device measures cortisol in sweat. Available from: https://news.stanford.edu/2018/07/20/wearable-device-measures-cortisol-sweat/
[21] Monitoring Your Health is No Sweat with Eccrine Systems' Sweatronics Wearable. Available from: https://www.medgadget.com/2018/05/monitoring-your-health-is-no-sweat-with-eccrine-systems-sweatronics-wearable.html/
[22] OBELAB. fNIRS devices. Portable NIRS. Available from: http://obelab.com/
[23] Huth A G et al. 2016 Decoding the semantic content of natural movies from human brain activity *Front. Syst. Neurosci. Frontiers Media S.A.* **10**
[24] Home - Affectiva : Affectiva. Available from: https://www.affectiva.com/
[25] BrainWaveBank. Available from: https://www.brainwavebank.com/
[26] Biomarkers & Digital Therapeutics for Mental Health. Available from: https://www.myfeel.co/
[27] Keum D H et al. 2020 Wireless smart contact lens for diabetic diagnosis and therapy *Sci. Adv. American Association for the Advancement of Science* **6(17)** eaba3252 (*Preprint* gr-qc/eaba3252)

Minimalism as a strategy for overcoming information overload of interface design for interactive city devices

V E Ryabinina-Zadernovskaya

Saint Petersburg Stieglitz State Academy of Art and Design, Saint Petersburg, Russia

E-mail: zadernovskaya@gmail.com

Abstract. The article is devoted to the consideration of the issues of forming a comfortable urban information environment using graphic design, namely, minimalism as a strategy for organizing visual information obtained through interaction with interactive interfaces in the urban environment. In modern megacities, devices with a touch screen increasingly provide the organization of the process of information interaction between a city and a person. From a visual point of view, they provide ample opportunities for the implementation of a graphic form. However, at the same time, the question remains to what extent the visual design of the software of urban information carriers affects the efficiency of their use, and how their aesthetic characteristics form the information and graphic ecology of the city. In recent years, the information space has expanded significantly; therefore, the organization of an environmentally friendly urban graphical environment has turned out to be associated with the need to apply approaches to organizing data that can reduce information and visual noise, as well as reduce the level of information pollution of the surrounding urban environment. This circumstance, on the one hand, enhances the importance of interaction design as a primary stage in creating an effective information design. On the other hand, it promotes a conscious restriction in the use of graphical tools, reflecting an ethical approach to interface design.

1. Introduction

The central issue of creating functional graphic design is the specificity of the visual interaction between man and any mechanical system. This item has been on the agenda since the middle of the 20th century when graphic design became an independent professional field. However, for the past time, the change of social paradigm from industrial to post-information type [4; 12] has predetermined fundamental changes of understanding graphic design as an instrument of human interaction with material environment, especially urban environment.

Therefore, visual designers had relied on the achievements of their predecessors and used principles of functionalism in the practical aspect of organising human interaction with mechanical systems by designing interfaces of different devices. This basis allows them successfully creating their projects in the post-war context of industrialisation, economic, social and information development. They based on the idea of using symbolic language as a tool of mitigating language difficulties and predetermined the development of graphic design as an instrument of overcoming problems of globalisation and informatisation [13, p. 48].

Design projects of that period, especially created by designers connected with The Ulm School of Design, as well as the principles they developed during the design process, including the semiotic approach to creating interaction sign systems, have been still applicable with the advent of the information age [14]. Moreover, they have become an inspiration source for many modern design solutions driven by the idea of creating functional interfaces. Even though inner contradictions school itself provided the impetus to the understanding the design as an instrument of improving quality of life through humanisation of the relationship between man and machine.

The approach of human-centred design [5] is closely connected with the development of interaction design in the second half of the 20th century as the professional field, which concerns the creating of the interactive digital systems. However, widespread increasing informatisation and the crisis of information hyper-consumption provide nowadays the gradual shifting the focus of designers attention towards humanisation of mutual relations not so much of a man and a machine (in global understanding), as of a man and information received using machine interaction [2]. This influences significantly on the features of visual representation of information, including designers, desire for minimalism in the use of graphic tools [15]. This trend should concern the visual design of city media devices. Today it is clear, that peculiarities of the organisation of visual information provide by these devices impacts not only on personal energy forces and time spending on step-by-step problem solving but also on the psychological comfort of citizens. It makes them an essential ecological factor of information and visual environment of the city.

In many cities, the interaction between different community members and city institutions (transport, museums, shops) drives towards distance communication assisted using interfaces of different digital devices. We can expect that because of the widespread use of city interactive devices minimalism strategy in the visual design of interfaces should help reduce the informational and graphic overload of the modern citizen.

2. Materials and Methods

According to Oleg Pigulevskiy, minimalism as a strategy of visual design is based on the importance of "sensory balance in conditions of information overload and commodity abundance» [7, p. 3]. In his words, minimalism as a component of functionalism "becomes one of the ways to create comfort, preserve the integrity of modern man" [6, p. 3-4]. Minimalism has long been a distinct phenomenon in graphic design. It is widely represented in different areas of work, from single samples of print design to a complex strategy of creating a corporate identity. Moreover, it is famous all over the world [16]. Modern UI-designers also frequently turn to minimalist aesthetics because of the fact, that combination of large and well-readable typefaces, colour coding of information, intuitive navigation and flat illustrations. These factors help them to create a design, which not only makes devices more beautiful and attractive to the customers but also solves the problems of functional perception of information [2, p. 356].

At the same time, the modern tradition of creating interfaces implies the separation of interaction design and visual design. Moreover, it is crucial in this case that understanding of user goals and wishes dominates over visual characteristics of the product and its aesthetic qualities. Hence, visual product design is the realization of users' behavioural models, based on learning and anticipating their goals and not vice versa [2, p. 11]. So minimalistic design as a product of visual design embodies in a visual way an intelligent system of interaction, as well as abuse of visual effects and graphic redundancies in the design of interfaces often indicates invalidity of the interaction design.

At the same time, monitoring processes of interaction with interactive city devices connected with social infrastructure and learning its interfaces in terms of human-centred design have demonstrated that evolution of functional and aesthetic characteristics of user experience is based on the increasing influence of graphic design as an instrument of visual communication.

Figure 1 demonstrates that such graphic components as illustrations, infographics, pictograms, elements of corporate identity become increasingly common in interfaces of different payment and service terminals [6], ticket machines, ATMs, information kiosks and other similar objects of urban infrastructure in St. Petersburg museums, shops and other public places.

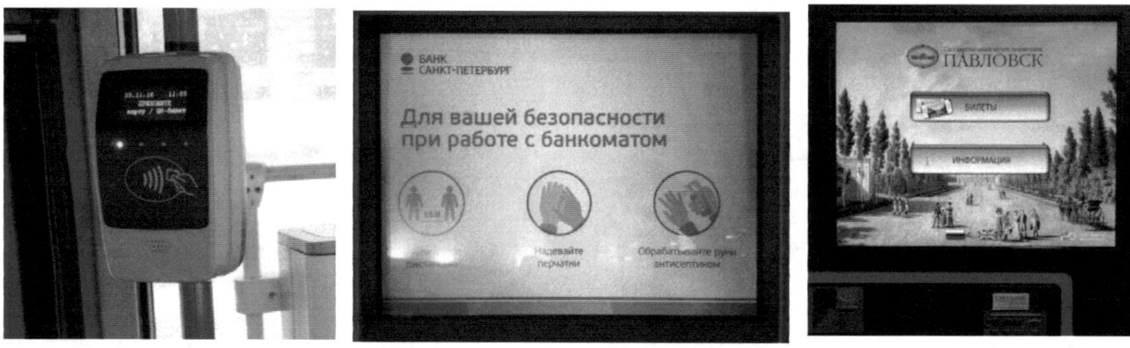

Figure 1. Interfaces of city interactive devices in St. Petersburg

According to Maxim Philippov the relevance of such graphic elements can be explained by the fact that the harmony of human interaction with computer systems "is in the area of product development of information based on perception, not text, but visual units" [10, p. 17]. However, it does not reduce the importance of the pre-project analysis, research of needs of potential users and making models of user interaction on this basis. Method of observation made it possible to reveal such problem: the more complex the system (that is the person is deeply involved in the interaction process with the device), the more often there is a shift of focus of attention of experts away from the design of interaction towards the implementation of the graphic form. Thus, visual effects replace the original logic of interaction. It significantly damages both to the communication with the device and, as a consequence, to the ecology of the informational and graphic environment of the city.

Such interfaces, which are toxic to the city environment, can be recognised by an abundance of textual information, confusing navigation, visual noise (illustrations, three-dimensional images, shadows). A significant example is a ticket machine in the St. Petersburg Metro (Fig. 2).

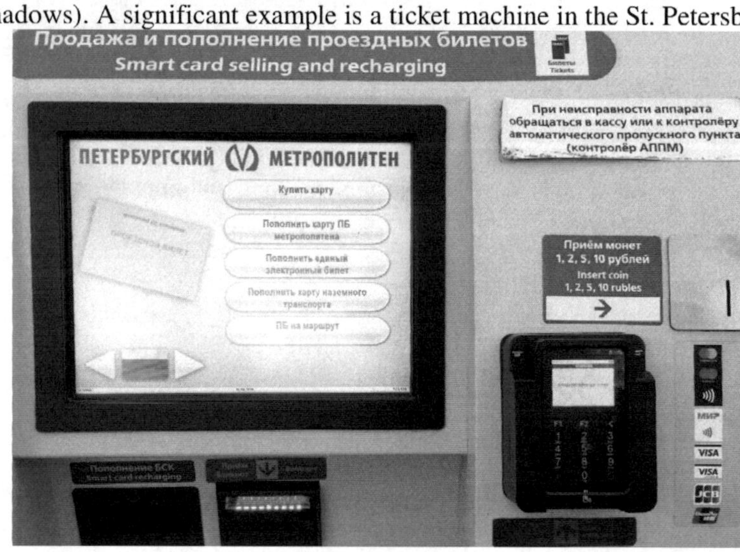

Figure 2. The interface of a ticket machine in the St. Petersburg Metro.

Display here provides complex support of a process of buying and recharging a travel ticket by connecting the person and the system. It directs the person in the process of receiving the service, determines its quality and speed, and as a result, it influences on global user satisfaction and his emotional state.

The algorithm of actions is provided by the touch screen, which offers excellent opportunities for visual design. Visual effects, however, mask interaction problems that significantly reduce the quality of service. Timur Repin, the designer of "Mobile Up" company, which is known as the expert in the developing mobile products and services, drew his attention to this problem and proposed an alternative design based on rethinking and greatly simplifying the system of user interaction with the device and as a result graphical simplification of the design of the interface (Fig. 3) [8].

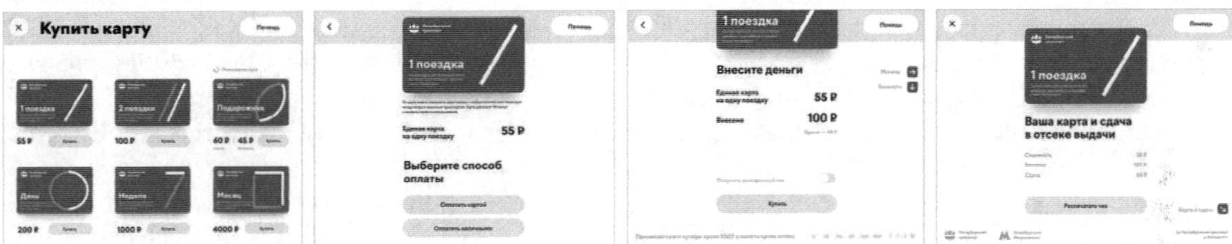

Figure 3. T. Repin's project proposal for redesigning the interface of a ticket machine in the St. Petersburg Metro.

3. Results

Following the strategy of minimalism in the user interface design of an interactive city, the device is an essential factor of the information environment arrangement. On the one hand, the limited use of graphic elements reflects a logical and functional approach to creating user-friendly information structure. On the other hand, it embodies an ethical approach to design because in the situation of total information overload, when the battle for attention is getting as intense as possible, aggressive design, which generates visual mess, is perceived not only as ineffective but also as disrespectful to the consumer.

Modern technical capabilities allowing to display on interfaces of interactive city devices many colours, various of typefaces, images and all kinds of visual effects produce an extra visual noise, which splits the focus of attention and facilitates uneconomical use of human energy, thereby preventing the receipt of important information. Thus, in the conditions of total information overload designers are additionally responsible not just to deliver a particular message by graphic means, but also to purposefully organise the entire interaction process considering the reduction of information load.

4. Discussion

In a general sense, information overload is a concept, the essence of which is that the presence of too much information leads to confusion, disorientation, and also causes difficulties in decision-making [9, 11]. As visual information dominates over other types of information, it is reasonable to consider the information overload from the visual point of view. Therefore, it is obvious the influence of the organization of the entire visual environment of the city on the level of information overload. Interaction with interactive city devices should be based on principles of respecting for a person's privacy, the specificity of perception and emotional comfort.

Visual pollution and visual noise together hurt the environment of modern cities. Not only medical experts, architects, psychologists, but also art critics and experts in visual communication pays their attention to its toxicity. According to Galina Kuryerova "the concepts of "mental pollution", "sign (semiotic) pollution", "visual noisiness" are moving from one article to another, and the "white noise" dominance which generates extreme exhaustion and satiety is recorded everywhere" [3, p. 505].

Alan Cooper also encourages avoiding visual noise. In his words "visual noise within an interface is caused by superfluous visual elements that detract from the primary objectives of communicating affordances and information" [2, p. 423].

Hence, it is not surprising that minimalism as graphic design strategy become nowadays increasingly popular. In particular, it concerns the process of information design of interfaces of interactive city devices. According to Jakob Nielsen's usability heuristics for user interface design, effective interaction design is based on principles of organization of information, which help to reduce information anxiety (visibility of system status, user control and freedom), improve understanding the content and promote better interaction with it (using uniform standards). All of these impacts on the visualization features, as it involves moving towards a minimalist design [15].

Swiss graphic designers have actively developed minimalism, as a graphic design concept, in the middle of the 20th century, and today it stays still popular. In the Oksana Vashchuk words, Swiss designers were in opposition to the mass pop culture, they "were guided by a socially-oriented worldview, choosing the categories "clarity", "responsibility" and "ethics" as the keywords" [1, p. 547]. Concise visual language not only contributes to the formation of a moving visual image but also helps to protect the consumer from the abundance of unnecessary, insignificant and distracting information. That is why minimalism strategy has gained new importance and become so popular nowadays. It helps to minimize stress, reduces the time of perception of information, promotes its better memorability, and forms ethics of design, which, due to information overload, extends to both the content and visual aspects.

5. Conclusion

The information revolution has significantly changed the quality of life of the modern man. Along with the apparent advances of the information age (mobility, data availability), we are witnessing an information explosion, which has led us to significant psychological problems: anxiety, information satiety. In the situation of information crisis, graphic designers have faced the new challenges, closely connected with the entire organization of the communication process between human and technic devices. For urban devices as a factor of comfortable city life, it is especially important, because they target multiply stakeholders.

If we consider the level of information pollution, it should be noted that the functionality of the graphic design is defined today by the quality of the information environment organization. It means a desire for information efficiency and reduction of graphic noise level simultaneously. Moreover, it makes further research focused on the abilities of minimalism as a strategy of visual design, especially crucial because it reflects a modern approach to the formation of comfortable information and visual city environment.

References

1. Vashchuk O A 2012 International graphic design style in the mirror of "Neue grafik" magazine (1958-1965) *Iskusstvoznaniye* Iss. 1/2 pp 546–564 (In Russian)
2. Cooper A 2014 *About Face: The Essentials of Interaction Design, Fourth Edition* (Indianapolis, Indiana: John Wiley & Sons, Inc.)
3. Kuryerova G G 2005 *What's in the front? Illustrated design textbook* (Tumen: Institut dizayna) pp. 504–533 (In Russian)
4. Matsevich I Y 2009 The phenomenon of "postinformation society" as an object of conceptualization *Mediaphilisophiya II. Granitsy distsipliny* (St. Petersburg) pp. 103–113 (In Russian)
5. Norman D A 2013 *The Design of Everyday Things* (NY: Basic Books; Revised edition)
6. Transportation organizer. URL: http://orgp.spb.ru/2016-335/ (In Russian)
7. Pigulevsky O V 2008 *Functionalism and minimalism in modern culture* (Rostov-on-Don) (In Russian)

8. Repin T Terminal interface in metro URL: https://medium.com/@gigapixel0000/интерфейс-терминала-в-метро-6773213e1afd (In Russian)
9. Toffler A 1970 Future Shock. (NY: Random House)
10. Philippov M V 2003 *Computer means of visual information in modern graphic culture (screen design: evolution, basic characteristics, structure)* (St. Petersburg) (In Russian)
11. Gross B M 1964 The managing of organizations: The administrative struggle (NY: Free Press of Glencoe)
12. Hunt J 1994 The Post-Information Society *The Virginia Quarterly Review* Vol 70 **1** pp 38–50
13. Moravánszky A and Kegler K R 2016 *Re-scaling the Environment: New Landscapes of Design, 1960-1980 Birkhauser Architecture*
14. Neves I C and Rocha J 2013 The contribution of Tomas Maldonado to the scientific approach to design at the beginning of computational era. The case of the HfG of Ulm. Future Traditions *1st eCAADe Regional International Workshop Proceedings* (University of Porto, Faculty of Architecture) pp 39–50
15. Nielsen J 1995 10 Usability Heuristics for User Interface Design URL: https://www.nngroup.com/articles/ten-usability-heuristics (дата обращения: 28.06.2020).
16. Tolley S 2016 *Min: The New Simplicity in Graphic Design* (London: Thames & Hudson)

IOP Conf. Series: Materials Science and Engineering **944** (2020) 012031 doi:10.1088/1757-899X/944/1/012031

Features of the design of animation graphics of educational multimedia edition on the discipline "Draft Geometry"

R P Sergin[1], V R Nigmatullin[1], T V Pechenkina[1], D G Tkach[2]

[1]Ufa State Petroleum Technological University, Ufa, Russian Federation
[2] Kosygin State University of Russia (Technologies. Design. Art), Moscow, Russian Federation

info@rusoil.ru

Abstract. The article describes the theoretical foundations and practical methods of designing animation graphics for educational multimedia publications on the subject of descriptive geometry. The use of animated graphics as an effective means of visualizing volumetric-spatial models is considered. The effectiveness of the educational multimedia edition with elements of animated graphics in the educational process is investigated, a comparison is made with the traditional methods of teaching the discipline.

1. Introduction

The analysis of modern trends in teaching graphic disciplines in universities shows that there has been a qualitative shift towards the widespread use of computer visualization tools. In computer visualization tools, animation graphics occupies a special place, being, undoubtedly, the most effective means of demonstrating complex spatial models and visualizing information.

The article is based on the practical experience of the staff of the Department of Integrated Engineering and Computer Graphics of the Ufa State Petroleum Technical University, in the development of a multimedia publication using animated graphics, in the discipline "Descriptive geometry".

The study guide designed and created by the authors is a multimedia edition consisting of a series of thematic visualization lectures totalling one hundred and ten minutes. In each lecture-visualization, the viewer (student) is accompanied by a character (virtual teacher of the course, actor). All course material is broadcast via high definition video. The visual image of the teacher, captured in advance, was transferred by video filming into the three-dimensional virtual space of the video scene, in which the animation graphics are the connecting link of all elements of the composition of the multimedia publisher. In such a kind of virtual classroom, the character plays a double role - a teacher and an operator, interacting with the graphic images of the multimedia edition, methodically consistently revealing the content of the course to the recipient (student).

The volume of video scenes in which animated computer graphics were used exceeded 80% of the total time. It takes many resources to create and produce such an impressive amount of animated graphics. Moreover, since this directly affects such basic consumer properties as the ergonomics of the visual environment, the quality of functionality and aesthetic properties, all these challenges could not be resolved without an integrated interdisciplinary design approach.

2. Materials and methods

One of the most effective methods of teaching graphic disciplines is lecture-visualization, which is most consistent with the specifics of teaching these disciplines [1]. In traditional lectures, various kinds of demonstration material are used: posters, layouts, separate illustrated images, volumetric models, each of which performs its tasks and functions [2]. It is necessary to take into account many formative, compositional and stylistic features of creating a dynamic (unfolded in time) composition of a multimedia publication for integrating of all this material in a completely different format, where motion and sound are added to the usual, static graphic form. Thus, the authors were faced with the task of finding optimal solutions in matters of content-functional, design and figurative-aesthetic plan. Figure 1 shows a fragment of a scene from a multimedia educational publication on the topic "Axonometric projections".

Figure 1. Example scene from the course

3. Analysis of the content and functions of animation graphics of educational multimedia edition

The solution to these tasks is based on the use of the following three main types of animation graphics, which differ in technical means of production, functional purpose, and visual and aesthetic characteristics:

1) **Volumetric 3D**, which belongs to the category of animated images, the graphics of which are realized mainly in 3-dimensional space. This type of animation graphics was the primary tool for creating dynamic compositions for educational multimedia publications. Thanks to the means of three-dimensional animation, basic tasks of demonstrating volume-spatial models were performed, as well as the rotation of objects in space, visualization of sections and intersections of objects (Fig. 2).

2) **Plane 2D,** which represents animated graphic, digital images that have only two values in the visual space of the scene – height and width. This type of animation graphics based on production technology is more reminiscent of traditional frame-by-frame animation. A characteristic difference from three-dimensional animation is linear animation of the graphic elements of the composition when the movement of the graphic element occurs in one plane. When developing an educational multimedia edition, this type of animation graphics was used mainly for creating navigation and indication elements (pointers, arrows, highlighting and highlighting the necessary elements of the composition). The primary function of such animation is navigation, due to which the viewer's attention is transferred from one graphic element to another.

IOP Conf. Series: Materials Science and Engineering **944** (2020) 012031 doi:10.1088/1757-899X/944/1/012031

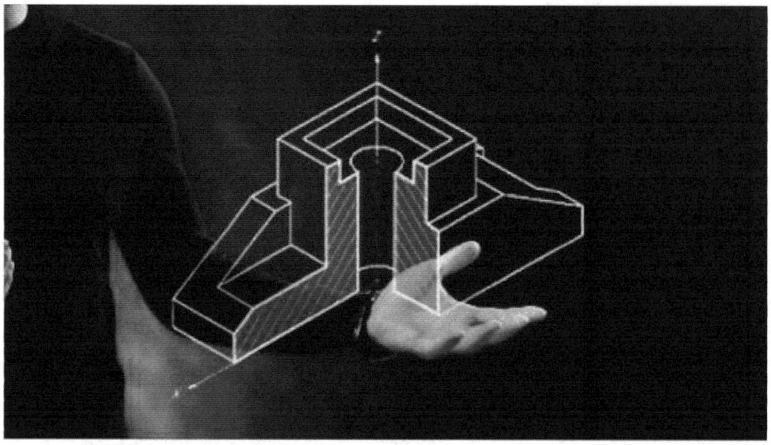

Figure 2. Animate an isometric section view of a part

3) Combined graphics, which combines visual, aesthetic and functional properties of both volumetric and flat two-dimensional animation, depending on the compositional tasks to be solved. In the video scene of the lecture-visualization, a comprehensive communicative complex of animated graphics is involved, illustrations, text, photographs, video and interactive interface elements are combined into a common plastic form and unfold in the dynamic composition of a multimedia edition (Fig. 3).

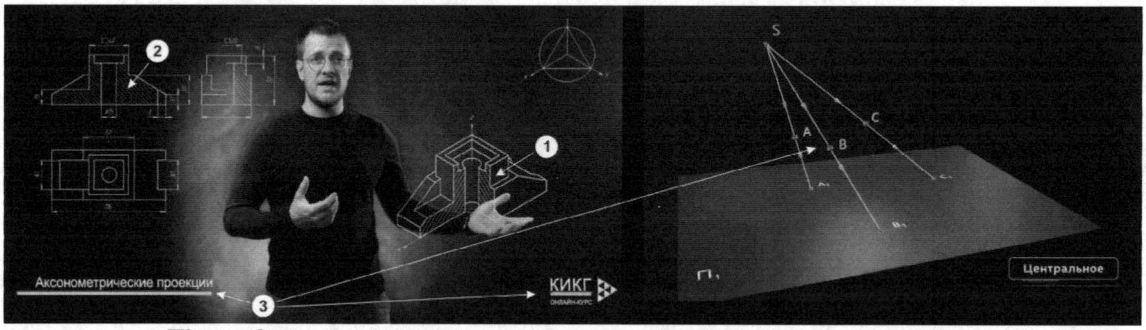

Figure 3. Types of animation graphics 1 - volumetric, 2 - flat, 3 - combined

4. Design principles and algorithm for creating animation graphics

Since the effectiveness of mastering the educational material of an online course depends on the quality of the communicative properties of the designed multimedia edition, then for the designer of animation graphics, the successful choice of graphic means and the construction of a sound-visual composition of the frame becomes extremely important. The success of this choice is determined by an integrated design approach – from the concept stage to implementation.

Project life cycles: sketching – at this stage, ideas of the concept of animation graphics were formed and formulated, the pedagogical experience of teaching graphic disciplines using "lectures-visualization" was studied; Prototyping – preparing a test version of an animated graphic model. Figure 4 shows a prototype of the "Lecture-Visualization" model made using animated 3D graphics. At this stage, the technical visualization tools were worked out; the Cinema 4D software package was chosen as the most optimal solution, it combined the convenience of modelling three-dimensional graphics with the ability to integrate tools and animation and video editing.

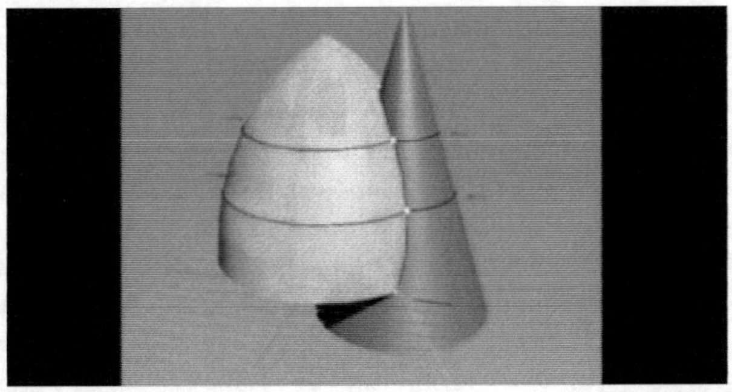

Figure 4. Lecture-visualization animation model prototype

Testing was carried out in a real educational process; at this stage, an experiment was carried out with the introduction into the educational process of fragments of lectures-visualizations using animated graphics. The results of the testing were recognized as successful and were detailed by the authors in the materials of the previous article [3].

The production is a project cycle divided into three main phases. The pre-production stage consists in preparing a calendar plan for the implementation of the project, a scenario of interaction with a working group of specialists (designers, animators, operator, director, scriptwriters, authors) was described. At this stage, a complete storyboard was prepared, which is a sequence of graphic frames presented in the form of pictures describing the content of the scenes, indicating the volumes of animation graphics, the number of combined scenes (video with a person in the frame and animation) and scenes with full animation visualization (without a person in the frame). At this stage, estimates were also made for video production. Figure 5 shows a drawing of a portion of a storyboard sheet.

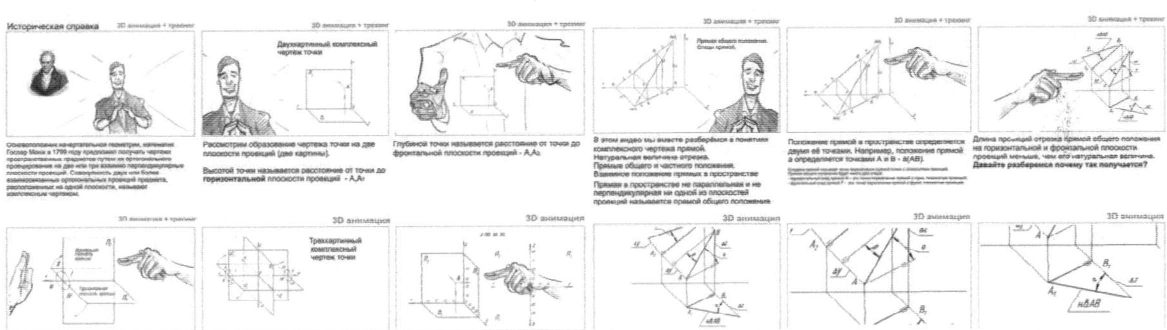

Figure 5. Fragment of project storyboard

The production stage is the second stage. At this stage, filming work was carried out; video footage was filmed, then a primary animatic was prepared, which is a sequence of video with a person in the frame and images of animation thumbnails. This made it possible to correct the timing (broadcast time), as well as to make corrections in the functional and aesthetic properties projected objects of animation graphics.

The post-production stage is the third stage. At this stage, work was carried out on computer animation tools; graphic source video files were processed using Motion tracking technologies (motion capture, the process of recording the movement of objects or people), the illusion of holography of animation graphics elements in the frame was created, as well as work on colour correction and creation of special effects in After effects software package.

IOP Conf. Series: Materials Science and Engineering **944** (2020) 012031 doi:10.1088/1757-899X/944/1/012031

5. Analysis of figurative and aesthetic properties

Since the primary expressive and pictorial means of building the composition was animation graphics, the authors paid much attention to the spatial-visual environment, in which the elements of the animation graphics were placed. For this, the parameters of the mise-en-scène construction were taken into account, combining the synthesis of direction, camera work and sound-visual effects [4]. In most of the scenes, the person (the course teacher) placed in the frame performed the functions of an operator, didactically guiding the viewer along with the internal elements of the scene, manipulating various graphic models in the virtual space of the stage. Pearson also performed the functions of an actor who introduced the necessary plastic accents into the composition of the frame. So, for example, for better contrast and readability of animation graphics elements, the structure of the frame's mise-en-scene was taken into account; colour and light, sound effects (noises) and musical accompaniment, word and movement. All these components were aimed at creating a harmonious visual and sound environment [5].

Colour and light were technically realized with the help of coloured soffit lighting, which was an important part of the composition; lighting schemes were selected in such a way as to give depth to space and to unite all the elements of the composition stylistically into a single harmonious connection. A backlight (key light) was used for the main cut-off picture. Modelling or fill light was used for softening the shadows created by the key light.

Sound is an essential part of dynamic composition, used for indicating various states. For example, to complete the construction of communication lines in a graphical plot, the technique of sound accompaniment was used, which noticeably increased the attention of students in the process of watching a lecture-visualization [6].

Word and movement is the integration of word and movement in the frame. The project actively used elements of kinetic typography (font animation) with the appropriate soundtrack. It should be noted that kinetic typography plays a vital role in animation design, combining a complex of communicative, artistic, and expressive means [7].

Stylistics of animation graphics. In this project, a synthesis of "line art" and "isometric" style was used, this combination allowed effectively combining the graphic aesthetics of drawings with volume-spatial forms of axonometric projections. These stylistic properties were also embodied in the visual packaging of the educational multimedia edition (design elements, splash screens, information windows, scene navigation elements, and font).

6. Results

In practical terms, thanks to the means of computer animation, the effect of realism were achieved. The effect of realism allows the teacher (operator) getting the opportunity to virtually move graphic objects in space, demonstrating their various properties.

In the synthesis of anthropomorphic, textual and graphic images of movement, animation graphics are the primary means of expression and a connecting link that unites all elements of the composition of an educational multimedia publication.

The following criteria for the quality of animated graphics were identified and achieved, affecting the speed of learning the material and visual ergonomics of the educational multimedia edition:

• Visibility – communicative properties when demonstrating volumetric-spatial models;

• Consistency – hierarchy of graphic elements of animated graphics that make up the overall compositional structure of an educational multimedia publication (3D animation, 2D animation, combined);

• Variability – the versatility of the use of animated graphic elements of the educational multimedia edition was taken into account for various areas of training (architects, designers, engineers);

• Aesthetic – the formation of a visual image of an online course with its unique style (splash screen, identity, packaging, font).

7. Conclusion

Visualization lectures with elements of animated graphics have a high degree of relevance and relevance at the present stage, as they are the optimal tool for distance work and online learning.

Acknowledgments

As part of the Ufa State Petroleum Technological University leadership grants competition 2020, the authors received an internal grant for the development of an online multimedia course, which is currently at the completion stage. This project will create a useful tool for the development of online education within the open online courses "Oil and Gas Education" in the Ufa State Petroleum Technological University.

We are grateful to the Rector of the Ufa State Petroleum Technical University Baulin Oleg Aleksandrovich.

References

[1] Stasenko M S and Ryazanova T N 2017 Specificity of teaching the disciplines "descriptive geometry" and "engineering graphics" as a multifactorial social process of ensuring a high level of professional training of future specialists *Young Scientist* **47** pp 227–232 URL: https://moluch.ru/archive/181/46739/ (date accessed: 18.06.2020).

[2] Pechenkina T V 2020 Personal experience in the use of workbooks and methods of teaching descriptive geometry at the university // *Modern science: theory and practice. Dushambe, Scientific Research Center of Science* p 131

[3] Pechenkina T V and Sergin R P 2019 Search for modern methods of teaching descriptive geometry and analyzes of their application // Materials of the International Scientific and Technical Conference **13** (Ufa: USPTU Publishing House) p 413

[4] Buntasanakul S and Dvorko N I 2014 Interactive documentary filmmaking: new possibilities of the digital environment and features of the director's activities *Historical and social educational thought* **3** URL: https://cyberleninka.ru/article/v/interaktivnaya-dokumentalistika-novye-vozmozhnosti-tsifrovoy-sredy-i-osobennosti-deyatelnosti-rezhissera

[5] Lugovtsev A Yu 2017 Design of animation objects and characters in the conditions of convergence of screen arts **10** **(2)** pp 93–98 URL: http://www.gramota.net/materials/3/2017/10-2/22.html

[6] Goryunova I E 2011 Problems of audiovisual synthesis in creative multimedia products (on the example of modern mass shows) *Bulletin of the Russian State University for the Humanities. Series: Literary criticism. Linguistics. Culturology* URL: https://elibrary.ru/item.asp?id=17356855 (date of access: 12.07.2020)

[7] Sitnikova L D 2013 The use of animation and computer graphics in the educational process *Humanitarian vedomosti TSPU named after L.N. Tolstoy* URL: https://cyberleninka.ru/article/n/ispolzovanie-animatsii-i-kompyuternoy-grafiki-v-uchebnom-protsesse/viewer (date accessed: 12.07.2020).

[8] Burkova S P, Vinokurova G F and Dolotova R G 2016 Analysis of the use of the electronic course "descriptive geometry and engineering graphics" *Modern problems of science and education* **3** URL: http://science-education.ru/ru/article/view?id=24536 (date accessed: 07/27/2020)

IOP Conf. Series: Materials Science and Engineering **944** (2020) 012032 doi:10.1088/1757-899X/944/1/012032

The concept of sustainability management of the ecosystem of cities and small settlements

N Sevryugina[1], A Apatenko[1], E Voitovich[2,3], N Kozhukhova[4]

[1] Russian State Agrarian University – Moscow Timiryazev Agricultural Academy, Timiryazevskaya St., 49, Moscow,127550, Russia
[2]National Research Moscow State University of Civil Engineering, Yaroslavskoye Shosse 26 Moscow, 129337, Russia
[3] Research Institute of building physics of the Russian Academy of architecture and building sciences, Lokomotivny passage, 21, Moscow, 127238, Russia
[4] Belgorod State Technological University named after V G Shukhov, Kostyukov St., 46, Belgorod, 308012, Russia

E-mail: kozhuhovanata@yandex.ru, e.voitovich@mail.ru

Abstract. Critical risks of ecosystem identity violation in the structure of landscape-recreational formation of the environment of cities and small settlements are identified. The dependence of the functional load of urban areas with industrial and civil facilities and the ability to restore the ecosystems of the adjacent small rivers and reservoirs is established. The necessity of modeling urban areas based on the “attraction zone” and “exclusion zone”, the verification of technical solutions for engineering systems of water protection complex harmonized with the natural environment is substantiated. A conceptual model for controlling of the risks of developing urban areas is proposed, that is focused on the mechanism for managing water body renovation programs taking into account risk factor analysis. A technique has been developed for restoring water bodies to the level of harmonized functioning by means of anticipatory diagnosis of risk-failures of the danger of environmental disruption. The introduction of digital indicator that determines the level of harmonized stability of functioning of engineering systems of water protection complex in the environment as an additional indicator of the economic and environmental activities of the regions will improve the accuracy of assessing environmental benefits and losses in planned projects.

1. Introduction

The landscape topography of urbanized territories includes various water bodies, technical and engineering structures that deharmonize the state of the natural environment. Given the globality of such deharmonization, it is clear that the creation of a mechanism for environmental sustainability management systems is required.

Don Goldstein [1] reported options for adaptive strategies at the regional level that can be identified and evaluated using a scenario-building methodology that includes the interaction between several variables and the actions of decision-makers over time.

Kerstin Hötte [2] studied the problem of slow involvement of innovative "green" technologies. Using the ecotechnological extension of the macroeconomic agent-oriented model Eurace@unib, it is shown how the effectiveness of various climate policies depends on the type and strength of diffusion barriers.

The investigations [3] present the results of empirical study on theoretical relationship between innovation in the field of "green" technology and efficient use of natural resources. The mechanism of influence of "green" technological innovations on efficiency of the use of natural resources is clarified based on the method of stochastic border analysis; positive and negative factors of "green" technological innovations are analyzed.

Andreas Schäfer [4] shown the relationship between population dynamics, technological change and the depletion of natural resources. The factors were studied in evidence-based form, that allows taking into account signs of the intertemporal secondary effects of R&D with respect to existing technological knowledge. Also, it is shown that an increase in the relative productivity of research or educational subsidies contributes to the long-term growth of technology.

Chris William Callaghan [5] proposed four technological scenarios for identifying and determination of key issues associated with the spread of hazardous technologies.

The research of Xiang Xiang Sun et al. [6] was devoted to study the market segmentation and mechanism of its influence on the environmental performance of electric power industry from point of view of technological innovation. It has been proven that market segmentation hinders the way to technological innovation. It has been established that this fact manifests itself more significantly in provinces with weak institutional quality.

It was found by [7] that technological innovations benefit society by improving communication and productivity in the supply chain. Although, technological innovations provides more opportunities for environmentally conscious choice and better infrastructure, its availability is ambiguous. Further research suggests considering efficiency as a key factor to decide on the promotion of certain technologies to add value to the social good. Similar studies have been conducted in related fields [8-12].

2. Materials and Methods

The analysis of regulatory acts for a ten-year period was aimed at finding out achievements and problems when the implementation process, as well as setting tasks for the future, taking into account existing technological innovations (The Decree of the Government of Moscow №1004-PP edited of October 28, 2008 «About the city target medium-term program for the remediation of small rivers and reservoirs in Moscow city 2009-2011»; The Resolution of the Government of the Moscow Region of 25.10.2016 N 795/39 (edited of April, 24, 2018) «On approval of the state program of the Moscow region «Ecology and the environment of the Moscow Region» for 2017-2026»).

The key aspect of these programs is the task of scientific and methodological support of a complex of actions for restoration and remediation of water bodies as well as to improve the management, operation and protection of water bodies.

Technically, this problem can be solved by taking the following actions:

- reduction the load on water bodies by developing programs that exclude changes in the water regime and water quality;
- rehabilitation of water bodies and coastal zones by clearing channels, dredging, designing the coastline, reconstruction of hydraulic structures, etc.;
- optimization of drainage networks through structural reconstruction and modernization of water bodies.

To achieve the proposed result, it is required to develop a scientifically and technically justified complex of actions focused on the formation of urban wastewater for an indefinite period of implementation; with a designated period of remediation; harmonized activity and advancing risk factor consideration of the danger of environmental disruption, i.e. long-term stable improvement of water quality in water bodies.

Hypothesis of the study: due to a large number of water bodies in the structure of landscape-recreational formation of the environment of cities and small settlements, taking into account the

IOP Conf. Series: Materials Science and Engineering **944** (2020) 012032 doi:10.1088/1757-899X/944/1/012032

complexity of the engineering systems that arise during reconstruction, it takes a long time and significant costs to complete the remediation of water bodies water system, the remediation program requires the simultaneous resolution of many related programs. This requires a structural modernization of the system, which provides targeted regulation of tasks for the remediation of water bodies and coastal areas.

Study goal: harmonization of ecosystems and landscape-recreational urban areas through the conservation of green planting and the remediation of small rivers and water reservoirs.

Methods: the studies are focused on the integration of the theory of large systems and probability theory with digital regulation mechanisms, which allow composition a control path adapted to correction in case of deviation when creating a mechanism for controlling the ecosystem of functioning of water bodies burdened by the urban environment

3. Results

3.1. Justification of modeling of urban areas based on the «attraction zone» and «exclusion zone»

The algorithm of engineering measures for the reconstruction of urban water bodies characterizes the systematic use of all available theoretical and practical innovative resources of the object formation of a sustainable environment with improved technical and environmental conditions [13-18].

The systematicity of the problem involves the use of all theoretical and practical, innovative resources. Integration of the theory of large systems and the probability theory with digital control mechanisms allows creation of controlled path with possible correction of it in case of deviation (Fig. 1).

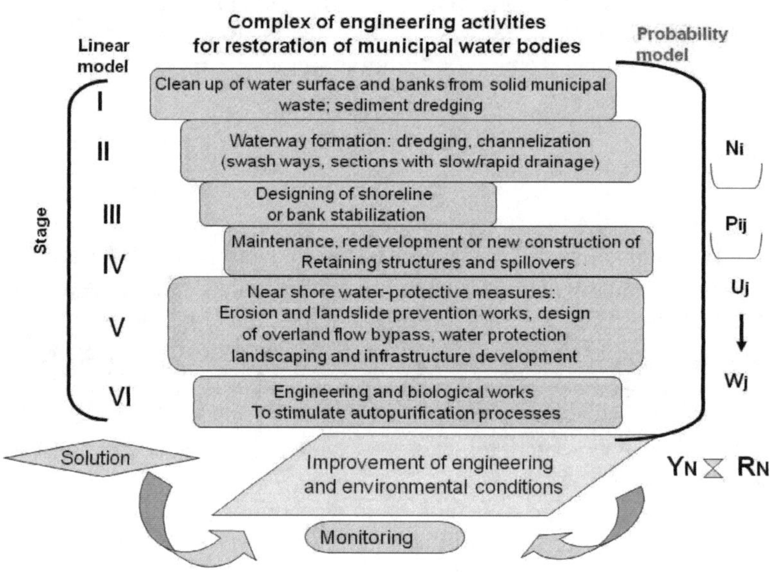

Figure 1. Modeling algorithm of rehabilitation measures for restoration of water bodies

The modal analysis of algorithm presented in Fig 1, includes the following parameters:

- probability of achieving the objectives;
- control activities;
- corrective actions;
- stage of the program realization.

The fixed state of the system during the design period of remediation is presented as the initial A stage, with the idealization of the linear translation to the key B stage.

Linear model allows description for implementation of each stage. The probability model of risky failures R_n for development of the programs $\overline{Y}_n$ is expressed by the equation (1):

$$R\{Y = 0\} = R_1 + R_2(1 - R_1) + R_3(1 - R_1)(1 - R_2) + ... \quad (1)$$

Modeling of probability hypotheses of development

To increase the accuracy of the model convergence, some hypotheses of development scenarios A_n were discussed.

For each hypothesis significance level, a weight characteristic w_n calculated discretely with a step, limited by the gradation $h = \frac{1}{n}$ of the number of hypotheses were assigned.

The numerical value of the weight indicator was calculated with the following equation:

$$\left\{0, \frac{1}{n}, \frac{2}{n}, ..., \frac{(n-1)}{n}, 1\right\}. \quad (2)$$

3.2. Verification of technical solutions for engineering systems of water protection complex, which are harmonized with the natural environment

Normally, risk theory offers an assessment of acceptable risk from the point of view of the mathematical expectation of damage when implementing a scenario of dangerous situations. Quantitatively, the value of acceptable risk is determined by the p^z parameter, and the actual result (p_i) of the hazard level, that is compared with the normalized value and indicates on the criticality of the situation. To implement the program of remediation of water bodies at all stages, the adaptation of the basic model of risk assessment, calculated by the following equation was proposed:

$$R_i = \frac{M[U]}{M[A_n] - M[C_n]} = \frac{\sum p_i U_i}{M[A_n] - M[C_n]}, \quad (3)$$

where U_i is physical damage in case of a negative scenario for–implementation of the program stage;

$M[A_i]$ is the mathematical expectation of a positive result when implementation of the program stage, the monetary equivalent of economic efficiency;

$M[C_i]$ is mathematical expectation of resource investments (costs) for implementation of the program stage.

It is a commonly accepted design parameters, estimated in monetary terms, to be presented over a period of time per year, by introducing an indicator of the differentiated vulnerability of objects V_{ij} from the complex realization of hazards at various stages of the life cycles of objects.

The socio-ecological and natural-ecological risk of the jth damage caused to the ith water body in the spatial boundaries of the territories from a hazard situation is mathematically described by the following model:

$$R^p_{ij} = V_{1i} V_{2i} V^p_{ij}, \quad (4)$$

where $V_{1i};V_{2i}$ are probabilities when implementation of a complex of hazards during the annual cycle; V_{ij}^{p} is differential vulnerability of objects from the integrated implementation of hazards during the remediaton of water bodies.

Considering the importance of the objects, development of a model for advanced development of situational hazard elimination when implementing programs to prevent factor risks was used as an alternative to the situation of eliminating consequences in the case of a negative scenario.

Calculation methodology for assessment the degree of damage caused to water bodies consists of the cost of restoring the disturbed state. It should be noted that this value increases with prolonged and intensive exposure to pollutants over time.

Using this assessment methodology (Ministry of Natural Resources and Ecology of the Russian Federation. Order of April 13, 2009 N 87 "On approval of the methodology for calculating the amount of harm caused to water bodies as a result of violation of water legislation" (amended on August 26, 2015)), an algorithm to control the level of contamination by introducing a multiplex digital indicator was applied.

There is a number of parameters in this technique that serve as input indicators:

K_{season} is a varied calculated coefficient that accounts for climate and environmental conditions within a given season;

K_{w} is a varied coefficient that determines environmental factors as well as conditions of water bodies;

K_{in} is an indexation coefficient that determines an inflationary component of economical development;

$K_{i.p.}$ is a coefficient that accounts for intensity of negative impact of pollutants to a water body. A coefficient with the values of 1, 2 and 5 defines an exceedance of pollutants with III-IV class of hazard up to 10 times, from 10 to 50 times and more than 50 times, respectively;

K_{dur} is a coefficient that defines duration of negative impact of pollutants to water bodies without taking any remedial actions. A coefficient value of 5 describes a solubility of pollutants in water;

K_{pollut} is a coefficient that stands for a level of contamination of water body polluted with municipal and industrial waste;

Q_{w} is a volume of water required for remediation to prevent an atrophy of water body and taken as a double volume of non-recoverable take out of water from the water body;

Amount of pollutants in total volume of water body can be determined based on the equation (6):

$$M_i = Q(C_{fi} - C_{Ni})T \cdot 10^{-6},\ \text{t.}\quad M_i = Q(C_{fi} - C_{Ni})T \cdot 10^{-6},\ \text{t.} \tag{6}$$

Q is a sewage flow rate, m³/hour;

C_{fi} is an average effective concentration of a substance in sewage water for a calculated time period, mg/dm³;

C_{Ni} is a permissible concentration of a substance within a limit of allowed volume of discharge for a calculated time period, mg/dm³;

T is a duration of sewage discharge with elevated amounts of contamination, from the moment of discharge indication to its complete discontinuation, hours.

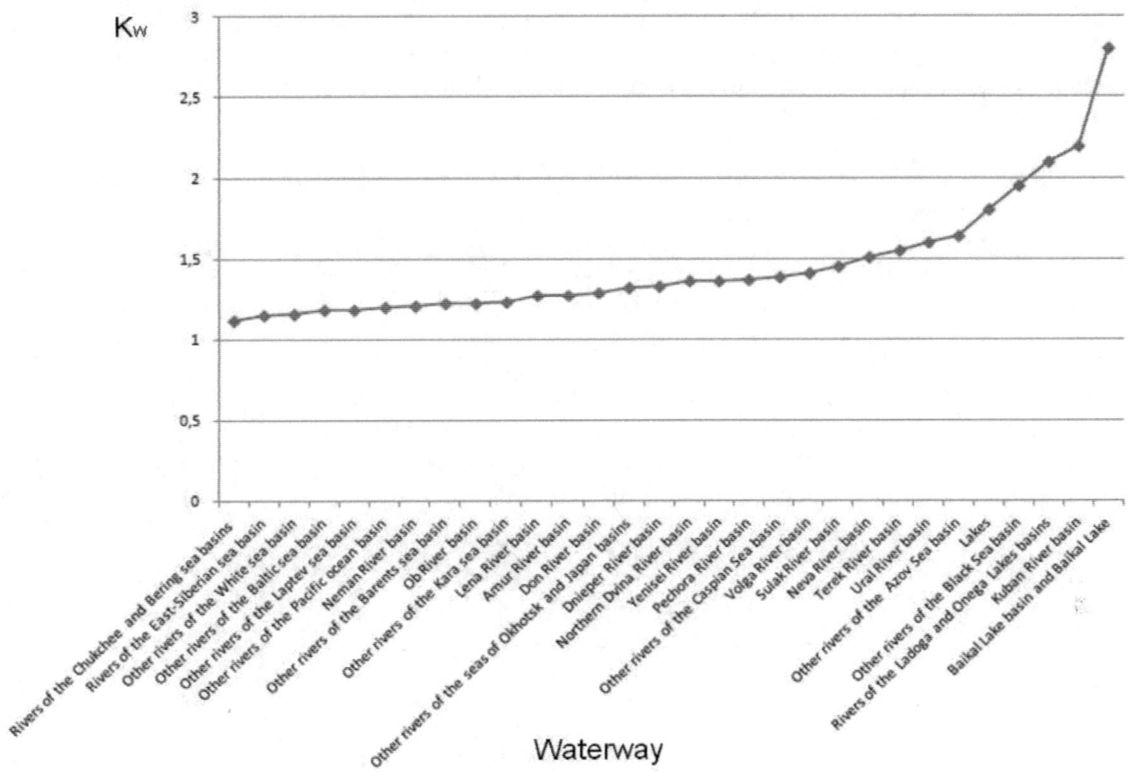

Figure 2. Coefficient that determines environmental factors (K_w) (state of water bodies) of modeling remediation measures for the restoration of water bodies

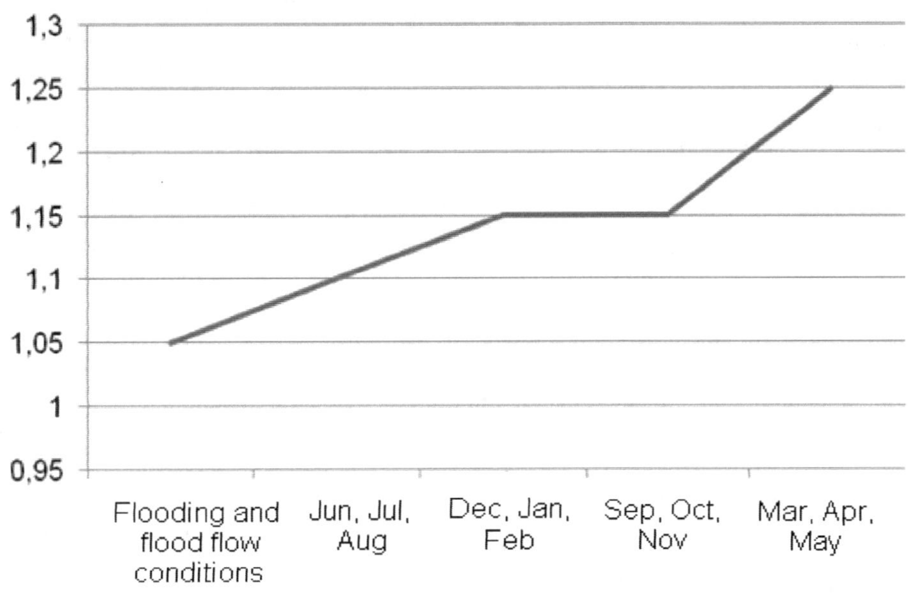

Figure 3. Coefficient that accounts for climate and environmental conditions within a given season (K_{season})

The forecast model is created with the correction of the result taking into account the temporary factor technospheric impact on the ecosystem. It is also necessary to carry out analytical adjustments

IOP Conf. Series: Materials Science and Engineering **944** (2020) 012032 doi:10.1088/1757-899X/944/1/012032

of the complexity of the impact of related areas and infrastructures on water bodies, i.e. it is necessary to solve volumetric problems with the maximum allowance for the factor influence of the state-forming system of water bodies and adjacent territories (construction of probability-determined and probability-statistical models of mathematical analysis).

3.3. The mechanism of scientific and technical support activities focused on formation of sustainable urban ecosystems, including coastal zones and municipal effluents

Nowadays, specialists are faced with the situation of the already implemented negative scenario of the impact of territorial development on water bodies. In this regard, it is necessary to assess the environmental damage and establish the level of shift from the initial natural state of the water body.

Development of a forecast remediation program includes the selection of instruments for a technical assistant with a "zero" additional environmental load during the program implementation.

It seems clear that the predictive model designed by the information-weighted analog algorithm will not allow for consideration of many specific factors for a particular object.

With respect to the complexity and significance of this task, it is proposed to apply an individual remediation simulation programs to restore the environment. Algorithm of the designed model is shown in Fig. 4.

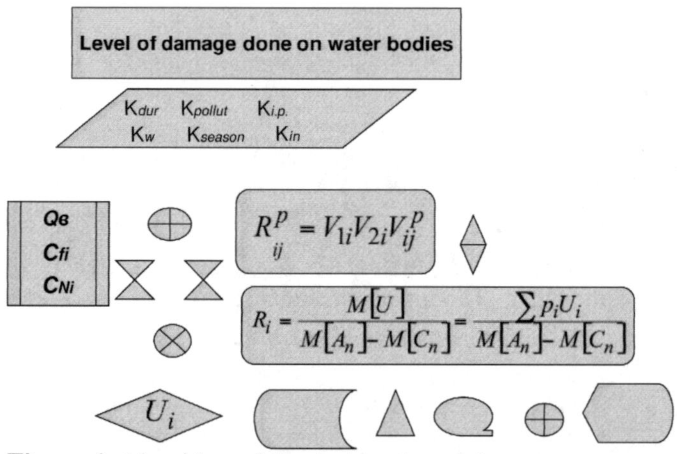

Figure 4. Algorithm of the designed model

The «road map» of the mechanism of scientific and technical support for activities focused on providing the sustainability of the ecosystem of the urban environment, including coastal zones, municipal effluent consists of combination of deterministic-probabilistic modeling and probabilistic-statistical analysis.

3.3.1. Deterministic-probabilistic modeling [18]. The model is considered as a complex of predictive elements which are grouped into deterministic and probabilistic category. Predictive variability is described by periods of advance control of parametric change in the value factor. Probabilistic-statistical processing of the obtained results of assessing the criticality of the state of a water body is carried out; the relationships and the nature of the influence for a time period and a factorial feature are established. Then an extrapolation of a continuous series of values is performed, including the identification of the level of influence and degree of hazard for the future period. Deterministic-probabilistic modeling provides with the study of multifactorial correlation between the rate of occurrence of a hazard and its identifying factors at different monitoring periods.

3.3.2. Probabilistic-statistical analysis [18]. The research is carried out by combining multifactorial correlation models, which allow for establishment of relationship between hazard indicators and its determining factors. It should also be noted that the effectiveness of developing the theory of graphs model, which allows for representing the implementation of a program for the remediation of water

IOP Conf. Series: Materials Science and Engineering **944** (2020) 012032 doi:10.1088/1757-899X/944/1/012032

bodies as a controlled dynamic space-and-time system, that is influenced by human activities as a hazardous factor.

Further studies suggest a detailed research of the digital twin model of a personified water body when landscape-recreational formation of the environment of cities and small settlements with a scenario module of augmented reality for managing ecosystem sustainability.

4. Discussion

The long-term program provides for the restoration of water bodies up to the level of harmonized operation of them. It can be achieved by proactive diagnosis of failures risk of an environmental contamination.

The intensive development of digital technologies allows an introduction of the calculated value of a digital indicator that determines the level of stability of the harmonized operation of engineering systems for the water protection complex in the natural environment as an additional indicator of the economic and environmental activities of the regions. This helps to improve the assessment accuracy of the environmental benefits and damages in planned projects.

It is necessary to give an assessment of ecosystem services in Russia as soon as possible and to use it as one of the indicators of the economic and environmental activities of the regions. This mechanism will allow more complete consideration of environmental benefits and damages in planned projects [19-22].

5. Conclusion

The necessity of modeling of urban areas based on the «attraction zone» and «exclusion zone» principle, as well as the verification of technical solutions for engineering systems of the water protection complex, which are in a harmony with the natural environment was demonstrated.

A conceptual model for control of the risks of developing urban areas was proposed, focusing on the mechanism for managing by water body renovation programs taking into account risk factor analysis. A technique for remediation of water bodies up to the level of harmonized operation by proactive diagnosis of failures risk of an environmental contamination was developed. Introduction of a digital indicator that determines a level of stability of the harmonized operation of engineering systems for the water protection complex in the natural environment as an additional indicator of the economic and environmental activities of the regions allows to improve the assessment accuracy of the environmental benefits and damages in planned projects.

References

[1] Goldstein D 2015 Climate-adaptive technological change in a small region: A resource-based scenario approach *Technological Forecasting and Social Change* **99** pp 168–180

[2] Hötte K 2020 How to accelerate green technology diffusion? Directed technological change in the presence of coevolving absorptive capacity *Energy Economics* **85** 104565

[3] Miao C, Fang D, Sun L and Luo Q 2017 Natural resources utilization efficiency under the influence of green technological innovation *Resources, Conservation and Recycling* **126** pp 153–161

[4] Schäfer A 2014 Technological change, population dynamics, and natural resource depletion *Mathematical Social Sciences* **71** pp 122–136

[5] Callaghan C W 2018 Surviving a technological future: Technological prolife ration and modes of discovery *Futures* **104** pp 100–116

[6] Xiang xiang S, Xiao liang Z, Zhang wang C and Yuping Y 2020 Environmental efficiency of electric power industry, market segmentation and technological innovation: Empirical evidence from China *Science of The Total Environment* **706** 135749,

[7] Mao C, Koide R, Brem A and Akenji L 2020 Technology foresight for social good: Social implications of technological innovation by 2050 from a Global Expert Survey *Technological Forecasting and Social Change* **153** 119914,

[8] Heckman J J and Leamer E 2002 *Hand book of Econometrics* **5**
[9] Solozhentsev E D 2012 Risk management technologies (with logic and probabilistic models). *Springer* p 328
[10] Karasev V V 2014 Risk Management Technologies with Logic and Probabilistic Models *Issues of* risk analysis 11(3) pp 14–21.
[11] Sheshenya N L Mathematical models of forecast estimates of the geoecological state of urban areas (elibrary_25719970_45769784).
[12] Gagarin V G, Lushin K I, Kozlov V V and Neklyudov A Yu 2016 Path of optimized engineering of HVAC systems Procedia Engineering 146 pp 103–111.
[13] Apatenko A S and Sevrugina N S 2019 J. Machine and equipment for the village **7** pp 35–38
[14] Kochetkova T V and Aleinikova N V 2019 Comfort of the urban environment Bulletin of BSTU named after V.G. Shukhov **11** pp 66–72
[15] Shchukina T V, Vorob'eva Yu A, Karoli M A and Logachev A V 2019 Improving the quality of the air environment through the greening of urbanized territories on the example of improvement of social facilities *Bulletin of BSTU named after V.G. Shukhov* **10** pp 93–101
[16] Sevryugina N 2017 *EMMFT2017 Advances in Intelligent Systems and Computing* **692** pp 273–281
[17] Sevryugina N 2017 *MATEC Web of Conferences* **117** UNSP00151
[18] Penteado I M, Nascimento A C S, Corrêa D, Moura E A F, Zilles R, Gomes M S, Pires F J, Brito O S, Silva J F, Reis A V, Souza V and Pacífico A C N 2019 Among people and artifacts: Actor-Network Theory and the adoption of solar ice machines in the Brazilian Amazon *Energy Research & Social Science* **53** pp 1–9
[19] Gulsrud N M, Raymond C M, Rutt R L, Olafsson A S, Plieninger T, Sandberg M, Beery T H and Jönsson K I 2018 Rage against the machine? The opportunities and risks concerning the automation of urban green infrastructure, *Landscapeand Urban Planning* **180** pp 85–92
[20] Abba Ari A A, Gueroui A, Titouna C, Thiare O and Aliouat Z 2019 Resource allocation scheme for 5G C-RAN: a Swarm Intelligence based approach *Computer Networks* **165** 106957
[21] Prenkert F, Hasche N and Linton G 2019 Towards a systematic analytical framework of resource interfaces *Journal of Business Research* 100 pp 139–149

IOP Conf. Series: Materials Science and Engineering **944** (2020) 012033 doi:10.1088/1757-899X/944/1/012033

Architecture of medium- and multistoried residential buildings with steel framework

T V Shamaeva[1], E S Zinkevich[1]

[1] National Research Moscow State University of Civil Engineering (NRU MGSU), 26, Yaroslavskoye Shosse, Moscow, 129337, Russia

E-mail: ShamaevaTV@yandex.ru

Abstract. In this article, the object of research is multistoried buildings and medium-rise buildings with a steel framework as load-bearing structures, which have been built over the past 13 years. Advantages and disadvantages, characteristic features of the architectural design of these buildings have been specified. As a result of the work it has been revealed that architectural solutions depend directly on the structural solutions of the metal frame when the number of floors is from 3 to 21; the space-planning type is used as sectional and corridor, etc. A number of disadvantages have been identified such as: the choice of a structural system depends on the number of stories in the building; a narrow range of steel products is for the civil construction; there are no typical architectural units, etc. As for the advantages, there are various solutions for facade systems; short installation times are compared with the "all-season" construction; the load on the foundation from a metal frame is less than that from a monolithic one. There is a possibility of using prefabricated standard elements of the frame-modular construction of the building, adaptation of the building to another function. Galvanized steel structures have turned out to be more resistant to external factors.

1. Introduction

The steel framework is being used in buildings for various functions in modern design. We will consider the use of the steel frame in the residential construction of medium-and multi-stored buildings. We will identify the advantages and disadvantages of the steel-framework and will investigate feasibility of the design.

The percentage of steel frame buildings in our country does not exceed 15% of the total number of completed projects. For example, in Norway the percentage is 48%; in Sweden, the UK and the USA – 65-68% [1]. Mainly, this scale is achieved through the multi-stories housing construction. In the Russian design of residential buildings, the primacy is taken by the monolithic reinforced concrete construction. But the architectural solutions of residential buildings based on a steel frame are not inferior to monolithic ones.

Architectural solutions are directly related to the choice of the structural scheme of the steel framework. There are more than 10 structural frame schemes for the design of residential buildings on a metal framework, including for residential buildings. The most common frame is frame-tie, precast-monolithic. High-rise buildings are frequently frame-cable-stayed, frame systems with suspended floors, with reinforced concrete diaphragms and stiffeners, trusses and frames with a floor height shell. In the high-rise design, the possibilities of using steel frames are most fully revealed.

Let us determine which buildings are considered mid-rise and multi-storey. The classification of buildings by the number of storeys changes over time. Indicators of the number of floors increased with the general trend of increasing the height of buildings, development of design solutions, the technology of construction, development of new metal alloys. Today, in accordance with the norms, mid-rise buildings have 4-8 floors; multi-store – 9 floors and more [2].

2. Materials and methods

An analysis of the architecture of residential buildings with a steel frame, built in Russia (7 pcs.), in the United States, Canada (6 pcs.) was illustrated by table 1.

The tasks were to determine the number of stories, space-planning type; characteristic features of the appearance of buildings. It is necessary to identify the disadvantages and advantages of using a steel frame in residential buildings over 4 floors, to raise the question of the feasibility of using a steel frame for residential multi-apartment buildings.

Table 1. Objects for analysis, part 1

1. The appearance of the building. Title. Country, city. 2. Space-planning type of building. Number of floors. Type of facade. Year.			
Russia		**USA, Canada**	
1. "Moskovskiy" Residential complex, Yekaterinburg. 	- Sectional. - 25 floors. - Ventilated, suspended facade. Glass, panels. - 2012.	1. Macallen Building, Boston, USA. 	- Corridor-type. - 14 floors. - Ventilated, suspended façade. - 2007.
2. Residential building for employees of the Ministry of Defense of RF. Murmansk, Murmansk region, Gadzhievo. 	- Sectional (2 sections). - 4 floors. - Curtain facade. Ceramic granite tile. - 2015.	2. The Canopy Lofts, Lincoln, Nebraska, USA 	- Sectional, corridor-type. - 6 floors. - Hinged panels, glass. - 2013.
3. Residential complex "Fregat-Neo", Krasnoyarsk. 	- Corridor-type, sectional. - 14-16 floors. - Ventilated, hinged facade. Glass, panels. - 2015.	3. Troy Boston, Boston, USA. 	- Corridor-type. - 9-19 floors. - Hinged panels, glass. - 2015.

IOP Conf. Series: Materials Science and Engineering **944** (2020) 012033 doi:10.1088/1757-899X/944/1/012033

Table 2. Objects for analysis, part 2

4. Residential complex "River Park", Moscow, Nagatinsky Zaton. 	- Sectional. - 17-18 floors. - Gas-concrete blocks. Ceramic granite panels on a steel frame. - 2017-2020.	4. 100 Norfolk Street, New York, USA. 	- Sectional. - 12 floors. - Glass. - 2016.
5. Residential neighborhood "Novomarusino", Novosibirsk. 	- Sectional. - 10 floors. - External walls. Hinged thermal panel. - 2017-2020.	5. Highline 23, New York, USA. 	- Sectional. - 10 floors. - Ventilated, suspended façade. - Glass.
 6. Residential complex «Tridesyatoe», Bogorodsk, Nizhny Novgorod region	- Corridor-type - 3-5 floors - thermal panel «Rusteh+» - 2016-2017		
7. Residential complex "Shakhtar", Chegdomyn, Khabarovsk Krai. 	- Sectional. - 4 floors. - Thermal panel «Rusteh+». - 2016-2017.		

3. Results.

In the course of the analysis, the following results were obtained:

- architectural solutions depend directly on the structural solutions of the metal frame. The formation of the architectural appearance of the facades is due to the "external" frame ("lattice facades"); the use of hinged facade systems; ventilation elevation systems; wall - "screen", curtain walls made of glass, panels, large stained-glass surfaces;
- the number of floors is from 3 to 21. In the Russian design of residential buildings with a steel frame, it is used in an equal ratio in buildings of medium height (3-5 floors) and in multi-stored buildings (10-18 floors). In foreign practice, multi-stored construction prevails (10-21floors);

IOP Conf. Series: Materials Science and Engineering **944** (2020) 012033 doi:10.1088/1757-899X/944/1/012033

- the space -planning type is used in two versions: sectional and corridor. This is the advantage for sectional houses (90%). In Russian examples there is a tendency to create standard projects, in foreign examples - in individual design;
- elevations and appearance of buildings in foreign examples are not always aligned with the function of housing. At the same time, the appearance is more vivid, memorable, and the steel frame is visible in the appearance and in the interiors;
- the categorization of housing is different: from social, economy housing (RF), apartments to housing with increased comfort (mainly in foreign practice).

In the course of the research on this topic, additional results were obtained.

Disadvantages of using a steel frame for residential buildings:

- The choice of a structural system depends on the number of stories in the building; the higher the number of stores of the building, the more varied the choice of the structural system. For medium and high-rise buildings, frame, cased-frame systems are cost-effective. The frame creates restrictions on the layout of the building; therefore, it is mainly used in office and administrative buildings, in rare cases – in residential buildings [3]. It also creates a more rigid planning structure in comparison with a building made of monolithic reinforced concrete.
- At present there is a low level of unification of frame elements, the predominance of manual operations when creating the "frame" of nodes [3]; a narrow range of steel products for civil construction.
- There are no typical architectural units, design solutions in comparison with monolithic design.
- Metal prices have increased compared to previous years; the benefit from using a metal frame compared to reinforced concrete has decreased [3].
- High labor costs of highly professional assembly workers.
- To increase the fire resistance of structures, it is necessary to carry out structural protection of the load-bearing elements of the frame made of metal.

The advantages consist in the following:

- there are various solutions for facade systems;
- installation times are short compared to that of the monolithic construction;
- this is the “all-season” construction, independent from weather conditions in comparison with the monolithic construction;
- the load on the foundation from a metal frame is less than that from a monolithic one (above 25 floors, it is better to use a steel frame);
- there is a possibility of using prefabricated standard elements;
- the frame-modular construction of the building is possible, which is more suitable for the construction of social housing, hotels and dormitories;
- the building may be adapted for another function; there is a variability of the building with large spans. This allows you to create free spaces; at the same time, kitchens, bathrooms, toilets can be created as modular elements;
- galvanized steel structures are more resistant to external factors, but also more expensive.

4. Discussions

The question remained open: how many floors are there in a multi-stories building? The classification of buildings in terms of the number of stories in Russian standards is ambiguous.

There is a definition of a high-rise building. A high-rise building is the height of a residential building over 75 m [4], and the height of the building is determined by fire regulations. At the same time, the designers (especially for fire systems) have a concept of "high-rise" buildings, but there is no official definition. It is necessary to introduce clear limits on the number of floors for multi-stories buildings. There are rules for designing reinforced concrete frames for residential multi-stored buildings [5], but there are no rules and regulations for steel frames.

IOP Conf. Series: Materials Science and Engineering **944** (2020) 012033 doi:10.1088/1757-899X/944/1/012033

The feasibility of using a metal frame for apartment buildings is ambiguous in terms of architectural solutions. The conclusions are as follows:

- For residential buildings of medium stores (from 4 to 8 floors), it is more economical to use a steel frame with reinforced concrete floors. Most likely, these are economy class or social housing, as well as apartments, hotels and hostels. Facades are without bulk plastic. It is possible to use standard elements (for example, reinforced concrete floor slabs). The typical modular design is possible.
- For multi-stories buildings, the variability and choice of structural systems are higher. This is related to the possible variety of facade solutions. When using large-span structures, "free" spaces are created; there is an increase in various planning techniques, adaptation to different functions.

5. Conclusions

Much depends on the tasks set and on the capabilities of designers, builders, customers, climatic conditions, the raw material base and other factors. The design and construction of steel-framed apartment buildings deserve special attention and development in the future.

6. References

[1] Mironov V Iron arguments for and against steel construction. RBC+. [Electronic resource]. URL: https://plus.rbc.ru/news/

[2] SP 42.13330.2016 urban planning. Planning and development of urban and rural settlements. Updated version of SNiP 2.07.01-89* (with Changes №1, 2). Date of introduction 2017.07.01

[3] Razumova O V 2012 Principles of forming architectural buildings, structures and architectural complexes using steel frames *Visnik Pridniprovskoi derzhavnoi Akademii budivnitstva TA architekturi* **1** pp 53–66

[4] SP 267.1325800.2016 High-rise Buildings and complexes. The rules of design. Date of introduction 2017.07.01

[5] SP 356.1325800.2017 Structures of frame reinforced concrete prefabricated multi-stories buildings. The rules of design.

IOP Conf. Series: Materials Science and Engineering **944** (2020) 012034 doi:10.1088/1757-899X/944/1/012034

Using computer graphics in design

E V Shopina[1], **A A Stativko**[1], **M A Markova**[1]

[1] Belgorod State Technological University named after V.G. Shukhov, Russia, 308012, Belgorod, ul. Kostukova, 46

E-mail: lena_shopina@mail.ru

Abstract. In the modern world, with the development of technology, it becomes possible to create design art objects from the developed sketch to the implementation of the project in material using computer graphics. Thanks to vector graphics programs and the emergence of laser and milling equipment with numerical software (CNC), designers were able to transfer the drawing to the material using material cutting technology. The object of the study is vector graphics programs, laser and milling cutting technology. As a result of the study, the authors conclude that thanks to software products and the development of technologies and equipment, designers were able to simplify the process of making decorative products.

1. Introduction

The ability to transfer the pattern along specified lines and dimensions using laser and milling equipment with CNC and vector graphics programs allowed designers to create unique decorative products, minimizing the subsequent process of processing edges and surfaces of blanks, due to the high quality of the cut.

The purpose of the study is to analyze the technology of making a plywood product by transferring a vector pattern and performing laser and milling cutting using the example of a decorative panel "Cranes".

2. Materials and Methods

Designers use different methods when creating products. With the advent of computer graphics and specialized equipment, the technology of making art objects has been significantly simplified. Thanks to computerization, designers were able to create a product flow using the same drawing. This became possible with the advent of vector graphics of specialized software products. CorelDraw is usually used as programs. Separately, it is worth highlighting the technology of laser and milling cutting of various materials. Materials that designers choose for future products may be glass, plastic, plywood and metal.

3. Steps of making a decorative product from a sketch to an embodiment in a material

Let us consider the technology of making a designer product using laser and milling equipment using the example of a decorative panel "Cranes". In most cases, plywood is the most common among the materials used to make decorative panels. The reasons for its wide distribution are not only the trends in fashion and the popularity of styles such as minimalism, loft and modern, but also in its environmental friendliness, strength, wear resistance, easy workability. Plywood is also easily cut, engraved, glued and decorated.

IOP Conf. Series: Materials Science and Engineering **944** (2020) 012034 doi:10.1088/1757-899X/944/1/012034

The production of an art object begins with the idea of the designer. After viewing similar options, obtaining various kinds of information, a sketch is created (Fig. 1). A sketch is an example sketch of an idea. For the best result, you usually make several sketches of a product that differ from each other in style, design, and creative solution. After the best option is chosen, it is finalized; all the smallest details are carefully drawn. Further, since the panel consists of several layers, you must draw each one separately. For these purposes, tracing paper is used. It is applied to the sketch and the outlines of each layer on a separate sheet are drawn, and new parts are added as needed.

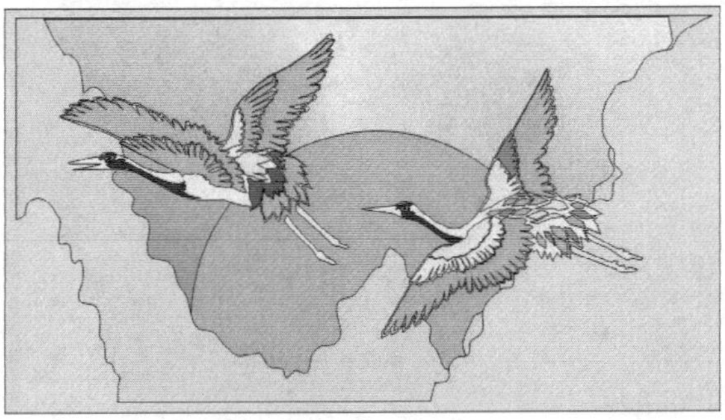

Figure 1. Sketch of a panel

At the second stage, the layers drawn on the tracing paper should be photographed and transferred to the program. To create a picture in the CorelDraw program, it is convenient to use a tool such as the "Bezier Curve", using it to indicate the main lines that can be edited later [1]. If necessary, you can delete or draw lines again to simplify the sketching process and optimize time. A program can also be used to draw a future product AutoCad.

It is at this stage that the overall dimensions of the product are finally established. Each panel layer is drawn separately (Fig. 2).

Figure 2. Example of a drawing for laser cutting

At the third stage, the layers drawn in the program are cut from plywood using a milling cutter. Feathers and other small parts for panels are cut with the help of laser equipment (Fig. 3). The design is assembled. Fasteners are installed on the background layer from the back.

After the article is cut out, the plywood panel must be refined and prepared for subsequent application of paint and varnish.

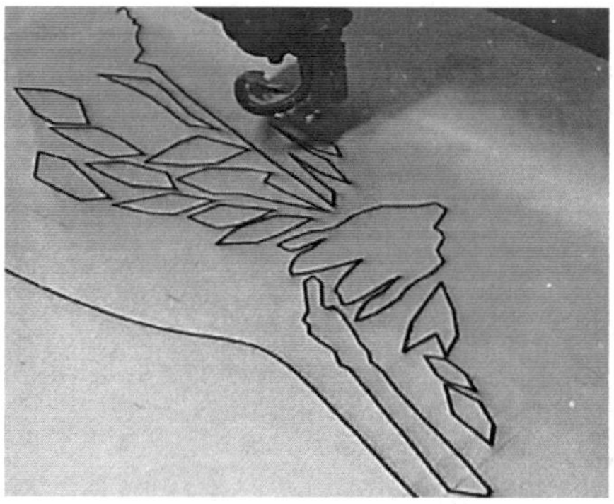

Figure 3. Exercise of a laser-beam cutting

A sandwich web is used for this purpose. Despite the fact that laser cutting is characterized by accuracy and high cutting quality, it is still necessary to grind the plywood product (Fig. 4). The cut quality after the cutter is much worse, so it is necessary to clean more. It also depends on the material itself, namely on plywood and its variety. After grinding, the layers are assembled into a single structure and glued together (Fig. 5).

Figure 4. Preparations for a panel

Figure 5. The glued product

Then protective and decorative coating is applied to the article (Fig. 6). This operation is performed in order to protect the product from rotting, insects, atmospheric effects, and in some cases from burning (Fig. 7).

IOP Conf. Series: Materials Science and Engineering **944** (2020) 012034 doi:10.1088/1757-899X/944/1/012034

Figure 6. Putting acrylic paint

Figure 7. Completed art object - decorative panel "Cranes"

4. Use of laser and milling equipment for cutting plywood

The complexity of the drawing and the overall dimensions of the future product depend on the choice of equipment. If the project includes many small parts, complex intersections, angles and it is small in size, then cutting on a laser machine is used [2].

The laser cutting technique was specially designed to give wood a decorative effect. Laser cutting can produce small parts with complex bends and breaks. With the help of such equipment, it is possible to obtain a high-quality cut. With all the advantages, laser cutting has the disadvantage of limiting the thickness of the material. On average, it is possible to treat material from 0.8 mm to 60 mm using a laser. A feature of this method is that all parts or holes cut by a laser have a black edge. If such edging is not desired in the finished product, the plywood material can be painted [2].

Laser cutting technology consists of directed action of laser beam on plywood, temperature at cutting point increases, after which wood material is removed by evaporation [2].

Milling equipment has less accuracy, but thanks to a cutting tool-milling tool, it is possible on average to process workpieces having a thickness of 1 mm to 14 mm (and more depending on the machine used). The most popular method of processing plywood as a derivative of wood material on an industrial scale is plywood milling. In such machines they are cut not only in a straight line, but also in any given trajectory. Such machines help cut holes, make recesses, slots, and create a volumetric pattern [3].

Milling cutting is carried out using a cutter, which is essentially a cutting tool [4]. Unlike laser cutting, milling does not leave a black edge, but is not able to cut small holes and too sharp angles. Despite this, thanks to the mill you can make a chamfer, get a neat not only rectangular, but also shaped cut [4].

When performing laser cutting and transferring a drawing from Corel Draw to Laser Work, the lines should be indicated in red, since most often it is the cut color. The laser machine can be used and for engraving the material the color for the lines is black then.

IOP Conf. Series: Materials Science and Engineering **944** (2020) 012034 doi:10.1088/1757-899X/944/1/012034

5. Results and discussions

As a result of the study on the example of the manufacture of the "Cranes" decorative panel, the authors conclude that computer graphics influenced the creation of design products, making it possible to transfer lines to the material and carry out cutting. Thanks to the developed technologies, the cutting of various materials was simplified, it became possible to cut small parts, obtain a chamfer and a curly cut, cut to a certain thickness and according to given sizes and parameters, and apply engraving. Computer graphics allows you to create various design solutions and edit them. On the basis of one sketch, you can produce an entire batch of products without significant labor.

6. Conclusions

Computer graphics made it possible to create a picture from vectors that can be changed, deleted and created again. This made it possible to simplify the process of creating a drawing and it is possible to cut according to given vectors by obtaining a drawing exactly corresponding to the drawing.

Thanks to laser cutting technology, it is possible to process various materials, ranging from paper to metal, to obtain small size and complex shapes. Milling made it possible to process large parts and create a curly cut. It is mainly used in the furniture industry.

7. Acknowledgements

This work was realized in the framework of the Program of flagship university development on the base of the Belgorod State Technological University named after V G Shukhov, using equipment of High Technology Center at BSTU named after V G Shukhov.

References

[1] Gursky Y, Gurskaya I and Zhvalevsky A 2015 *Computer graphics: Photoshop CS, CorelDRAW 12, Illustrator CS* (Moscow: SPb: Peter) 812

[2] Antsiferov S I, Lyutenko A O, Sychev E A, Sivachenko L A and Karachevtseva A V 2019 Digital design using generative design *Scientific journal Technical aesthetics and design research* **1(4)** 38

[3] Cecil J 2001Computer-aided fixture design - A review and future trends International *Journal of Advanced Manufacturing Technology* **18(11)** pp 790-793 doi: 10.1007 / s001700170004

[4] Horvát I and Vroom R W 2015 Ubiquitous computer aided design: A broken promise or a Sleeping Beauty? *Computer-Aided Design* **59** pp 161-175 doi: 10.1016 / j.cad.2014.10.006

[5] Antsiferov S I, Karachevtseva A V and Sivachenko L A 2019 DESIGN AND DESIGN Products in CAD / CAM / CAE NX system controlled by PLM system Teamcenter *Scientific journal Technical aesthetics and design research* **1(2)** 45

[6] Norenkov I P 2002 *Basics of computer-aided design: textbook. for universities* (Moscow: Publishing house of MSTU im. N.E. Bauman)

[7] Kukhta M S, Kumanin V I, Sokolova M L and Goldshmidt M G 2013 *Industrial design* (Tomsk: Tomsk Polytechnic University) **12**

[8] Grigoryants A G 1989 *Fundamentals of laser processing of materials* (Moscow: Mashinostroenie) 13–84

[9] Avrutin S V 2018 *Osnovy milling business* (Moscow: YoYo Media)

[10] Kopylov R 2010 *Work on milling machines* (Moscow: Lenizdat) pp 3–10

IOP Conf. Series: Materials Science and Engineering **944** (2020) 012035 doi:10.1088/1757-899X/944/1/012035

Features of designing small architecture objects in the design of environmental space

E V Vasilenko[1], P G Vasilenko[1], O S Goltseva[1], T E Belyakova[1]

[1] K.G. Razumovsky Moscow State University of technologies and management (the First Cossack University), Russian Federation, Moscow, Zemlyanoy Val, 71

E-mails: e.vasilenko@mgutm.ru, p.vasilenko@mgutm.ru, o.goltseva@mgutm.ru, t.belyakova@mgutm.ru

Abstract. The article deals with research on the organization of modern design of small architectural objects in the city environment. Analysis of the architecture of the environment of modern cities has shown that design and architecture are not given due attention. The artistic and constructive design solution of small architecture objects in the urban environment should be appropriate in the specific conditions of the districts, functionally justified, convenient to use-ergonomically, economically, practically, and should also meet modern aesthetic requirements. Although elements of the external equipment of the environment are not the main means of composing urban architecture, they significantly enrich planning and architectural and spatial solutions, and contribute to the formation of an integral architectural appearance of a modern city.

1. Introduction

The modern city, consisting of large and small architectural structures-simple and complex forms, open spaces, highways-wide avenues and narrow streets, is characterized by the absence in these large volumes and spaces of elements that are commensurate with the human scale, adversely affects the mental and emotional state of a person. In such an environment, a person feels uncomfortable, depressed. However, the purposeful use of various types of green spaces and small architecture objects in the territories adjacent to residential buildings allows these huge spaces to be divided into a number of smaller spaces, enrich them, and make them attractive. Since the size and proportions in the design of small architecture objects are always linked to the human scale, regardless of the distance from which we perceive them, they allow us to feel the size of the surrounding buildings and structures and make the environment understandable, cozy, human.

2. Materials and methods

As an object of research, the modern organization of design of small architectural objects in the city environment is considered. The analysis of methodological and special literature on the problem is carried out, General theoretical research methods are used. The design methodology is based on a systematic approach and the development of new methods for solving problems, based on various theories and recommendations for organizing the creative process. The analysis of all stages of design in design is structured and studied by many authors in this field: S. M. Mikhailov, A. S. Mikhailova, V. S. Kulikov, A. G. Naumov, V. N. Tkachev, A.V. Efimov, G. B. Minervin, V. T. Shimko, B. G. Barkhin

IOP Conf. Series: Materials Science and Engineering **944** (2020) 012035 doi:10.1088/1757-899X/944/1/012035

et al.) Methodological findings, interesting and thoughtful methodological techniques that reveal certain provisions in the design of small architectural objects of the urban environment (art objects) includes the formation of the structure of industrial and residential objects, squares, highways, etc.

3. Results

Objects of small architecture (art objects) can perform a utilitarian, practical function and also a very important aesthetic function that contributes to the formation of the aesthetic appearance of modern urban space and the education of aesthetic taste in the city's residents. Analysis of the architecture of the environment of modern cities has shown that design and architecture are not given due attention. The main reason for many unsuccessful decisions is that the organization of the modern urban environment is not sufficiently studied. The basic rules and guidelines for the design of design in this area do not contain calculation standards and recommendations for the nomenclature and quantitative application of small architectural forms there are no instructions on the content and volume of the relevant design documentation6 in some cases, standard designs of small architecture objects are incorrectly developed and applied.

Table 1. Style-forming factors

Factors for the design of					
Compositional unity and interaction with the surrounding architecture	Unity of design solutions and materials used	Combination of proportions and volumes, consistency of shapes and textures	Modularity of elements, Assembly with unification of nodes and parts, dimensions	The optimal range of objects that correspond to the ergonomics and aesthetics of the environment is set and quantity	Unity of the color solution, the relationship with the overall color solution of the architectural environment

The intensive growth of cities, the increase in areas and territories requires that landscaping and design of territories, as well as small architecture, are designed today on a scientific basis. It is necessary to work on the design of small architecture design in two directions [4]:

Determine an acceptable approximate nomenclature of small architecture objects. Develop the basic principles of architectural and artistic design design and on this basis prepare manuals for designers.

Develop scientifically based standards for quantifying the need for small architecture objects in certain environmental territories, taking into account population density, territory size, sanitary and hygienic conditions, local natural and climatic features, which would complement the SP (Construction Rules) and instructions for the composition of project documentation. Analyzing the shortcomings in the design of small architectural objects, we see that in most cases the design does not contribute to the creation of a perfect architectural and artistic urban environment. Often, each object is designed separately, without compositional and constructive linking it with other architectural objects. Therefore, as a rule, objects installed side by side do not harmonize with each other, destroy the compositional unity, and create an unpleasant impression for the aesthetic perception of space. When designing small architecture objects in an urban environment, the designer must cover them in a complex – this is the only way to create a full-fledged environment that is in good harmony with large architectural structures and the entire infrastructure of the city [3]. After analyzing designer-the designer should identify: tectonic (symbol color of the constructive elements) features of interaction of color and form (mismatched patches of color with form elements); the nature of the color composition (single color, using related hues, contrast); the color character of the area as a whole, whether the colors of objects in the population or there are surfaces that stand out; all colors, gathering them in a single table and divided into main groups: basic, additional, accent. You should also choose the finishing materials taking into account the color and texture, physical characteristics, cost, and specify the specific places where they are used.

4. Discussion

Objects of small architecture, depending on their functional purpose, can be divided into several complexes: small architectural structures, engineering and economic equipment, equipment for sports and recreation, visual information and decorative forms. The artistic and constructive design solution of small architecture objects in the urban environment should be appropriate in the specific conditions of the districts, functionally justified, convenient to use-ergonomically, economically, practically, and should also meet modern aesthetic requirements.

An artist-designer must necessarily have three qualities-his own vision, a high professional level of performance of analogs, as well as knowledge of the basic principles of composition. An artistic image is an artist's view of the world through their own consciousness, that is, the expression of themselves through other objects or phenomena. Form is one of the main means by which an artistic image can be expressed. According to Paul Klee, a German-Swiss artist, form is born from a point that forms a line when moving [2, 10].

The form is created by several means: rhythm, contrast, scale, nuance, and proportions. Form in combination with color enriches the created works, however, it is worth noting that there is no form without color, light and texture. You also need to take into account that the color must be selected in accordance with the intended meaning to correctly embody the meaning of the artistic image. Another means of expressing an artistic image is texture. Texture is the external properties of an object that are determined by the material of the composition. The characteristics of the texture include roughness, slipperiness, softness, etc. Texture and color can be expressed differently depending on the lighting, another means of expressing an artistic image.

The composition has many rules, following which the artist risks losing his own originality – this is the paradox. The correct composition can beautify the work, but make it boring, so Golubeva in a sense refutes all the material presented and encourages you to follow your own feelings. However, without knowledge of the laws of composition, it is impossible to "not follow them correctly". The laws of composition must be observed in any work, but every artist, designer, architect has the right to depart from them, if such is the intention. Modern art has a tendency to violate General canons and rules. This was always the case on the threshold of the birth of a new style [9, 6]. The three main laws of composition: unity(integrity), separation of the main and secondary, balance. The composition can be called complete when you look at the image, it does not break up into separate parts, but is a single one. Balance can be called both symmetry and asymmetry. Balance is the basis of harmony in the work.

5. Conclusion

The concept of "small architecture" includes: pavilions, fountains, swimming pools, bicycle parking, yard furniture, sports and gaming facilities(playgrounds), household equipment of the territory, structural visual information, fencing, lighting, decorative sculpture, vases, flowerbeds. Thus, small architecture should be called objects of equipment and landscaping that are proportional to the human scale and meet the utilitarian and aesthetic needs of modern man [11]. When designing or standard small

IOP Conf. Series: Materials Science and Engineering **944** (2020) 012035 doi:10.1088/1757-899X/944/1/012035

architecture objects, you should create small architecture complexes that are characterized by unity of style [5].

The number of objects included in the complex is determined by their functional purpose and artistic intent. For example, a set of environmental design objects on a recreation area may consist of a pavilion, benches, and trash cans. However, in some cases, depending on the specifics and artistic intent, it is advisable to include other objects: a swimming pool, information stands, decorative sculpture (art object), which can become a compositional center and affect the artistic decision of all parts of the complex. In all cases, when creating objects of small architecture, you must simultaneously follow the laws of expediency and aesthetics [8,12].

Beauty cannot be accidental in relation to the functional features of small architecture. The task of an architect and designer is to correctly combine functional and aesthetic characteristics in small architecture objects that correspond to a wide range of utilitarian and artistic needs of a person. To create an artistically complete system of small architecture objects throughout the territory of the designed environment, it is necessary to adhere to a single style [13]. The main style-forming factors of small architecture can be considered optimal nomenclature, modularity and Assembly of elements, compatibility of objects, unity of design, composition and color solutions.

Small architecture objects get special expression in those territories where the architecture of buildings, landscaping and landscaping are solved taking into account modern aesthetic requirements [1]. Although elements of the external equipment of the environment are not the main means of composing urban architecture, they significantly enrich planning and architectural and spatial solutions, and contribute to the formation of an integral architectural appearance of a modern city [7].

References

[1] Andreeva E V and Voronov M Yu 1970 Equipment for city streets. Review (Moscow: VNIITE, Art construction abroad) **2**

[2] Vasilenko P G, Vasilenko E V and Pruzhina E S 2019 Features of light design in modern urban planning Modern trends in fine and decorative applied arts and design **1** pp 34–39.

[3] Vavilova I and Pirker V 1972 Social functions of small forms of architecture *Scientific notes of V. I. Mukhina lvhpu* **3**

[4] Dedyukhin V F 1970 On small forms in the architecture of cities *"Izvestiya vuzov". Construction and architecture* (Novosibirsk) **6**

[5] Efimov A V 1990 *Coloristics of the city* (Moscow: stroizdat)

[6] Lopasova E V 2014 Formation of an aesthetic image of the urban environment by means of modern monumental art In the collection: Modernization of humanitarian and art education: innovative development strategies *Materials of the International scientific and practical conference: in 2 parts. Edited by: Gritsenko V. P., Alekseeva I. V., Zengina S. S., Enikeeva A A* pp 354–361.

[7] Cerver F A 2000 *The World of Architecture* (Cologne: Konemann)

[8] Isozaki A 1988/89 *The international design yearbook* (Milan: Thames and Hudson)

[9] Pallotta V I, Dudnikova A A 2018 Art object as a modern form of artistic integration in design *In the collection: Education. The science. Culture. materials of the international scientific forum* pp 135–137.

[10] Lancaster M 1996 *Colourscape* (London, Academy Editions)

[11] Lenclos J-Ph, Lenclos D 1999 *Couleurs du monde* (Paris, Le Moniteur)

[12] Smirnova M A 2017 Shaping in fine art and design in the collection: Pedagogical education at the intersection of epochs: innovations and traditions in the field of educational technologies *Collection of scientific papers of the International scientific and practical conference. Institute of social and humanitarian technologies* pp 252–253.

IOP Conf. Series: Materials Science and Engineering **944** (2020) 012036 doi:10.1088/1757-899X/944/1/012036

Features of style shaping in modern design

E V Vasilenko[1], P G Vasilenko[2], T A Gordeeva[3], I I Gerasimenko[4]

[1] Federal State Budget Educational Institution of Higher Education K.G. Razumovsky Moscow State University of technologies and management (the First Cossack University), Russian Federation, Moscow, Zemlyanoy Val, 71, e.vasilenko@mgutm.ru, +79771836015
[2] Federal State Budget Educational Institution of Higher Education K.G. Razumovsky Moscow State University of technologies and management (the First Cossack University), Russian Federation, Moscow, Zemlyanoy Val, 71, p.vasilenko@mgutm.ru, +79771836012
[3] Federal State Budget Educational Institution of Higher Education K.G. Razumovsky Moscow State University of technologies and management (the First Cossack University), Russian Federation, Moscow, Zemlyanoy Val, 71, gordeeva07@mail.ru, 8-903-775-38-02
[4] Federal State Budget Educational Institution of Higher Education K.G. Razumovsky Moscow State University of technologies and management (the First Cossack University), Russian Federation, Moscow, Zemlyanoy Val, 71

Abstract. The article is devoted to the analysis of modern design, features of style formation and conceptual solutions in design. The dynamics of social development of our society, energy interactions, flows and mixing of social, cultural, aesthetic layers and groups contribute to an even greater variety of styles in the modern household environment. Eclecticism - free variation of images of any of the historical styles, their arbitrary combination, adaptation to needs-becomes the main feature of modern design. The organization opposes the desire for unlimited freedom of expression, from collectivism to individualism. Despite all the complexity and multilayered style-forming factors, in the end, the style is a system of externally perceived and easily recognizable features. Recent trends in design and world art are not at all without roots in the past and do not represent a decisive break with them. The use of retro style, pop art, hyperrealism and abstraction in design is generated by a modern over-urbanized culture. Modern style is a complex system of interactions in a single process of spiritual development of society. It can be understood as a conclusion, as a generalization of real experience. The existence of a single modern style with its unity of forms, features and attributes is impossible. Modern style formation involves the combination of different directions.

1. Introduction

The design of architectural structures and complexes in our time combines the strictness of forms and whimsicality, mannerism, constructiveness with decorativeness. Style does not freeze in place, in one position, it moves forward, using reinterpretation and self-denial.

Retro style in modern design is becoming more functional and constructive, and decorative and stylistic trends complement and give a special conceptual aspect to the formation of architectural design.

IOP Conf. Series: Materials Science and Engineering **944** (2020) 012036 doi:10.1088/1757-899X/944/1/012036

The dilemma that arises in modern design in the search for diversity within the framework of stylistic unity is what can be considered a single whole within the framework of eclecticism, and what is already a new style direction.

In the past, stylistic unity arose more or less spontaneously, and even in retrospect, it was not always realized by its creators. It seemed like a natural. It could only contrast the phenomena of other style series with normative ideas about taste. Only a re-evaluation of the values of culture and national color (characteristic of this area) opened the way to an unprecedented freedom of choice of taste and style in art, design and architecture. This was the beginning of eclecticism-free variation of images of any of the historical styles, their arbitrary combination, processing and reinterpretation, adaptation to the needs of modern society.

2. Materials and methods

Information analysis of sources included searching for source information in combination with a preliminary study of their content. Obtaining objective data on the problem under study allowed us to determine the goals and directions of our research work, as well as apply the method of project research. In relation to texts published in the Internet space, the method of Ethnography was used, which can be used in project research technologies. Method of analysis in the formation, creation of a certain synthesis in combining the stylistic features of various trends in art. Rationalistic, structural, and functional tendencies are contrasted and intertwined with irrational, surreal, expressive, and decorative aspirations. The organization of shaping in design engineering is faced with the desire for unlimited freedom of expression, from collectivism to individualism.

3. Results

The desire for a new modern style of shaping that meets the spiritual and physical needs of modern man, giving his way of life the most complete, visible expression and specially organized and meaningful spatial environment - all this has inspired many innovators since the modern era.

The social dynamics of our time, energy interactions, flows and mixing of social, cultural, aesthetic layers and groups contribute to an even greater stylistic diversity of the modern household and spatial environment. The unity of modern style is utopian [6]. Modern mass perception, which does not try to combine all possible phenomena of historical and new culture in any spiritual value, is not an ideal. Therefore, the search for unifying values of modern culture is very relevant today, but they are based on some other, not strictly stylistic, in the traditional sense of the word..

4. Discussion

The needs and fantasies of the new time will be imperceptibly embodied within the boundaries of one style, at least in the usual, traditional sense of it. It is impossible to imagine modern design in a single style (Renaissance, Gothic, etc.), the variety and interweaving of styles forms the overall picture of modern art culture. Streets of modern cities, interiors of public and private buildings, industrial design and fashion design, advertising and graphics-all this uses the heritage of styles from different eras. The design uses items that bear indelible traces of their origin from different, difficult to relate to each other spheres of creative activity – technical, folk, artistic and applied, etc [5]. There are many disagreements and judgments about what will become inconvenient and inconvenient in such an artistically unified world, whether there will not be a feeling that the removal of stylistically alien elements is unacceptable for us [12]. The ability of a person to navigate the world of familiar values, and in just surrounding objects, will be lost. the differentiation of images of the human subject environment that has occurred over the past century is irreversible.

The culture of the XX and XXI centuries is focused not on artistic Association, but on a sharp division of the spheres of creativity by their specialization. Within each of these areas of design, architecture, and decorative arts, there is a prevailing trend towards individualization of all artistic products. Even within a single artistic ensemble, some of its decorative elements often claim not to be synthesized, merged with the environment, but to be "solo" - an easel solution.

Of course, this does not mean that there are no unifying trends in modern design. We are used to living and navigating in an artistically complex, ambiguous and mobile environment, in conditions of constant intersection of various aesthetic principles [9]. The complex structure of public consciousness and its dynamism, the diversity of our ideas and knowledge about culture – modern and historical. There is an opinion that, despite all the complexity and multilayered style-forming factors, in the end, the style is a system of externally perceived and easily recognizable features. And further, in an effort to distinguish the entire ensemble or work from its stylistic component, it reduces it to several stylistic distinctive features - to the fact that it allows you to refer the subject to this, and not to another style. Such an operation with signs replacing the phenomenon itself would be quite appropriate in a technical work. This approach can be called attribution. But it is not enough to study the aesthetic nature and historical component of the artistic style [7]. Objects can be selected, analyzed, and finally repeated. Obviously, we will define such repetition as a phenomenon of stylization, but not of style, simply because the style-forming factor in this case is imitation. In order for the concept of "style "to become an indisputable tool of historical and artistic analysis, it is advisable to understand it less formally-not as a" system of signs", but as a certain image of a decorated, artistically expressed worldview of the epoch [10].

The eclecticism of the XIX th century was inspired by the idea that you can use the entire artistic world, the spiritual treasures of antiquity or Ancient Russia, the Renaissance or the court of Versailles, along with other techniques [8]. The attitude of the modern "retro" to the art of the past is significantly different and more complex. Eclecticism should not be confused with retro style. Eclecticism is an attempt to remake cultural heritage; retro style is a play on certain cultural and historical situations. Playing in the beautiful past, keeping the distance between the past and the present that separates us. We clearly define this distance and introduce it into the structure of the style itself [11].

Another example of such a typical European art of the early XVII th century stratification into the high pathos of the Baroque and the naturalism of caravaggism is also, apparently, the polarization of trends within a single form, the superstructure organization of artistic consciousness.

Examples of the XX th century-the beginning of the century) - primitivizing trends of modern art, the search for "wisdom of ignorance", closely associated at that time with the most radical trends of the artistic avant-garde. In them you can see not only a variant of the "retro style", but also an important feature of modern culture associated with its anti-rationalistic aspect [3].

World art culture and design contain many examples of stylistic intersections and interactions throughout the history of existence that permeate all spheres of art.

Design in the field of leisure gives the greatest scope to retro stylizations [15]. After all, a restaurant, bar, or cafe is likely to serve as an outlet for every day, business-like existence. People leave their worries for a while, change the rhythm, style and manner of behavior, forms of communication with each other. All this is facilitated by a change of scenery, a special saturation and style of the decorative structure of the interior. The restaurant is decorated in a simple, functional retro style. He was formed and lives in a special field of decorative and applied art, similar to scenography. His field of activity is interior design, which has gradually evolved from a holistic art of design and architecture into a special genre that has developed its own language and design methods [14]. As a rule, such an interior fit into this architectural structure, old or new, without merging with it, without becoming an organic part of it. The suspended ceiling is separated from the real ceiling, and decorative panels and facings demonstrate their independence from the tectonics of walls and supports.

Inside a rigid architectural " box " everything merges into an elegant ensemble that forms a certain action and gets all its spatial, decorative, plot and style qualities from the inside – from what is closed in itself, and not from the outside, from what is suspended and attached. Retro style, forming the building as a whole, extends to its appearance the same idea of object, structure-decor. First of all, the plot motivation of the retrospective course is preserved [1]. That is why modern retro stylizations should not be interpreted as a total stylistic superiority in modern art culture. They easily get along with the most rigid structural forms of architecture and design. The same decisions can be made with the help of folk art.

Such stylistic counterweights are often used in modern interiors. This style balances within the framework of a single artistic culture existed in the past. This is how once fantastic and stylized pseudo-Gothic got along with strictly regulated and severe classicism in its stylistic purity. It served as some compensation for this regularity and seriousness, but not as a stylistic alternative [13].

5. Conclusion

Thus, the latest trends in world design and art are not at all devoid of roots in the past and do not represent a decisive break with them. The use of pop art and hyperrealism, abstractions in design are generated by a modern over-urbanized culture. Technology and the entire human-engineered environment are seen as rigidly drawing the world around them. It is essential that the very role of the technical component (negative or positive) is significant in the formation of modern world perception. It is necessary to see the modern art of design in its ambiguous relations with the problems of modern times, in the diversity of its social functions, in its contradictory and continuous development, in relation to this development of style periods and creative directions – these are not real borders, but rather coordinates for orientation in a complex process [2].

Modern art of design in its most objective directions requires the designer to make a conscious choice among the many means of image and expression available to him to create a conceptual image of the project. Hypotheses of style dictate the search for an unattainable unity of handwriting of artistic culture. They require normativity, and an all-encompassing stylistic unity is impossible without a generally significant scale of aesthetic assessments [4]. The real modern style, in its broad sense — is a complex system of interactions in a single process of spiritual development of society. It can be understood as a conclusion, as a generalization of real experience. A single modern style does not exist, the unity of forms, features and features, as in the styles of the past, in our time in design and art culture is not destined to be realized.

References

[1] Vasilenko E V and Mareeva Yu S 2019 national culture in the design of Russia *Bulletin of the Institute of world civilizations* Vol 10 **1 (22)** pp 44–49

[2] Vasilenko P G and Kozlova U A 2020 Application of sgraffito technique in modern interior design *In the collection: Science. Culture. Art: actual problems of theory and practice. collection of reports of the all-Russian (with international participation) scientific and practical conference Belgorod* ***3*** 265–268

[3] Glazychev V 2006 Design as it is *2nd Ed publishing house "Europe"*

[4] Lopasova E V 2014 Formation of an aesthetic image of the urban environment by means of modern monumental art *Modernization of Humanities and arts education: innovative development strategies* pp 354–361

[5] Pallotta V I 2019 Ideal and material in environmental design as the basis of intellectual landscape *In the collection: Russia and the world: the development of civilizations. The transformation of the political landscape during the period 1999-2019. Materials of the IX scientific and practical international conference: in 2 parts* pp 432–437

[6] Pruzhina E S, Demidova E Yu and Pallotta V I 2019 Features of formation of open urban spaces by design means *In the collection: Problems and trends in the development of the socio-cultural space of Russia: history and modernity. Materials at the VI scientific and practical international conference. Under the editorship of T I Ryabova* pp 446–456

[7] Stepanova S A 2006 Evolution of the visual image of the city *Architecton: proceedings of higher education. — Ekaterinburg: publishing house of the USAAA: (http://archvuz.ru.)* **5**

[8] Shulgina D P 2010 Regional features of eclectic architecture in the Russian province D P Shulgina (Moscow: LENAND)

[9] Almquist E, Senior J and Bloch N The elements of value *Harvard business REV*

IOP Conf. Series: Materials Science and Engineering **944** (2020) 012036 doi:10.1088/1757-899X/944/1/012036

[10] Muchazondida M 2011 alienation of food in tourist Eatertainment: in nitcography *tourism research* **11 (3)** 1–18
[11] Simon H A 1969 The science of the artificial (M I T Press, Cambridge, Mass)
[12] Vaus de D (2006) Editor's Production: Research Design – a Review. In D. De Vaus (ed.) ResearchDesign.Vol.1. London: Sage: XXIII-LV
[13] Kozinets R V 2010 Netnografiya: Ethnographic Research On The Internet London 211
[14] Muchazondida M 2011 Alienation of food in tourist Eatertainment: in nitcography *tourist research* **11 (3)** 1–18.
[15] Gold R L 1958 Roles in sociological field work *social forces* **36** 217–223.

IOP Conf. Series: Materials Science and Engineering **944** (2020) 012037 doi:10.1088/1757-899X/944/1/012037

Truss model data analysis in Autodesk Revit

A A Vedernikova[1], **R A Shishmarev**[2]

[1] Saint Petersburg State University of Architecture and Civil Engineering, 4 Vtoraya Krasnoarmeiskaya ul., Saint Petersburg, 190005, Russia
[2] weblog "InfoBIM", http://infobim.ru/

E-mail: parallelogram13@gmail.com[1], mr.shruslan@gmail.com[2]

Abstract. The article presents the process and results of truss model data analysis using the Revit API. Methods for obtaining model data are presented, as well as an example of their application: calculation of the truss using the finite element method. Fragments of the program code for the plug-in are given. The calculation results in the plug-in correspond to the calculation results in specialized programs. This article is useful for architects and designers using the Revit API and Dynamo.

1. Introduction

Working with BIM often requires data analysis to perform particular design tasks [1]. For example, technical expertise on tower cranes [2], mining implicit 3D modelling patterns [3], daylighting simulation and analysis [4], organization of Autodesk Revit software interaction with applications for structural analysis [5], analysis of the overall sustainability of the building [6]. Engineers usually use plugins for such tasks.

The article considers an analysis of the model of the truss frame (Fig. 1): obtaining and preparing geometry and finite element analysis to obtain forces in the truss elements. C# programming language and the Revit API were used for this. Many engineers use programming tools in their work. This article shows that the capabilities of BIM are not limited to modelling. Engineering calculations can also be performed without the use of additional software.

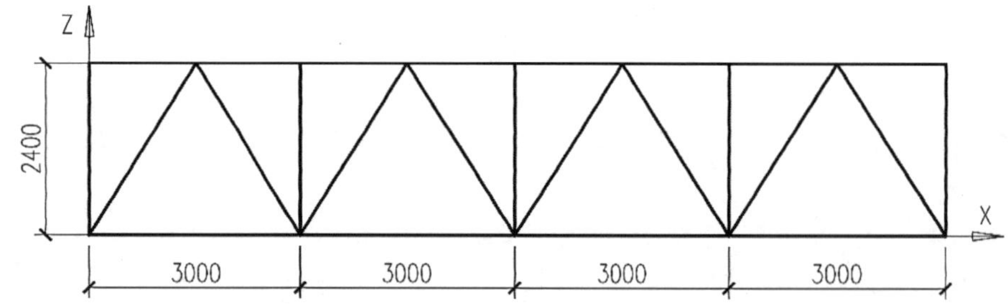

Figure 1. Truss frame.

IOP Conf. Series: Materials Science and Engineering **944** (2020) 012037 doi:10.1088/1757-899X/944/1/012037

2. Materials and methods

The plugin was based on the calculation of truss elements by the finite element method for bar systems described in [7], briefly outlined in part 3. C# and the Revit API were used. The data for the calculation was obtained by special methods for objects in the Revit API. The plugin methods code is presented in part 3.

3. The plugin creating process and results of its work

The plugin extracts initial data (truss geometry) from the model in Revit, filters necessary data, and sets missing parameters (loads, boundary conditions).

The plugin framework consists of three stages. The first stage gets the geometry of the truss elements which contain lines in base (Fig. 2). Every line has coordinates of the end and start points. Therefore, all truss frame lines are obtained using two methods *GetLine* and *GetLines* (Fig. 3). After that, a list of all points is formed using *GetXYZs* (Fig. 4) to obtain truss joints based on it. Then an additional *GetJoints* method sorts the list of all points to get unique points (Fig. 5).

One of the goals is dividing the truss cords into finite elements because often they are drawn with one line. The lines that need to be divided into finite elements are found using the obtained nodes. The result is a list of 25 lines.

The second stage is sorting. The nodes are sorted by three coordinates: first by Z, then by X and Y (Fig. 4). After that elements are sorted. An alternately scanning nodes and finding correspondences between nodes and lines method was developed for that. The order of the lines corresponds to the minimum sum of the node numbers. The first line is the line between nodes 0 and 1, the second between 0 and 2, the next between 1 and 2, and so on. As a result, the necessary sorted list of lines and nodes are obtained.

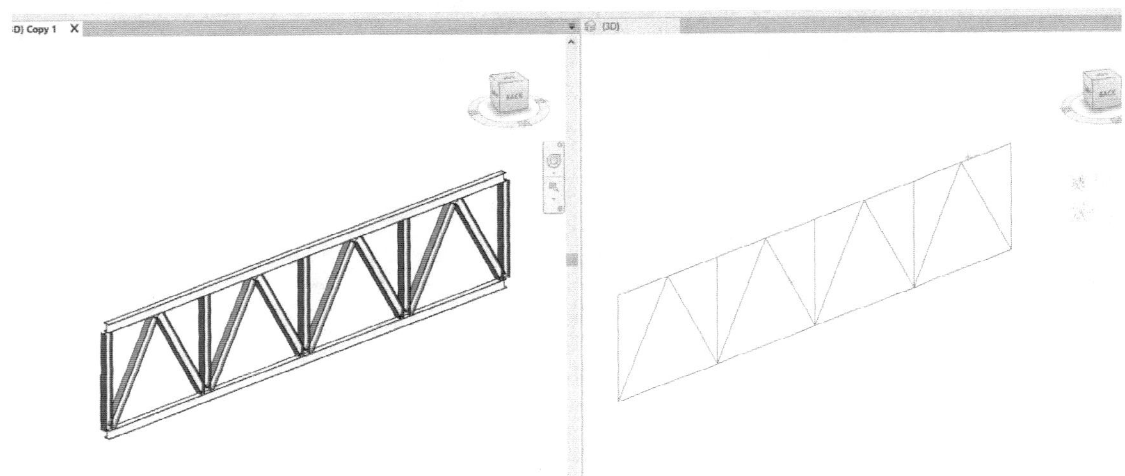

Figure 2. Truss frame and line base.

IOP Conf. Series: Materials Science and Engineering **944** (2020) 012037 doi:10.1088/1757-899X/944/1/012037

```
class FrameLines
{
    //Get the lines from element
    public Curve GetLine(Element el)
    {
        Curve beam_c;
        LocationCurve loc = el.Location as LocationCurve;
        beam_c = loc.Curve;
        return beam_c;
    }

    //get the lines from elements
    public List<Curve> GetLines(List<Element> elements)
    {
        List<Curve> truss_curves = new List<Curve>();

        foreach (Element el in elements)
        {
            Curve curve = this.GetLine(el);
            truss_curves.Add(curve);
        }
        return truss_curves;
    }
```

Figure 3. Class *FrameLines* and method *GetLine, GetLines*.

```
    //get all points from all curves
    public List<XYZ> GetXYZs(List<Curve> all_curves)
    {
        List<XYZ> points = new List<XYZ>();
        foreach (Curve c in all_curves)
        {
            XYZ p_s = c.GetEndPoint(0);
            XYZ p_e = c.GetEndPoint(1);

            points.Add(p_s);
            points.Add(p_e);
        }
        return points;
    }
```

Figure 4. Obtaining truss frame nodes coordinates.

```
public List<XYZ> GetJoins(List<XYZ> all_xyz)
{
    List<XYZ> joins = new List<XYZ>();
    joins.Add(all_xyz[0]);
    foreach (XYZ p in all_xyz)
    {
        int count = 0;
        foreach (XYZ n_p in joins)
        {
            if (this.IsCompared(p, n_p))
            {
                count++;
            }
        }
        if(count == 0)
        {
            joins.Add(p);
        }
    }
    return joins;
}
```

Figure 5. Obtaining truss frame nodes coordinates.

The third stage is calculation. Finite element method is used to calculate the truss frame [7]. It is assumed that the truss has pin-joints. Loads are nodal.

The array of forces:

$$\left[F_{x1}, F_{z1}, F_{x2}, F_{z2}, \ldots F_{xn}, F_{zn}\right] \tag{1}$$

Here F is the load along the axis indicated in the index, n is the node number. Zero is set if there is no load in a particular direction and a particular node. Each node of the truss top cord is loaded with a concentrated load of 10 *kN* for the calculation example.

The list of boundary conditions is set similarly to the list of loads but instead of the values of the loads, the positions for the fixed nodes are set to 1, for not fixed – to 0. The nodes of the truss lower cord are assumed to be fixed so the boundary conditions list contains [1, 1] and [0, 1] respectively for these nodes.

The stiffness of each element is defined as $A \cdot E$, where A is the cross-sectional area, E is the elastic modulus of steel. The area can be extracted from the model or set conditionally. $E = 2 \cdot 10^5\, MPa$.

The result is several data arrays: the coordinates of the truss nodes, the length of the elements, the lists of loads, boundary conditions.

The vector of internal nodal forces of a structure in the general coordinate system $X^0Y^0Z^0$ [7]:

$$\left\{S_k^0\right\} = \left[K_k^0\right] \cdot [A] \cdot \left([A]^T \cdot \left[K_k^0\right] \cdot [A]\right)^{-1} \left\{P_0\right\} = \left[K_k^0\right] \cdot [A] \cdot \left\{Z^0\right\} \tag{2}$$

$\left\{S_k^0\right\}$ – vector of internal nodal forces of the structure, consisting of blocks (cells) of vectors of internal nodal forces of finite elements (FE);

$\left[K_k^0\right]$ – quasi-diagonal design stiffness matrix, consisting of stiffness matrix blocks FE in the general coordinate system;

$[A]$ – correspondence matrix (nodes) of the structure, consisting of blocks of correspondence matrices FE;

$\left\{P_0\right\}$ – nodal load vector;

$\left\{Z^0\right\}$ – vector nodal displacements of the construction.

This part of script gives truss element stresses. It multiplies matrices using the Matrix library (C#) or Numpy library (Python). A calculation was made in another program (SCAD 21.1) to confirm the calculation as fig. 6 and 7 shows.

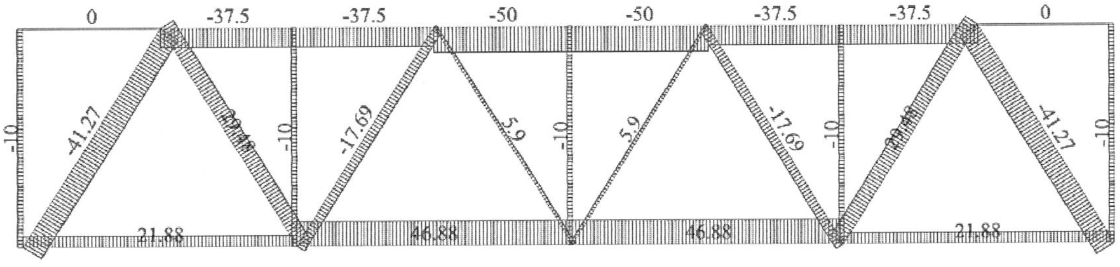

Figure 6. SCAD 21.1 calculations results [kN].

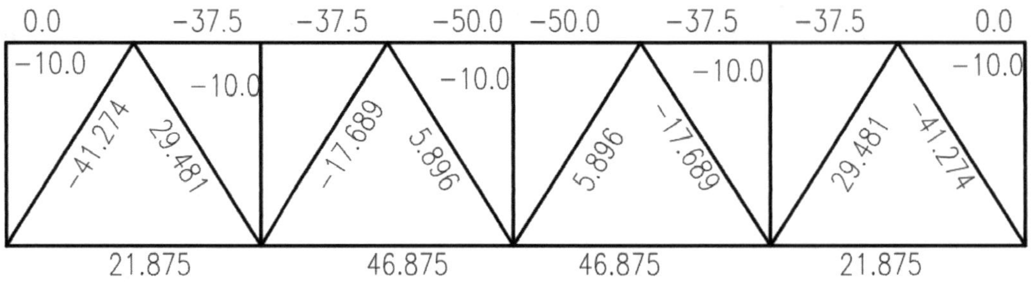

Figure 7. Plugin calculation results [kN].

4. Conclusion

This article will be useful to engineers and designers working with the API and looking for ways to speed up and simplify their workflow. The Revit API has special methods for retrieving data, and this article provides examples for its application. It is possible to extract data for structural analysis as a result of the analysis of the Revit model data. For example, FEM calculation that sets the cross-section of truss elements.

Perhaps in the future such plugins will be fully developed by large software vendors, but today there is a need for designers to have tools for solving particular problems and to be able to adapt it to their needs.

References

[1] Mladzievskiy E P 2020 *Academy* **2(53)** pp 9–10
[2] Sharmanov V V, Mamaev A E and Simankina T L 2019 *Vestnik nauchnogo tsentra* **2** pp 76–86
[3] Yarmohammadi S and Castro-Lacouture D 2018 *Automation in Construction* **93** pp 91–111
[4] Kota S, Haberl J S, Clayton M J and Yan W 2014 *Energy and Buildings* **81** pp 391–403
[5] Zotkin S P, Ignatova E V and Zotkina I A 2016 *Procedia Engineering* **153** pp 915–919
[6] Xiongjue W 2016 *Perspectives in Science* **7** pp 317–322
[7] Chernyy A N and Chernov S A 2010 *Truss finite element method calculation* (Ulyanovsk: UlSTU) p 30

IOP Conf. Series: Materials Science and Engineering **944** (2020) 012038 doi:10.1088/1757-899X/944/1/012038

Modern embankments of minor rivers on the example of the Belgorod region's cities

T S Yarmosh[1, 2], **P V Grigoreva**[1]

[1] Belgorod State Technological University named after V.G. Shukhov, Russia, 308012, Belgorod, 46 Kostyukova st.

*E-mail: polina9598@yandex.ru

Abstract. The article considers the problem of creating a comfortable recreational space in degraded riverside territories. The author analyzes the experience of the embankments' reconstruction of the Belgorod region, identifies priority directions in the development of the territory improvement projects. The article provides examples of new attractive, comfortable, and functionally developed embankments in the Belgorod region's cities. The creation of which was and is being carried out on previously degraded riverside territories, which contributes to improving the level and quality of the population's life, as well as attracting visitors to the city.

1. Introduction

A city is a complex structure in which the hydrosphere, atmosphere, flora and fauna, relief, soils, etc. are currently undergoing significant changes due to poor environmental conditions. They affect the main indicators of the natural environment characteristic of a given region, making various changes, and generating natural elements that are not typical for this terrain or, on the contrary, destroy them. As a result, a whole range of environmental problems arise, one of which is the problem of minor rivers and riverside territories in cities. After all, the active development and building of riverside territories in many cities affect almost all components of the natural environment. Foreign experience shows how riverside territories, turned into parks and biotopes, contribute to the maintenance of the natural ecological environment [1].

Many Russian cities have watercourses of various sizes, profiles, and character on their territory. Most of them experience man-made impacts. There are often contradictions over the use of such territories [2]. If major rivers and lakes have a sufficiently large potential and a significant degree of stability and self-regulation to various types of impacts, then along with them there are small watercourses, that are completely dependent on the operational and urban planning policies of cities. At the same time, small rivers, lakes, and ponds play an important role in forming the structure, image and microclimate of a quarter, residential group, or urban area [3]. The riverside areas of such water reservoirs can become wonderful new recreational areas for residents and guests of the city.

Minor rivers are one of the main landscape components of great ecological significance. Water resources of minor rivers, as part of the total water resources, are important factors in the placement of productive forces and the socio-economic development of society [4].

Almost all minor rivers in cities have significant resource potential. On the riverside territories, the policy of improving recreational areas, green, walking and sports zones, places of mass recreation and tourism is currently actively developing [5]. Improving the aesthetic components of the urban environment and creating its unique image can develop the tourism sphere, increase the level of the economy and incoming investment in the city and its facilities from the outside [6].

2. Materials and methods.

The theoretical and methodological basis is an ecosystem approach focused on process management based on the biosphosphere compatibility principles [7]. The main principle of the ecosystem approach to environmental management is predetermined by the interconnection of processes and phenomena in nature. This makes it possible to identify the interest and need of participants in urban planning activities, draw up legal requirements for preventing and avoiding any harm to natural objects and the environment.

3. Results and discussions.

Analysis of the riverside territories' condition, as well as steps for their development and restoration, were carried out on the example of three rivers in the Belgorod region: Vezelka, Vorskla, Nezhegol. Several acute problems were found in all three territories: the lack of well-equipped public spaces along the rivers, connections between green spaces, and there were practically no approaches to the rivers. This is affected by the impact of degrading industrial zones, as well as parking and garages.

Since 2017 to 2020 these coastal zones were reconstructed as part of the federal program for the embankments' improvement.

In 2017, the reconstruction of part of the Vezelka river embankment in Belgorod city was included in two federal programs at once: "Formation of a comfortable urban environment" and a project to organize public spaces for watching World Cup matches in 2018.

The chosen territory is located in the central part of the city. Important social, cultural and historical sites are located nearby: Victory Park, museum-diorama "Battle of Kursk. Belgorod direction", local lore museum, art museum, drama theater named after M. S. Shchepkin, etc.

"Strelka" consulting bureau was engaged in the upgrading project's development. The main objectives in developing this project were not only the creation of fan zones for the World Cup, but also the creation of a comprehensive improvement of the Victory Park's land, taking into account the interests of those groups that visited it. Before the reconstruction, the embankment consisted of several separate and uncomfortable parts. As a result of the project implementation, these territories were merged into a single whole, and walking routes and bike tracks were created connecting different banks of the river. The embankment was divided into several thematic blocks:

1. The zone of mass public visiting with the focus on family and children's recreation. The central section of the Victory Park has been allocated for it, as well as the territory near the children's park "Kotofey".
2. The zone located near Belgorod State University, focused on students.
3. A calm recreation zone with aesthetically attractive landscapes, located near the museum-diorama's territory.

Also, the following functional zones were identified in this project: 1- student embankment; 2- central embankment; 3- memorable embankment; 4- industrial embankment (figure 1).

IOP Conf. Series: Materials Science and Engineering **944** (2020) 012038 doi:10.1088/1757-899X/944/1/012038

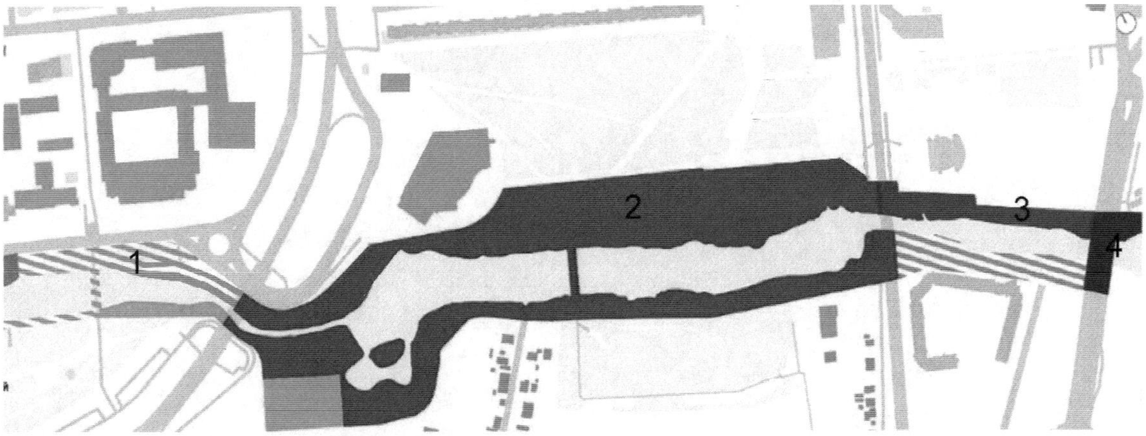

Figure 1. Functional zoning scheme of the Vezelka river embankment in Belgorod. The scheme is prepared by CB "Strelka".

In addition to these zones, taking into account the interests of citizens, special dog walking grounds, children's and sports grounds were designed and equipped. In picturesque places located in close proximity to the water, "contact" zones that resemble fishing bridges were equipped. Along the river, where there was once a steep descent to the water reservoir, equipped seating areas have appeared, competently distributed in several tiers up the relief. Organizing this recreational area, the following principle was strictly adhered to: recreation and tourism objects should be designed as objects that are organically included in the natural landscape, but in no case changing it [8]. Mandatory for implementation were: planting additional greenery (coniferous trees, various shrubs and herbs), creating bright species flowerbeds, cleaning water areas and installing a new drainage system (because in recent years, due to ground water, the territory of the Victory Park began to turn into a swamp). The improvement of the embankment in the Victory Park was completed before the 2018 FIFA World Cup.

At the moment, the reconstruction of the Vezelka river embankment in Belgorod has already been completed. All conducted measures were able not only to transform the embankment and form a new community center, but also to have a great positive impact on the river's state and coastal areas.

In 2019, Graivoron city of the Belgorod region won the all-Russian competition "Towns and historical settlements" with the project of reconstruction of the Vorskla river embankment in the park "Petrovskaya steep".

The territory with an area of 4.6 Ha was improved in 4 months. During the renovation works, several asphalt rings (with a total length of more than 3 km) were laid, which are ski-roller, bicycle and walking routes. An amphitheater 53 m long and 13 m wide was installed above the river. The main structures installed in Petrovskaya steep correspond to the eco-trend. The amphitheater, playground, benches and even urns are made of natural materials.

The Vorskla river embankment is divided into several functional zones: 1-walking zone, 2-beach, 3-festival zone, 4- entrance zone, 5- active recreation zone, 6- calm recreation zone, 7- islands, 8- forest park zone (figure 2).

More than 1 400 different trees and shrubs were planted on the territory of the new embankment. In order to admire green spaces on the park's territory all year round, experts planted more than 200 different tree species: linden, birch, spruce, pine, etc.

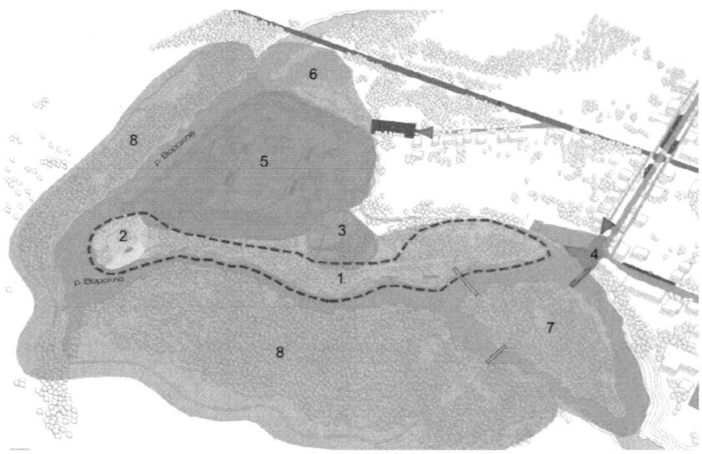

Figure 2. Functional zoning scheme of Vorskla river embankment in Graivoron.

The well-organized embankment of the Vorskla river in the park "Petrovskaya steep" has become a new public and recreational center of Graivoron.

In 2020, the project for the improvement of the Nezhegol river embankment in Shebekino won the all-Russian competition for the best projects for creating a comfortable urban environment in towns and historical settlements.

Shebekino has a complex functional structure due to the presence of a large number of industrial facilities, and storehouse and public utilities zones located dispersed throughout the city [9]. In the ccentral part, there is a lot of badly organized zones for long or short term stay of citizens. The most common problems of riverside territories include: ecological, geological, degradation of part of natural territories, lack of a clear communication framework, divergence of interests of different user groups, incongruity of the functional organization of riverside territories with the modern context of the city [10].

The administration of Shebekino decided to put in order the central part of the city and the Nezhegol river embankment, thereby solving many of the urgent issues. Reconstruction of the coastal zone will be made from Likhachev St. to 50th anniversary of October St. This is the central part and the landmark of the city. The improvement will affect the territory with a length of 694 m and an area of about 3 ha, and the residents themselves chose a site for it. The project creators considered all the wishes of local residents. Trees and shoreline that exist now will be preserved as much as possible, but green spaces will be added. The main idea of the project according to the authors' idea will be the creation of a public space surrounded by picturesque nature using a unique landscape.

The objectives of this project are: 1) creation of attractive urban spaces that contribute to improving the level and quality of life, as well as attracting visitors to the city; 2) development of the service industry by stimulating the development of small and medium-sized enterprises; 3) improvement of the positive image of the urban district.

Nezhegol river embankment is divided into the following functional zones: 1- entrance zone with cafes, 2- landscape park with walking areas and equipped recreation areas, 3- promenade with bike paths and viewing platforms, 4- mass event zone (festival park), 5- city beach, 6- boat pier, 7- park of generations, 8- youth cultural center square, 9- forest park, 10- arboretum (figure 3).

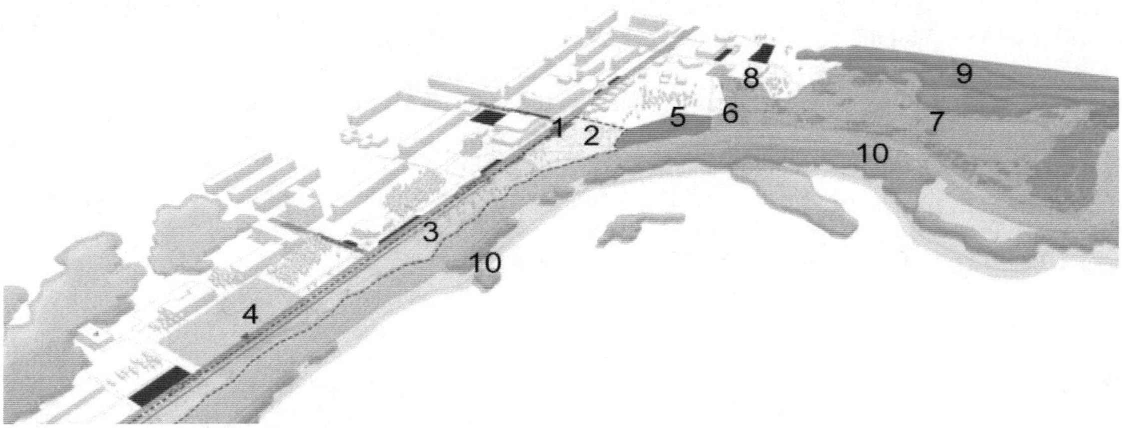

Figure 3. Functional zoning scheme of the Nezhegol river embankment.

Today, social and environmental factors of the formation of an architectural environment are of particular importance, which meet the tasks of "green" construction and are currently being actively developed all over the world [11]. In this regard, during the design and construction of the considered objects for the reconstruction of river embankments: the Vezelka river (Belgorod), the Nezhegol river (Shebekino) and the Voskla river (Graivoron), much attention was paid to the citizens' opinion and their wishes for the improvement of territories.

All the studied embankments are located in an urban environment, but despite the differences in their location and the initial degree of improvement, it is possible to identify functional zones characteristic of all and compare them. The comparison results are shown in table 1.

Table 1. "Comparison of functional zones".

Functional zone	Belgorod	Shebekino	Graivoron
Entrance zone	yes	yes	yes
Transit zone	yes	yes	insufficient
Calm recreation zone with bike and walking paths	extended	insignificant	insignificant
Active recreation zone	yes	yes	yes
Mass event zone	included in another zone	separate	separate
Beach zone	yes	yes	yes
Water objects (bridges, piers, etc.)	yes, but not enough	yes, but not enough	sufficiently
Accessibility (availability of parking, stops, etc.)	insufficient	good	good
Shoreline	extended	extended	extended

4. Conclusion

Trends of development in many Russian cities are currently aimed at creating favorable conditions for the eco-reconstruction of urban spaces [5]. Without a competent approach to the organization of an ecologically friendly environment, without a rational arrangement and implementation of green zones in the planning schemes of cities and suburbs, there can be no "smart" and, most importantly, safe city for the human life [12].

Riverbanks are extremely valuable both ecologically and for the organization of recreational areas. Due to the large number of economic and residential developments near rivers, there are many conflicts over the use of coastal territories. Open public spaces are the most important element of the city's planning structure, as they primarily characterize the quality of the population's life as a whole, reflect the level of development of social and cultural infrastructure, and form the overall city image.

The article provides examples of new attractive, comfortable, and functionally developed embankments in the Belgorod region's cities. The creation of which was and is being carried out on previously degraded riverside territories, which contributes to improving the level and quality of the population's life, as well as attracting visitors to the city.

References

[1] Perkova M V, Blagovidova N G and Tribuntseva K M 2015 Features of Design of Ecovillages in Depressed Areas in the City *Research Journal of Applied Sciences* **10** pp 608–619

[2] Perkova M V, Roshupkina O E, Kolesnikova L I and Vaytens A G 2019 The urban conflicts identification and resolution on the Belogorie reserve territory *IOP Conference Series: Materials Science and Engineering*

[3] Kulikov A S and Luchkova V I 2010 Khabarovsk city small rivers problems and suggested solutions *New Ideas of New Century* **1** pp 79–85

[4] Minor rivers of Russia. Available at: URL: http://www.microanswers.ru/article/malie-reki-rossii.html (accessed: 22.04.2020).

[5] Perkova M V and Krushelnitskaya E I 2016 Recreation and tourism organization issues in modern urbanization conditions *International Journal of Pharmacy & Technology* **8 (4)** pp 24919–24929.

[6] Yarmosh T S and Mikhailova I D 2019 Landscape urbanism - a new direction of modern concepts of urban space development on the example of Russian cities *V.G. Shukhov BSTU Vestnik* **7** pp 72–80

[7] Perelet R A 2006 Ecosystem approach for environmental management and protection // *Environmental economics* **3** pp 3–19

[8] Perkova M V and Krushelnitskaya E I 2014 Environmental problems of harmonization of landscape and recreational environment of resting and tourism facilities *V.G. Shukhov BSTU Vestnik* **5** pp 11–15

[9] Perkova M V and Zaikina A S 2016 The way to slove the problems of degraded areas in Shebekino town *V.G. Shukhov BSTU Vestnik* 6 pp 58–63

[10] Perkova M V Baklazhenko E V and Vaytens A G 2020 A Method for Identifying and Resolving Conflicts in Urban Riverside Development *IOP Conference Series: Earth and Environmental Science*

[11] Perkova M V, Drebezgova M U, Chernysheva N V, Perkova A U and Stoikovich N Option design of "green" office for SMBs based on additive modular technologies *Additive manufacturing technologies* **1 (4)** pp 22–34

[12] Yarmosh T S and Ivanilova E I 2017 The formation of a system of green areas of the city, as a means of improving the quality of life of the urban population *V.G. Shukhov BSTU Vestnik* **12** pp 109–112.

AUTHOR INDEX